Science and technology of building seals, sealan
TH9031 .S36 1992 33361

DATE	ISSUED TO

STP 1200

Science and Technology of Building Seals, Sealants, Glazing, and Waterproofing: Second Volume

Jerome M. Klosowski, editor

ASTM Publication Code Number (PCN)
04-012000-10

ASTM
1916 Race Street
Philadelphia, PA 19103

Library of Congress

ISBN: 0-8031-1857-0
ASTM Publication Code Number (PCN): 04-012000-10
ISSN: 1062-967X

Peer Review Policy

Each paper published in this volume was evaluated by three peer reviewers. The authors addressed all of the reviewers' comments to the satisfaction of both the technical editor(s) and the ASTM Committee on Publications.

The quality of the papers in this publication reflects not only the obvious efforts of the authors and the technical editor(s), but also the work of these peer reviewers. The ASTM Committee on Publications acknowledges with appreciation their dedication and contribution to time and effort on behalf of ASTM.

Printed in Fredricksburg, VA
November 1992

Foreword

This publication, *Science and Technology of Building Seals, Sealants, Glazing, and Waterproofing: Second Volume*, contains papers presented at the symposium of the same name, held in Ft. Lauderdale, FL on 5–6 Feb., 1992. The symposium was sponsored by ASTM Committee D-24 on Building Seals and Sealants. Jerome M. Klosowski of the Dow Corning Center in Midland, MI presided as symposium chairman and is the editor of the resulting publication.

Contents

Overview

The ASTM symposium which produced the papers presented in this book (Second Symposium on Science and Technology of Building Seals, Sealants, Glazing and Waterproofing held Feb. 5-6, 1992 in Ft. Lauderdale Florida), was a landmark event . Most of these are highly significant papers which will be the stimulus for many more research projects and could be the stimulus for radical changes in the specifications for sealants.

All the papers in this book should be of some interest to those interested in sealing and waterproofing. Of particular importance is the first set of papers on durability. These papers ask one to rethink the old thoughts on what durability means and how it should be tested. The papers on adhesion are also quite unique in approach and the conclusions to some these are controversial. While the papers are controversial, they are worth reading. Each stimulates thinking about adhesion in a new way. Some of the papers on joint design are review papers but they do bring out important-to-remember concepts. There is a concept paper on insulated glass and a series of papers on new products. All of this is here for the interested reader.

The Fedor paper is one more in a series where artificial weatherometry is compared to actual outdoor exposure of sealants. The idea from the earlier paper is reinforced in this paper is that it takes in the range of 1000 hours to 1200 hours in an artificial weather machine to simulate approximately 1 year in south Florida. That it takes something less to simulate a year in more temperate climates, perhaps as little as 500 hours, can also be concluded. This data, and many of the things that can cause variations in this data, are addressed. The enormous significance of this is that some specifications in ASTM indicate that sealants that are tested and survive 150 hours or 250 hours in such weathering machines are called weather resistant, when this paper indicates such short exposures have no way of determining the long term weather resistance. Thus we expect that many will, and should, study the Fedor paper when writing new test methods and specifications which address climatic durability. With the Fedor paper in mind, perhaps the scope of several test methods could be rethought as could the significance in use sections.

The Fedor paper also gives added significance to the papers on sealant durability by Bridgewater and Carbary, and by Beech and Beasley. These papers use similar artificial weathering to study the deterioration of various sealants and both papers study the mode of deterioration of the various generic sealants. The length of time in the weathering machines in these papers extend to 4000 hours and 5000 hours and thus could be perhaps reminiscent of between 5 years and 10 years of outdoor exposure, depending on climate. Thus the papers are not to be considered as presenting extremely taxing data on the sealants studied, but perhaps representative of real life. These papers, and the Fedor paper, only test sealants in the absence of movement and all the history of sealants would indicate that even more severe deterioration would be expected if the sealants were moved (extended and compressed) while weathering. Thus from the static nature of the tests, the papers describe conditions that are actually more mild than real life.

There would seem to be a discrepancy in the two similar papers on weathering of sealants in that one shows more deterioration than the other in similar conditions of the artificial weathering, on similar sealants. In the Bridgewater - Carbary paper there is more severe deterioration of some sealants with moderate heats, in normal humidities with only UV radiation as the significant stressing mode. This editor discussed this with both sets of authors and found that they used different color sealants and that they do expect color to be significant. They expect that the black and white have significant pigmenting fillers while the concrete colors have little of these pigments to act as U.V. blockers. Thus both papers have very interesting and pertinent data and both are worth of serious study.

The Pernisz - Klosowski paper shows that the radiation on a sunny day is between 15 and 21 times greater on a Southern exposure, in full sun, than on a northern exposure. This paper gives possible reasons why all sealant joints don't deteriorate at the rate suggested in the papers on weathering. Different joints, even on the same building, get different amounts of the sun's radiation. Thus with many sealants, variations in durability of the sealant are expected in the surfaces of a given building, depending on the orientation of the joints to the sun. The sidewalks and roadways are expected to be yet different, even with the same sealant installed in the same way.

The Tanzer - Frankel work shows a outdoor movement device which is easy to use and inexpensive to build. This device would be a handy evaluation means in any development lab or on an architects building (architects should indeed study the materials they use). In addition the paper shows a comparison of a variety of latex acrylic sealants and two silicone sealants which are tested in the movement device. Of special significance is the argument for the hourglass design for all sealant beads, to get the maximum movement ability and how the water borne sealants naturally shrink to achieve this while the non-water borne materials should be tooled to achieve this. All of this is interesting to almost all in the sealants area and this could and perhaps should influence our future test designs, especially if all premium quality sealants will be compared on the same test.

Professors Minkarah and Weisgerber each co-author a paper on highway sealants, which also have applicability to plaza decks and parking garage sealants. The Minkarah - Reiser paper looks at the damage to a number of highway sealants with various exposures. Their work considers 600 hours in an accelerated weather machine and with most of the sealants they consider there is already significant damage. This is especially interesting in light of the earlier mentioned Fedor paper who suggest this exposure is probably less than or equal to one year's exposure.

The Weisgerber and Wang paper discuss the stresses in highway joint sealants and helps to explain some of the stresses that cause the sealant failure.

The O'Connor - Panek paper is especially interesting to those who are involved with structurally glazed curtain walls. This paper too is a study of durability. This paper shows the deterioration, or more correctly, the lack of it in a well studied structurally glazed building. They report on the strength and adhesion of an aged silicone system and give a great deal of confidence to those who have used such systems.

An opposite of durability is early joint failure. One aspect of this is discussed in the paper on early joint movement by Flackett. In this paper silicones are studied as a function of cure rate. This paper makês a good follow-up to papers presented at the 1991 symposium the discussed movement during cure. All of this work emphasizes the need for the completion of the ASTM C-24 committee work on exactly this topic.

Silicone sealant adhesion technology is reviewed in a series of 3 papers by Gutowski, Cerra and Russel. Their studies at an Australian national lab, study the all important property of adhesion and methods to predict this. Their conclusions are interesting and perhaps controversial, but give the silicone technologist something to think about.

The paper by Klosowski and Dent also addresses adhesion. In this paper, a modification of the constrained blister test which was first pioneered by Moet at Case Western Reserve and Dillard of VPI, the test method is detailed. In addition to the test method is a discussion of the other sealant test methods that test adhesion and these are compared to the constrained blister adhesion test.

Primers are a fundamental aspect of adhesion to some surfaces with most sealants. The paper by Berg and O'Brien discuss a new primer concept in an unusual situation where silane coupling agents are put in a silicone solvent and are shown to be useful with both silicone and non-silicone sealants.

The paper by Webb addresses yet another aspect of adhesion. In this paper the preparation of the concrete blocks and factors that lead to variable adhesion to these surfaces is detailed. This too is valuable information.

There is more to successful joint sealing than a good sealant. There is also a good design. The design aspect of unique fillet joints is detailed in the paper by Hoigard and Kudder. Several special joints designs and problems with them are described in the paper by Nicastro and a general overview of many joint designs and building designs are shown in the paper by Yanoviak.

Insulated glass sealants are also discussed in this book. A very well done, comprehensive paper by Wolf details the various factors effecting life expectancy of an insulated glass unit and prioritizes them. To those who study this paper a whole new way of considering the factors effecting longevity of an insulated glass sealant are revealed. Following this is a paper about a modified polybutadiene sealant for insulated glass by Hubin-Eschger.

New materials are addressed in papers by Elfring and Rosano on new acrylic coatings; a new elastomeric silicone wall coating that is water borne and has an anti-graffiti nature is described in the paper by Lefler; a new solvent based acrylic for bedding applications in glazing is in the paper by Potts; the formulation and performance of a new water borne silicone sealant (silicone latex) is described by by Liles, and Fiorillo describes new polysulfide sealant technology.

In essence the papers in this book touch on almost every aspect of sealing and waterproofing. While the durability papers stimulated the most discussion at the symposium, all the papers were interesting. To some, there will be other interpretations of the data than that presented. Perhaps some readers will want to do some similar studies to those presented here and report them at other symposiums or publish them in subsequent STP's. That would be a very good product from all of this work, for it's the new research and testing that furthers the science and technology of building seals and sealants and the world can only benefit from continuing research and publication of that work.

Jerry Klosowski

Jerome (Jerry) Klosowski
editor

Thomas F. O'Connor[1] and Julian R. Panek[2]

EXTENDED LABORATORY TESTING FOR TWO STRUCTURAL GLAZING SILICONE SEALANTS

REFERENCE: O'Connor, T. F., and Panek, J. R., **"Extended Laboratory Testing for Two Structural Glazing Silicone Sealants"**, Science and Technology of Building Seals, Sealants, Glazing, and Waterproofing: Second Volume, ASTM STP 1200, Jerome M. Klosowski, Ed., American Society for Testing and Materials, Philadelphia, 1992.

ABSTRACT: The owner of an eighty story building was concerned that insufficient extended laboratory durability testing information was available for two silicone structural glazing sealants proposed for use on the new building curtain wall. A two phase laboratory testing program was conducted to resolve these concerns.

Phase I involved testing for durometer and weight loss after extended heat aging by ASTM C661 and C792, cyclic movement after extended exposure in a weatherometer by ASTM C719, and adhesion-in-peel after extended water immersion by ASTM C794. Also, the pull-off strength of the organic coating, that the structural sealants were adhered to, was tested by ASTM D4541. Phase II used glazed components of the curtain wall for mock-up testing to destruction, to determine the ultimate strength of the structural silicone sealant joint by ASTM E330.

Phase I testing indicated that the extended laboratory exposures did not result in any detrimental changes to the properties of the sealants. Also, the pull-off strength of the organic coating exceeded the strength of the structural sealants. Phase II testing indicated that the ultimate strengths of the structural sealants exceeded manufacturer's published strengths.

KEYWORDS: adhesion-in-peel, curtain wall, cyclic movement, durability, durometer, heat-aging, silicone sealant, structural glazing, tensile strength, water-immersion, weatherometer, weight-loss.

The owner of an eighty story building was concerned that the American Society for Testing and Materials (ASTM) standard C920, Specification for Elastomeric Joint Sealants, did not adequately describe the material properties that would be required for water, ultraviolet radiation, and heat aging exposure for structural sealant glazing (SSG) sealants. There was also a concern that the SSG joint strength would differ from that established by small scale sample testing. To resolve these concerns a testing program was developed to answer the following questions:

[1]Mr. O'Connor is a Vice President and Consulting Architect at Smith, Hinchman & Grylls Associates, Inc., 150 W. Jefferson Avenue, Suite 100, Detroit, MI. 48226.

[2]Mr. Panek is President of Specialized Consulting Services, 1261 Madison Drive, Yardley, PA. 19067.

1. What was the effect of extended exposure of the structural glazing sealants to water, ultraviolet radiation, and heat?

2. Was the factory applied paint coating on the curtain wall extrusions, to which the structural sealant was to adhere, sufficiently bonded to the aluminum?

3. Was the strength of the structural sealants, as established by small scale laboratory samples, achieved in the actual sealant joint design?

The testing program was divided into two phases. Phase I consisted of small scale laboratory sample testing, to provide answers for the first two questions; while, Phase II tested to destruction, a full size curtain wall component to answer the third question. The test program structural silicone sealants included a two-part, factory applied, high modulus (2PHM) material and a one part, medium modulus (1PMM) material both from the same manufacturer.

TESTING PROGRAM

Phase I

The following ASTM test methods, as modified, were used:

1. C661-83 - Indentation Hardness of Elastomeric-Type Sealants by Means of a Durometer.

Heat aging was conducted at 70°C (158°F) for six months with periodic readings taken initially and monthly on two specimens for each sealant.

2. C719-86 - Adhesion and Cohesion of Elastomeric Joint Sealants Under Cyclic Movement (Hockman Cycle).

Test specimens were prepared in triplicate for each sealant by applying a bead of sealant between parallel block faces, as specified in the method, with a sealant profile of 12.7 by 12.7 by 50.8 mm (1/2 by 1/2 by 2 in.). One block was glass with a reflective coating and the other aluminum with a factory applied polyvinylidene fluoride (PVDF) paint coating.

After completion of the cycling test as outlined in paragraph 8.7 of the method, with the 2PHM sealant tested at a joint movement of +/- 12.5 percent and the 1PMM sealant at a joint movement of +/- 25 percent, the tensile strength and elongation of each of the specimens was determined at a rate of movement of 12.7 mm (1/2 in.) per minute.

Additional test specimens were prepared as described above; however, instead of being cycled they were exposed in an accelerated weathering machine (ASTM G23, Method 1 - light/water cycle) with the glass side of the specimen facing the light and water source for a period of up to six months. A group of three specimens were removed at intervals of one week, 1, 2, 3, 4 and 6 months and immersed in deionized water for seven days. The specimens were then immediately tested to determine their tensile strength and elongation at a rate of movement of 12.7 mm (1/2 in.) per minute.

3. C792-75 - Effects of Heat Aging on Weight-Loss, Cracking, and Chalking of Elastomeric Sealants.

Heat aging was performed at 70°C (158°F) for six months with periodic readings taken initially and monthly on three specimens for each sealant.

4. C794-80 - Adhesion-in-Peel of Elastomeric Sealants.

Test specimens were prepared by applying both sealants to uncoated gray tinted glass, glass with a reflective coating, and aluminum with a PVDF paint coating. Four specimens for each sealant were tested after each of the following conditioning stages:

a. No water immersion.
b. Immersion in deionized water for one week.
c. One month immersion in deionized water.
d. One month immersion in deionized water followed by one week dry at 70°C (158°F), followed by one week immersion in deionized water.
e. Two months immersion in deionized water.
f. Three months immersion in deionized water.

5. D4541 - Pull-Off Strength of Coatings Using Portable Adhesion - Testers.

This method determined the adhesive strength of the PVDF coating to the curtain wall extrusions. An adhesive dolly was attached to each specimen using an epoxy adhesive. The adhesive was allowed to cure for seven days at 25°C (77°F) before testing the three specimens.

Phase II

A nominal 1016 by 508 mm (40 by 20 in.) component of the curtain wall system was selected for testing the structural sealant joint (Fig. 1). The component was glazed with the same 25.4 mm (1 in.) insulating tempered glass and fabricated with the same materials, finishes, and procedures as used for the production of the curtain wall system.

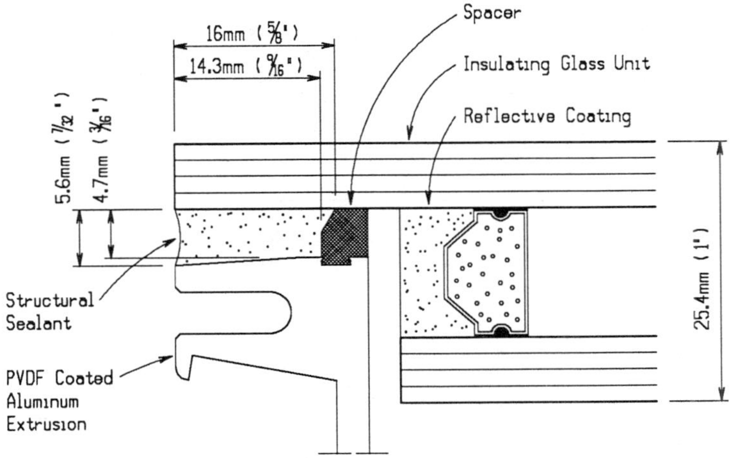

Fig. 1 - Curtain wall component structural sealant joint profile

The following ASTM test method, as modified, was used:

E330 - Structural Performance of Exterior Windows, Curtain Walls, and Doors by Uniform Static Air Pressure Difference.

The two structural glazing silicone sealants used for the curtain wall system were tested in the following configurations:

1. As glazed at the factory using the 2PHM sealant.

2. A simulation of the reglazing of a lite of glass using the 1PMM sealant, wherein about a 2 mm (1/16 in.) thickness of the 2PHM sealant remains on the curtain wall framing surface.

3. Glazed with the 1PMM sealant as was planned for some glazing at the job site to accommodate construction procedures.

Using, in general, Procedure A of E330 the following loads were applied. The specimens were preloaded to a negative 6.74 kPa (140.7 psf), which was the lower limit of the test apparatus, for 10 seconds and the load released. A continuously increasing negative load was then applied and held for 10 seconds at 30 kPa (630 psf), which is equivalent to a design factor of 4 or a structural sealant stress of 552 kPa (80 psi). Without interruption, increasing negative load was then applied to achieve failure of the sealant within 45 seconds after application of the pressure difference across the specimens.

Measurements were made and recorded of: the actual size of the structurally adhered glass panel; the width of structural glazing sealant at three locations on each side of the unit (at the glass surface and PVDF coated aluminum surface), and the mode and negative pressure at failure.

TESTING RESULTS

Phase I

1. C661 - The 2PHM sealant had an initial shore A value of 48 which increased to 59, and the 1PMM sealant 43 which increased to 47 (Fig. 2). Both sealants did not change in hardness after three months of heat aging. For comparison, specification C920 requires a value from 25 to 50 after 14 days cure and without heat aging.

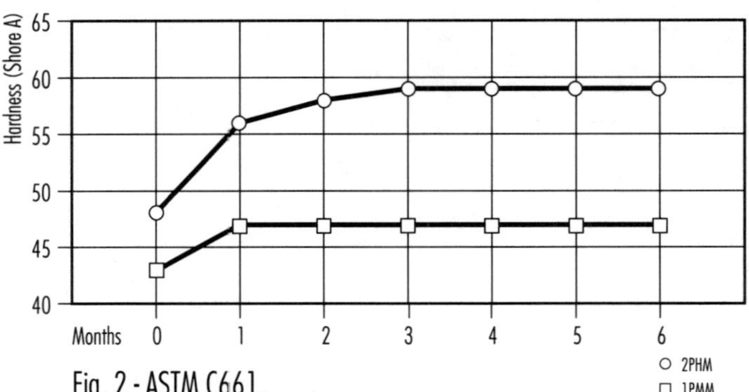

Fig. 2 - ASTM C661,
Increase in hardness after exposure at 70°C (158°F).

2. C719 - All 2PHM specimens exhibited 100 percent cohesive failure (Fig. 3 & 4). The tensile strength of the 2PHM sealant ranged from a high of 807 kPa (117 psi) at two months exposure to 627 kPa (91 psi) at one week exposure with 51 and 108 percent elongation respectively. The initial tensile strength and elongation is low since the glass side of the specimen broke during testing. The tensile strength after completion of the accelerated weathering was 669 kPa (97 psi) with 51 percent elongation.

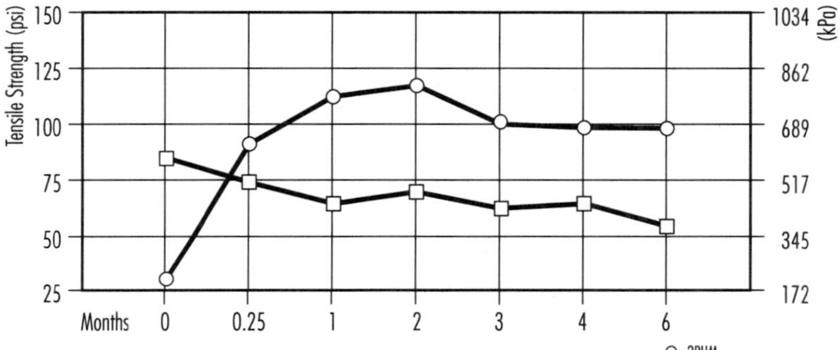

Fig. 3 - ASTM C719,
Tensile strength after exposure in weatherometer.

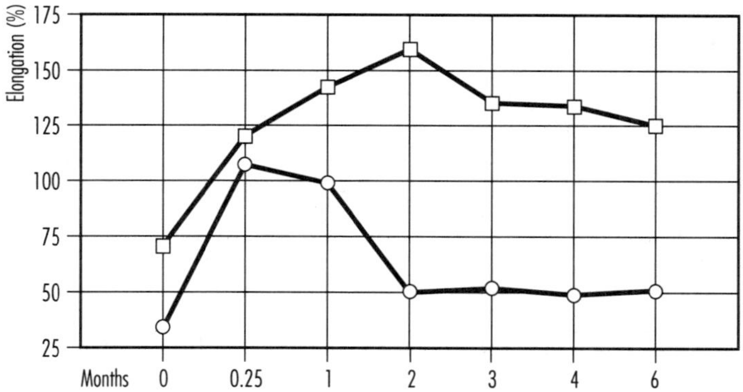

Fig. 4 - ASTM C719,
Elongation after exposure in weatherometer.

The 1PMM sealant exhibited 100 percent cohesive failure after one week and one and two months exposure in the weatherometer (Fig. 3 & 4). All remaining specimens exhibited at least 80 percent cohesive failure after 3, 4, and 6 months exposure. The 20 percent adhesive loss occurred at the reflectively coated glass surface. The tensile strength of the 1PMM sealant ranged from an initial value of 579 kPa (84 psi) to 372 kPa (54 psi) at six months exposure with 71 and 125 percent elongation respectively.

The above tensile strength and elongation values do not have any C920 comparable requirements. The goal of this test was to determine any effects of extended weatherometer exposure, with water and ultra-violet, on the adhesive characteristics of the sealants.

3. C792 - The 2PHM sealant stabilized at 1.3 percent weight loss after five months of heat aging. The 1PMM sealant weight loss after six months was 2.43 percent and was still increasing. The rate of increase after two months exposure was approximately 0.1 percent increase per month (Fig. 5). There was no evidence of any surface chalking or cracking of the sealants. For comparison, maximum weight loss by C920 is 10 percent after 21 days exposure.

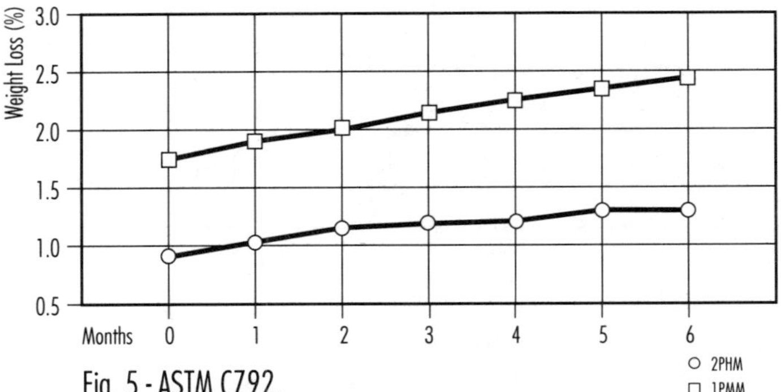

Fig. 5 - ASTM C792,
Increase in weight loss after exposure at 70°C (158°F).

4. C794 - All specimens failed cohesively (Fig. 6 & 7). The minimum peel strengths of the 2PHM sealant were 8.2 kg (18 lb.) at 2 months exposure on reflectively coated glass, 10.9 kg (24 lb.) initially on uncoated glass, and 6.4 kg (14 lb.) at three months exposure on PVDF coated aluminum. The minimum peel strengths of the 1PMM sealant were 5.4 kg (12 lb.) on reflectively coated glass and PVDF coated aluminum, and 5.9 kg (13 lb.) on uncoated glass all at three months exposure. For comparison, C920 requires a minimum peel strength of 2.3 Kg (5 lb.) after 7 days water exposure.

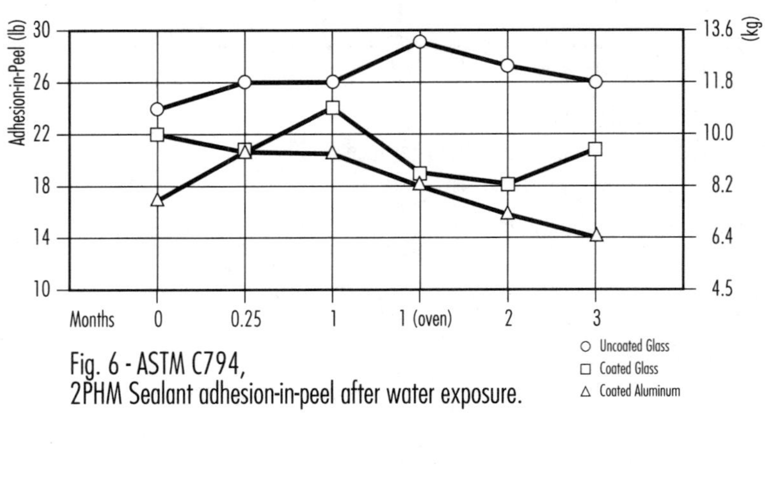

Fig. 6 - ASTM C794,
2PHM Sealant adhesion-in-peel after water exposure.

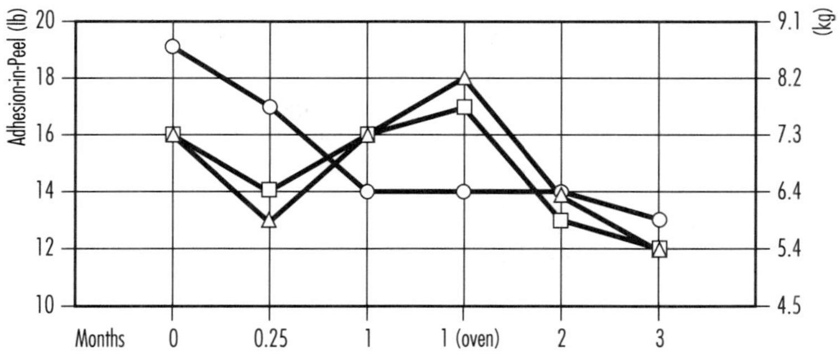

Fig. 7 - ASTM C794, 1PMM Sealant adhesion-in-peel after water exposure.

The most consistent adhesion values for the 2PHM sealant were obtained on uncoated glass with both the reflectively coated glass and PVDF coated aluminum showing some variability or decline in peel strength with exposure. The 1PMM sealant exhibited a consistent gradual decline in peel strength to all three surfaces.

5. D4541 - For the three tests, it was observed that none of the factory applied PVDF coating failed adhesively from the extrusion surface. The primary failure mode was 80 percent adhesive and occurred in the epoxy adhesive bonding the test device to the surface of the coating. The remaining 20 percent failure was delamination within the factory applied PVDF coating. The values obtained for the three tests were averaged resulting in a failure stress of 1482 kPa (215 psi) for the test dolly adhesive.

Phase II

1. E330 - No failure of the sealant occurred below the manufacturer's rated strength of the sealant. All sealant failures were cohesive with the sealant tear propagating for as much as 864 mm (34 in.) along one side of the unit until the load was relieved through the created gap in the sealant. The tearing did not continue the full length of the side. For unit G, the sealant, for no apparent reason, failed on both long sides with a 12.7 mm (1/2 in.) tear in one and a 25.4 mm (1 in.) tear in the other side, both of which did not propagate. For unit H the load was held for 5 seconds when the sealant failed. All units, except unit B, failed on the long side of the unit where deflection of the insulating glass would be expected to be the greatest.

The laboratory measured the structural sealant contact width for each unit except for unit C, which was removed for shipment to the owner. The actual size of the insulating glass was also measured. The data was analyzed to determine the average width and loading for each configuration of sealant usage. Table 1 was generated based on this analysis and the sealant manufacturer's published data.

From Table 1 the 2PHM and 1PMM sealants failed at average stresses of 1041 kPa (151 psi) and 565 kPa (82 psi) respectively, both of which are slightly greater then the manufacturer's published strengths of 965 kPa (140 psi) and 552 kPa (80 psi). The 2PHM/1PMM combination curiously failed at 814 kPa (118 psi) which is slightly above the average of the 2PHM and 1PMM sealant failure stresses [(1041 + 565)/2 = 803 kPa] [(151 + 82)/2 = 117 psi].

At the conclusion of the testing, it was noticed that the 2PHM sealant adhered to the plastic spacer (Fig. 1), that positioned the glass a constant distance from the frame, while the 1PMM sealant did not.

TABLE 1 - Curtain Wall Component Test Data.

Sealant Configuration	Failure Load kPa (psf)	Calculated Sealant Stress kPa (psi)	Manufacturers's Stress Rating kPa (psi)
2PHM (Ave)	56 (1174)	1041 (151)	965 (140)
Unit A	43 (900)	779 (113)	
B	62 (1300)	1207 (175)	
C	64 (1336)	... ...	
D	56 (1160)	1131 (164)	
1PMM/2PHM (Ave)	43 (906)	814 (118)	... ...
Unit E	41 (862)	772 (112)	
F	45 (950)	855 (124)	
1PMM (Ave)	30 (630)	565 (82)	552 (80)
Unit G	30 (630)	572 (83)	
H	30 (630)	558 (81)	

CONCLUSIONS

Phase I

1. No loss of volatiles or chemical changes which result in detrimental hardening or embrittlement of the sealants after heat aging for six months based on method C661.

2. No detrimental change in strength or adhesion after six months exposure in an accelerated weathering machine followed by seven day water immersion based on method C719.

3. Both sealants exhibit very low weight loss and, in general, reach stable configurations that are relatively unaffected by six months exposure to high temperature based on method C792.

4. No detrimental change in peel strength or adhesion after constant exposure to water for three months based on method C794.

5. The PVDF coating has an adhesive strength at least 2.6 times the average failure stress of the 1PMM sealant, 1.4 times the average failure stress of the 2PHM sealant, and 1.8 times the average failure stress of the 1PMM/2PHM sealant combination as determined by method D4541. These values, which are sufficiently higher than the sealant strengths, indicate that adhesive failure of the organic coating would not be expected before failure of the sealant

Phase II

1. For this particular structural sealant joint design, the tensile strength of the sealants, based on full scale testing by method E330, is in good agreement with the tensile strength values established by the sealant manufacturer using small scale laboratory specimens.

2. The typical reglazing configuration, a combination of the 1PMM and 2PHM sealants, had a sealant strength that was significantly above instead of at or near the 1PMM strength as was expected.

3. The adhesion of the spacer to the 2PHM sealant, which resulted in three-side sealant adhesion, does not seem to have affected sealant strength for this joint configuration.

4. At least under static loading conditions, a cohesive tearing failure of the structural sealant does not propagate to the other sides of a rectangular shaped specimen.

SUMMARY

The owner, after completion of the test program, was satisfied that both sealants had the potential capability to perform satisfactorily based upon the testing program.

ACKNOWLEDGMENTS

The authors would like to recognize the efforts of D/L Laboratories, New York, New York, and Construction Research Laboratory Inc., Miami, Florida for the performance of the testing.

Gregory R. Fedor

USEFULNESS OF ACCELERATED TEST METHODS FOR SEALANT WEATHERING

REFERENCE: Fedor, G. R., **"Usefulness of Accelerated Test Methods for Sealant Weathering"**, Science and Technology of Building Seals, Sealants, Glazing, and Waterproofing: Second Volume, ASTM STP 1200, Jerome M. Klosowski, Ed., American Society for Testing and Materials, Philadelphia, 1992.

ABSTRACT: Sealant test samples were exposed to accelerated and outdoor weathering to determine the usefulness of accelerated weathering test methods in predicting material degradation. One formula from seven different generic sealant types were exposed to three different outdoor environments for two years. Identical samples were exposed to three different accelerated test methods for 2000 hours. The degradation from outdoor exposure in terms of appearance, cracking, and change in hardness was duplicated in each case by at least one of the accelerated test methods.

KEYWORDS: weathering, sealants, Q-U-V, xenon arc, ultraviolet radiation, spectral power distribution

The usefulness of the various accelerated test methods for sealant weathering is a subject which is often debated. This can be attributed to 1) confusion regarding the numerous variables in accelerated testing and 2) poor correlation with outdoors in certain instances. The variables which cause confusion are the different types of test devices, light sources, cycles, exposure times, and evaluation methods. This confusion is

Greg Fedor is a research and development engineer at The Q-Panel Company, 26200 First St., Cleveland, OH 44145 often responsible for poor correlation because it leads to the use of improper test procedures which fail to simulate the climate, joint design, and installation of a natural exposure.

The objective of this study is to determine the usefulness of accelerated weathering tests by comparing the results of seven different sealant types exposed outdoors and in accelerated devices. For an accelerated test to be useful it must be able to duplicate the real time degradation of a material in an accelerated manner. If this can be demonstrated, the accelerated test method can be useful in evaluating and developing better sealants.

EXPERIMENTAL METHOD

Sealant test samples were prepared and submitted by volunteer companies from the C-24 ASTM committee (Building Seals and Sealants). The test samples were prepared by the sealant manufacturers in accordance with ASTM C-793 which specifies that the sample be a 3mm (1/8") thick slab of sealant applied to a clean bare aluminum substrate. The slab of sealant is 38mm (1-1/2") wide by 127mm (5") long and the substrate 76mm (3") by 152mm (6"). Eight materials from seven generic types were tested and are listed below:

- One Component Urethane
- Two Component Urethane
- Polyvinyl Acetate Latex
- Thermoplastic Rubber
- Silicone (black)
- Silicone (white)
- Acrylic - Solvent type
- Acrylic - Latex

Replicates were exposed to three different outdoor environments and three different accelerated test methods. The first phase of this study, which included one outdoor exposure for 1-1/2 years, was published in the May 1990 issue of Adhesives Age [1]. Two additional outdoor exposures (Ohio and Florida black box) have since been completed and the samples in all three outdoor exposures have been run for 2 years.

Natural Weathering

Ohio: Test samples were exposed in Cleveland, Ohio according to ASTM G-7 at 45° south with a plywood back mounting. Replicates were exposed for 1/2, 1, and 2 years beginning in March 1989. This exposure represents a northern, mild industrial climate.

Florida: Test samples were exposed in Homestead, Florida according to ASTM G-7 at 45° south with an open back mounting. Replicates were exposed for 1/2, 1, and 2 years beginning in March 1987. This exposure represents a hot subtropical climate.

Florida (black box): Test samples were exposed in Homestead, Florida according to ASTM G-7 at 5° south with a black box mounting as described in section 4.2.6 of G-7 . Replicates were exposed for 1/2, 1, and 2 years beginning in July 1989. This is the standard outdoor test method used to simulate an insulated, increased temperature application (i.e. a sealant applied to a building or automotive interior).

Accelerated Weathering

Q-U-V with UVA-340 Lamps: Test samples were exposed in a Q-U-V weathering tester per ASTM G-53 using UVA-340 lamps. The test cycle consisted of 8 hours of UV exposure at 65°C (149°F), alternating with 4 hours of condensation at 50°C (122°F). Specimens were exposed for 500, 1,000 and 2,000 hours.

Q-U-V with UVB-313 Lamps: Test samples were exposed in a Q-U-V weathering tester per ASTM G-53 using UVB-313 lamps. The test cycle consisted of 8 hours of UV exposure at 65°C (149°F), alternating with 4 hours of condensation at 50°C (122°F). Specimens were exposed for 500, 1,000 and 2,000 hours.

Xenon Arc: Test samples were exposed in a Ci 65 xenon arc device per ASTM G-26. The lamp used borosilicate inner and outer filters and was operated at .35 W/m2/nm at 340 nm. The test cycle consisted of continuous light with 18 minutes of water spray every 2 hours. The black panel temperature was 63°C (145°F) with a water spray temperature of 42°C (108°F). Specimens were exposed for 1,000 and 2,000 hours.

EXPOSURE VARIABLES

The damaging forces of UV light, moisture and temperature are different in each of the exposure methods. Each of these differences is examined to lend some understanding of the differences in test results.

Light Sources

Sunlight: The intensity and spectral distribution of sunlight is variable and constantly changing. The intensity and spectral distribution varies because of filtering from the air mass, cloud cover, pollution, etc. Because the sun is lower in the sky during the winter months, it is filtered through a greater air mass. Furthermore, the further from the equator the more filtering occurs. This filtering reduces the overall sunlight intensity, but more importantly, reduces the most damaging short wavelengths by a greater percentage than the longer wavelengths. For example, the intensity of UV at 320 nm changes about 8 to 1 from summer to winter. In addition, the short wavelength solar cut-off shifts from about 295 nm in the summer to about 310 nm in the winter. Consequently, materials sensitive to UV below 310 nm would degrade only slightly, if at all, during the winter months.

UVB-313: The UVB-313 lamp is the most widely used light source for the Q-U-V because it generally produces the fastest degradation. It produces fast degradation because it emits wavelengths shorter than those found in natural sunlight. These unnaturally short wavelengths, however, may cause unnatural degradation (i.e. degradation which does not occur outdoors). An example of this is yellowing of certain polymers which are known not to yellow outdoors. Special caution is necessary when using lamps which emit unnaturally short wavelengths when making comparisons between generically different materials. This is because different material types have different spectral response characteristics. Materials which are sensitive to these unnaturally short wavelengths may appear inferior in accelerated tests. When exposed to natural sunlight they may, or may not, be inferior. Thus using wavelengths shorter than natural sunlight to accelerate testing can lead to anomalous results.

UVA-340: The UVA-340 lamp was introduced in 1987 to improve correlation between Q-U-V testing and natural weathering. It has been tested extensively on both plastics and coatings and, in many cases, greatly improves correlation. This lamp is an excellent simulation of the most damaging wavelengths of sunlight and emits no unnaturally short wavelengths. Degradation from the UVA-340 lamp is generally slower than from the UVB-313 but the advantage is improved correlation with outdoor results.

Figure 1 shows how the spectral power distributions of the UVB-313 and UVA-340 compare to noon summer sunlight. As can be seen from figure 1, these lamps do not reproduce the entire solar spectrum, only the short ultraviolet wavelengths. This is effective because short wavelength UV causes almost all of the damage to exterior grade materials.

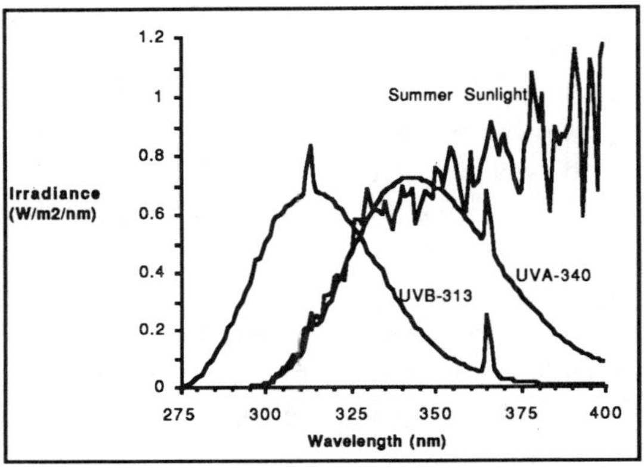

Fig. 1. Spectral power distributions of UVA-340 and UVB-313 lamps.

<u>Xenon</u>: Xenon lamps emit visible light and UV, including wavelengths much shorter than natural sunlight. Consequently, various optical filters are commonly used to eliminate these unwanted wavelengths. This study utilized the most common filter combination, borosilicate inner and outer filters. These filters eliminate most, but not all, of the unnaturally short wavelengths.

Figure 2 shows how the spectral power distribution of a xenon lamp with borosilicate filters compares to noon summer sunlight. As can be seen from figure 2, the irradiance increases throughout the UV portion of the spectrum and continues to increase well into the visible portion of the spectrum. Since the energy is spread across the entire spectrum, rather than concentrated in the UV region, a large amount of electrical power is required to drive the lamp in order to get reasonable intensity at any given wavelength. Electrical power constraints, therefore, necessitate lower intensity in the critical short wavelength UV region.

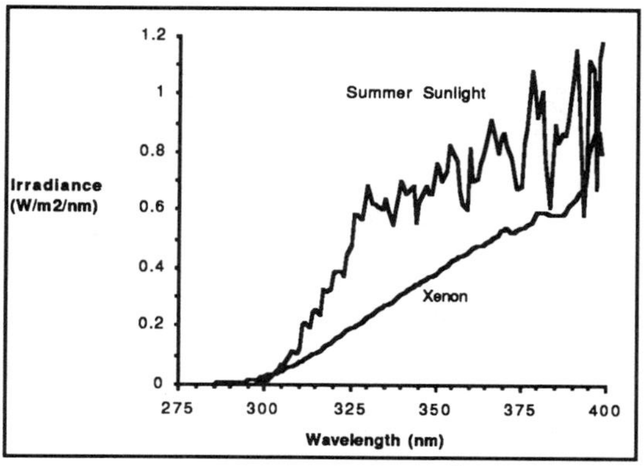

Fig. 2. Spectral power distribution of Xenon lamp with borosilicate filters.

Moisture

The second variable among the different exposure tests was the amount of time the samples were wet. As shown in figure 3 from data collected at the test site in Florida [2], materials were wet on the average of 45% of the time. Although accelerated weathering devices are able to produce such a high percentage of wet time it is common to use less moisture and spend a greater percentage of time in the UV cycle. This is based on the assumption that UV radiation causes greater degradation than moisture and therefore reduces test times. The Q-U-V test samples in this study were wet 33% of the time. The xenon test samples were wet 15% of the time.

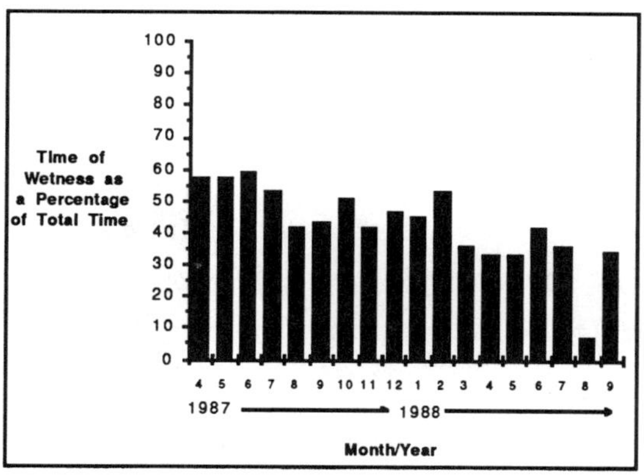

Fig. 3. Amount of time test samples were wet in Florida.

It is also significant to note that most of the wetness that a material encounters outdoors comes in the form of dew or condensation, not rainfall. Figure 4 shows the daily time of wetness compared to the amount of rainfall for a typical three week period in Florida. The fact that the days with the longest wet time had little or no rainfall indicates that rainfall has little to do with the amount of time a material is wet. This has significance for accelerated weathering tests because of the way moisture is produced. The Q-U-V uses relatively long (4 hour) cycles of condensation to reproduce the effects of naturally occurring moisture, whereas the xenon device uses short cycles (18 minutes) of water spray.

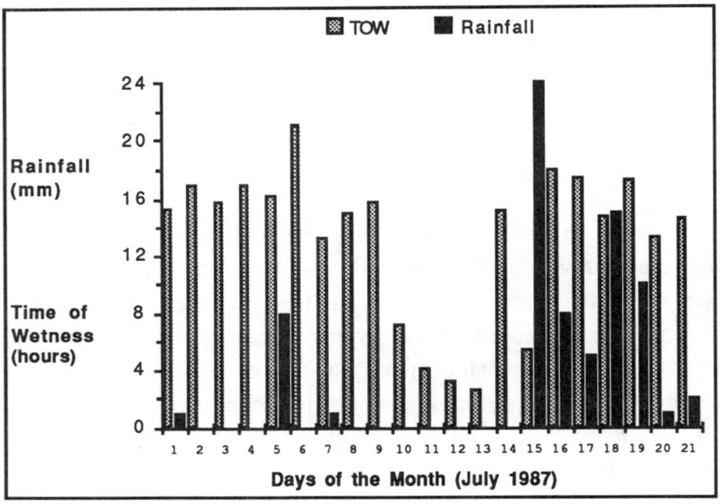

Fig. 4. Effect of rainfall on the time of wetness in Florida.

Temperature

Temperature is the third variable among the exposure conditions. There are several factors to consider when analyzing the thermal history of the test samples.

The most important consideration is the maximum temperature the test samples are subjected to. Accelerated testing is normally run at temperatures higher than the temperature a material normally experiences outdoors, but not so high as to cause abnormal degradation. This is done because higher temperatures usually increase the rate of degradation. The Florida samples mounted on an open rack (i.e. not insulated) reached a maximum surface temperature of 48°C (110°F). The Florida samples mounted on a black box (i.e. insulated) reached a maximum surface temperature of 54°C (129°F). These maximum temperatures were what the dark samples experienced. The lighter samples were as much as 10°C (18°F) cooler. The maximum temperature of the accelerated tests was 65°C (149°F). Consequently, the temperature of the test samples in the accelerated tests was significantly higher than the temperatures in the outdoor tests.

Another consideration is the rate and frequency of thermal cycling. The predominant temperature change for outdoor samples occurs gradually between day and night. The rate of change for the Q-U-V samples was also very gradual, because it takes about an hour for the chamber to reach equilibrium temperature after a cycle change. Xenon arc samples are exposed to direct water spray at the beginning of the moisture cycle, creating a high rate of temperature change.

The final consideration is the range of thermal cycling. Samples in the accelerated tests experienced a temperature excursion of 15 or 20°C between the UV and moisture cycle. Outdoor samples often have a daily temperature excursion of 30°C . This may have significance with respect to the amount of thermal expansion the test samples experienced.

RESULTS

The test samples from the six different exposure tests were compared by material type. The results include significant changes to the surface of the material and change in hardness. The outdoor test samples are compared to each other and the accelerated test samples are compared to each other. The correlation between the natural and accelerated weathering is then summarized for each material.

<u>One Component Urethane</u>

<u>Natural weathering</u>: Fine cracking was seen after 2 years in Cleveland but after only 1 year in Florida. The cracking became more severe after 2 years in Florida. No difference was found between the insulated and non-insulated Florida samples. (ref. fig.'s 5, 6, 7)

<u>Accelerated weathering</u>: All three accelerated test methods produced cracking in the test samples. The UVA-340 and UVB-313 tests produced fine cracking whereas the xenon tests produced a more coarse type of cracking. (ref. fig.'s 8, 9, 10)

<u>Correlation</u>: The fine cracking observed after 1 year in Florida was duplicated after 2000 hours using the UVA-340 and UVB-313 tests. The coarse cracking from the xenon test was not seen outdoors.

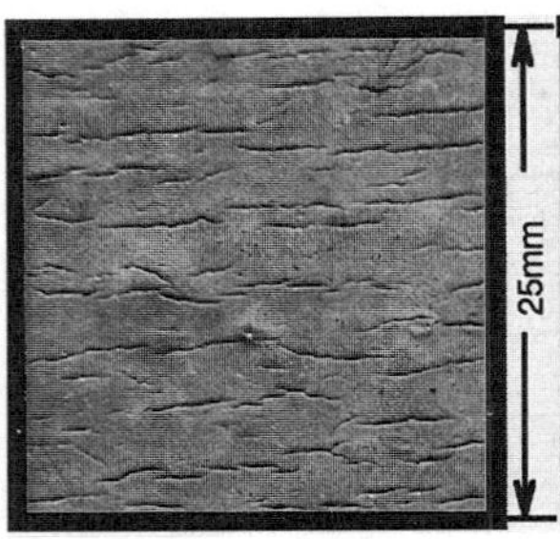

Fig. 5. Cleveland, 2 yrs.
One Component Urethane

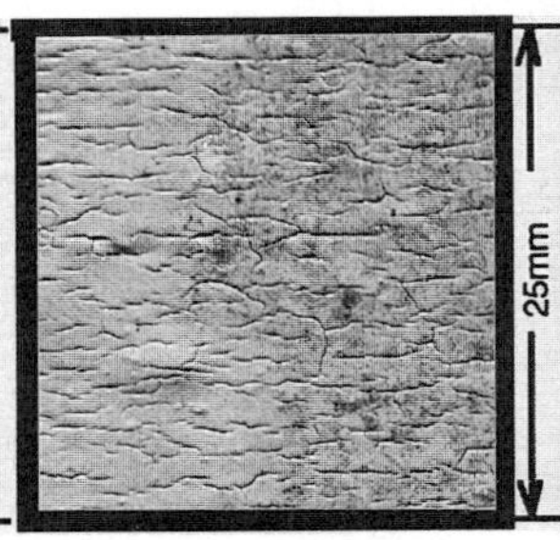

Fig. 6. Florida, 1 yr.
One Component Urethane

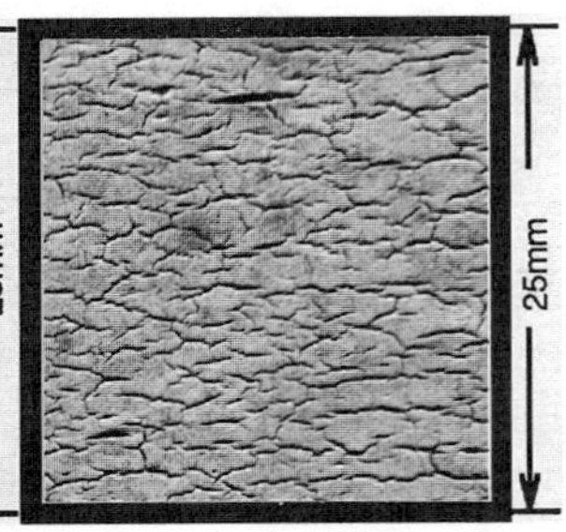

Fig. 7. Florida, 2 yrs.
One Component Urethane

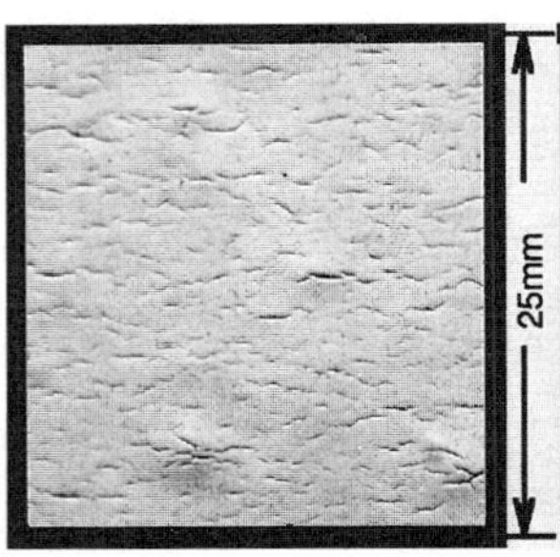

Fig. 8. UVA-340, 2000 hrs.
One Component Urethane

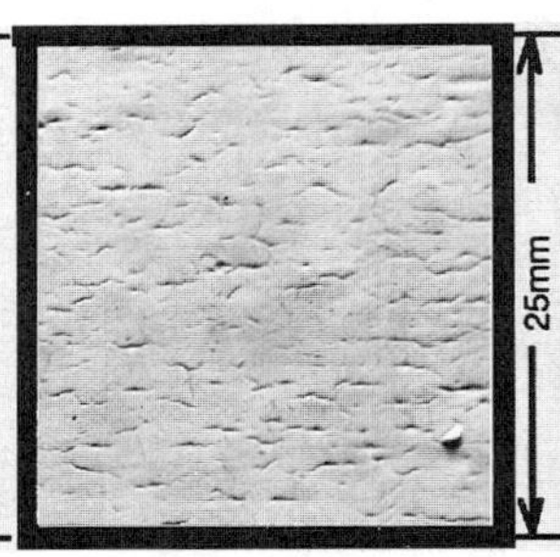

Fig. 9. UVB-313, 2000 hrs.
One Component Urethane

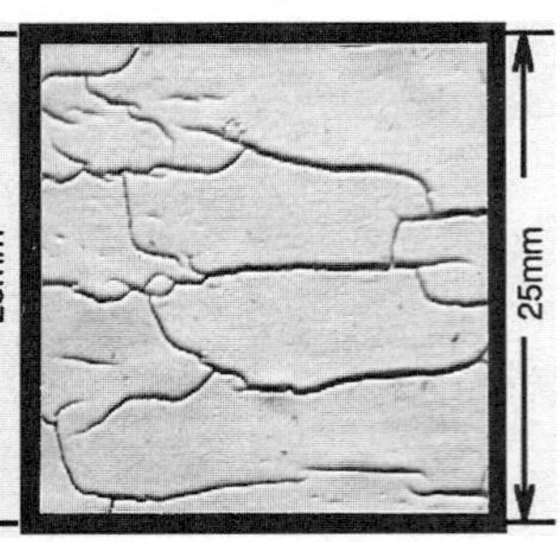

Fig. 10. Xenon, 2000 hrs.
One Component Urethane

Two Component Urethane

Natural weathering: Very minor cracking was seen after 1 year in Florida but severe cracking was seen after 2 years in Florida. No degradation was apparent after 2 years in Cleveland. The 1 year insulated samples experienced the same degree of cracking as the 2 year non-insulated samples. Blistering occurred on the insulated test samples but not on the non-insulated test samples. (ref. fig.'s 11, 12, 13, 14)

Accelerated weathering: All three accelerated test methods produced cracking and blistering on the test samples after 2000 hours. Very little cracking or blistering occurred after 1000 hours, except for the UVB-313, which showed some cracking but no blistering. (ref. fig. 15, 16)

Correlation: 2000 hours in the accelerated tests is slightly less severe than 2 years Florida exposure but much more severe than 1 year Florida exposure. UVA-340 and xenon tests should be run for 2000 hours to produce cracking. Cracking was visible after 1000 hours with the UVB-313 lamps.

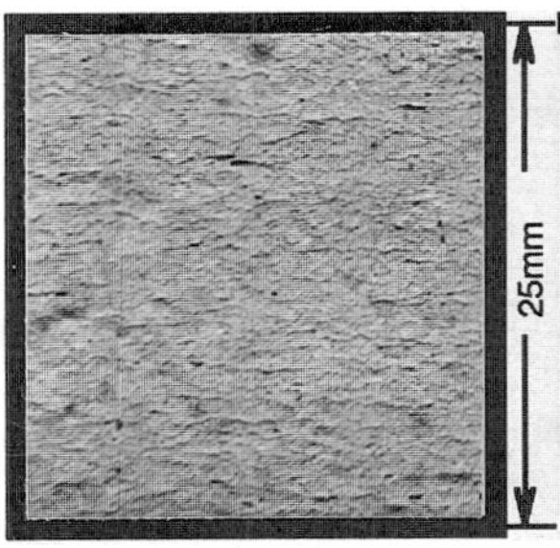

Fig. 11. Cleveland, 2 yrs.
Two Component Urethane

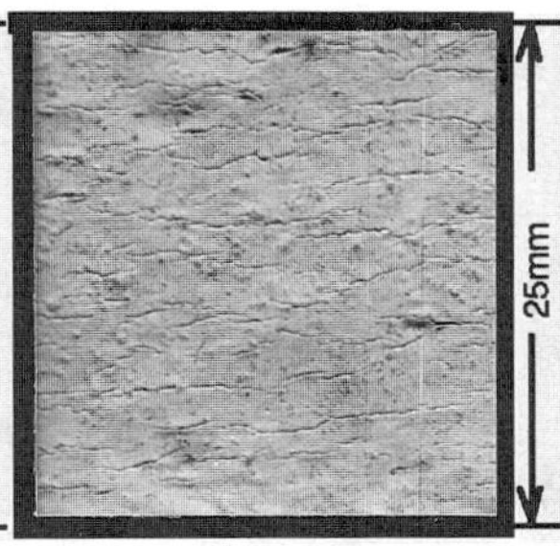

Fig. 12. Florida, 1 yr.
Two Component Urethane

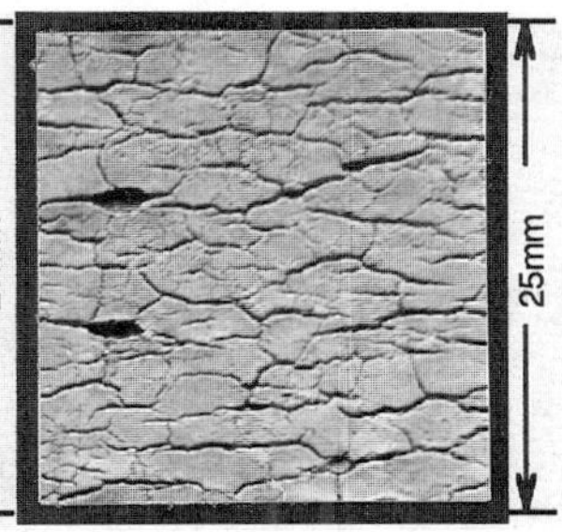

Fig. 13. Florida, 2 yrs.
Two Component Urethane

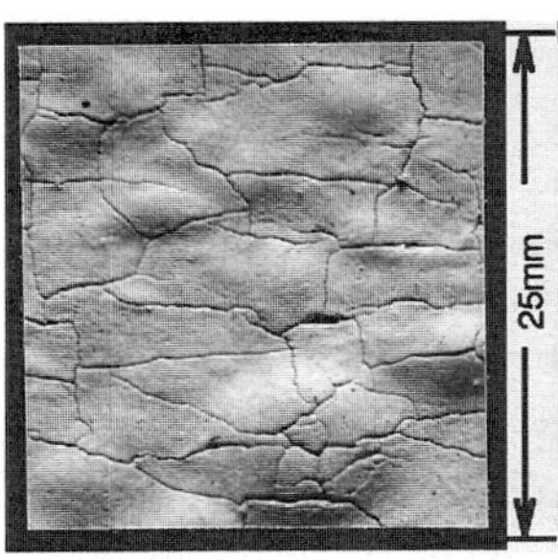

Fig. 14. Florida B.B., 1 yr.
Two Component Urethane

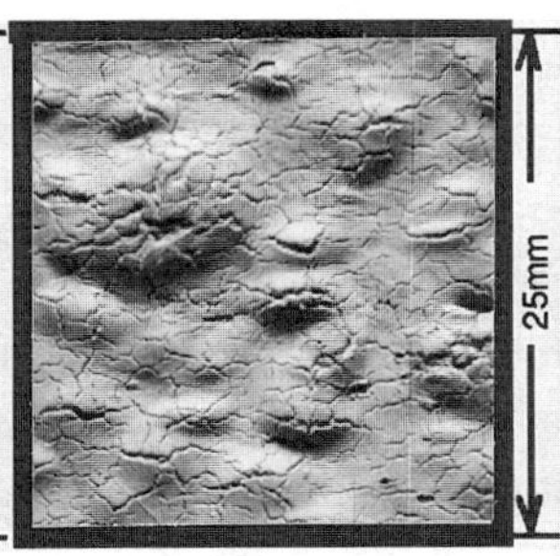

Fig. 15. UVA-340, 2000 hrs.
Two Component Urethane

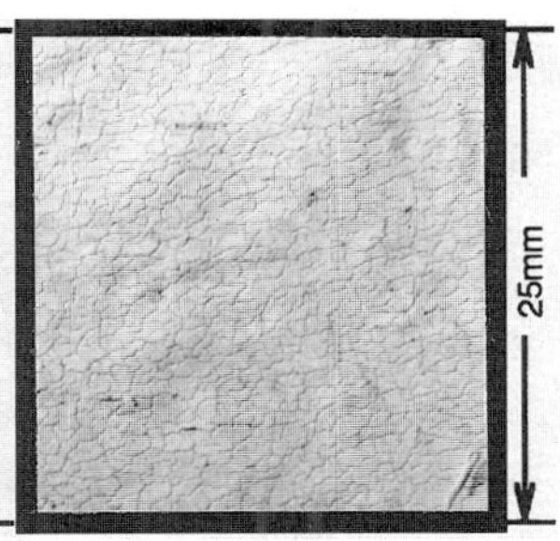

Fig. 16. UVB-313, 1000 hrs.
Two Component Urethane

Polyvinyl Acetate Latex

Natural weathering: The only visible change which occurred to the outdoor test samples was the growth of micro-organisms on the surface of the material. This occurred on both the Cleveland and Florida samples but much more so on the Florida samples. The samples also experienced degradation in the form of an increase in hardness. This increase in hardness was observed after 6 months with no further increase after that. There was no difference between the insulated and non-insulated samples. (ref. fig.'s 17, 18, 19)

Accelerated weathering: The only degradation due to the accelerated tests was a significant increase in hardness. The increase was observed after 500 hours with no further increase after that point. (ref. fig. 19)

Correlation: The increase in hardness found in the outdoor samples was duplicated in all of the accelerated tests. As in the outdoor tests, the increase in hardness occurred quickly and did not change after further exposure. The accelerated tests did not duplicate the micro-organism growth on the the outdoor samples.

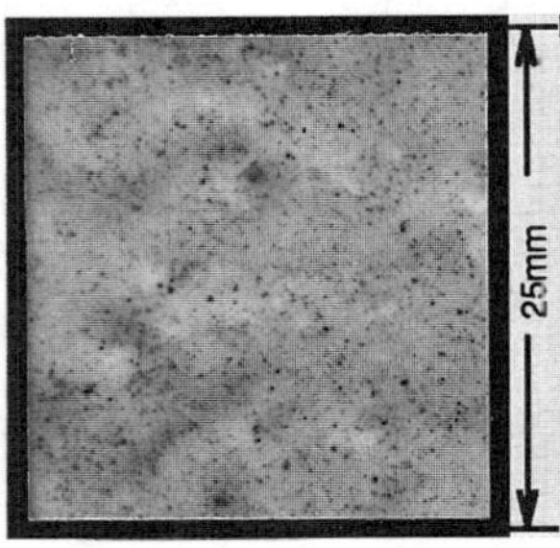

Fig. 17. Cleveland, 1 yr. Polyvinyl Acetate Latex

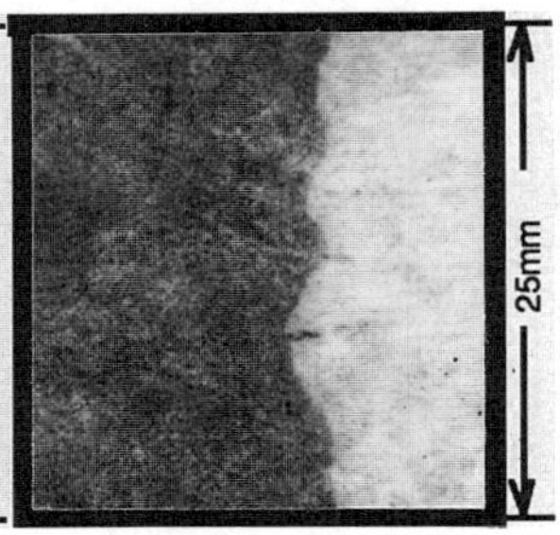

Fig. 18. Florida, 1 yr. Polyvinyl Acetate Latex

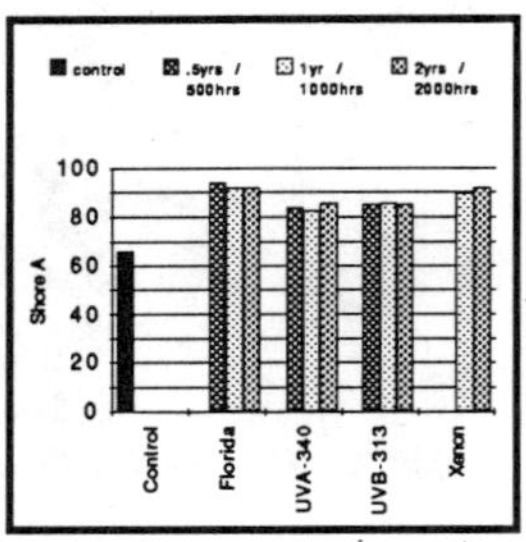

Fig. 19. Shore A Hardness. Polyvinyl Acetate Latex

Thermoplastic Rubber

Natural weathering: An alligator type skin formed on the surface of all the outdoor samples. The islands between the cracks started out fine but became more coarse with increased exposure time. The islands got darker while the cracks remained the original beige color. The Florida samples were somewhat more severe than the Cleveland samples. No difference was found between the insulated and non-insulated Florida samples. (ref. fig.'s 20, 21, 22)

Accelerated weathering: All three accelerated test methods produced cracking in the test samples. The UVA-340 and UVB-313 tests produced cracking which became worse with additional exposure time. The UVA-340 samples turned noticeably yellow and the UVB-313 samples turned severely yellow. An alligator type skin formed on the surface of the xenon samples. The islands between the cracks started out fine but became more coarse with increased exposure time. The islands got darker while the cracks remained the original beige color. (ref. fig.'s 23, 24, 25)

Correlation: The cracking observed outdoors after 2 years in Florida was duplicated after 2000 hours in each of the accelerated tests. The type of cracking produced by the xenon was more similar to the outdoor cracking. The UVA-340 and UVB-313 tests produced unnatural yellowing, the UVB-313 more so than the UVA-340.

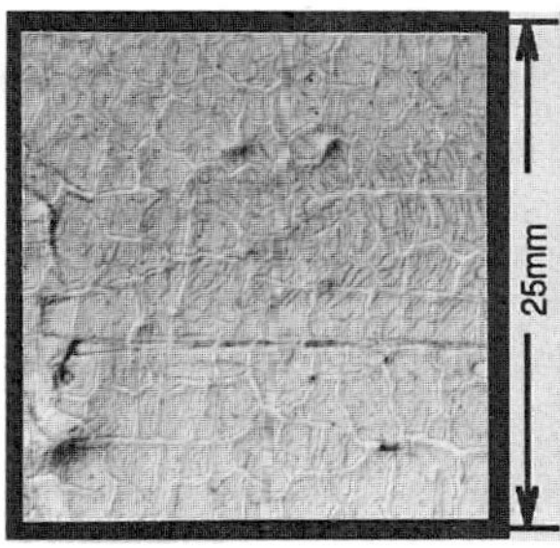

Fig. 20. Cleveland, 1 yr.
Thermoplastic Rubber

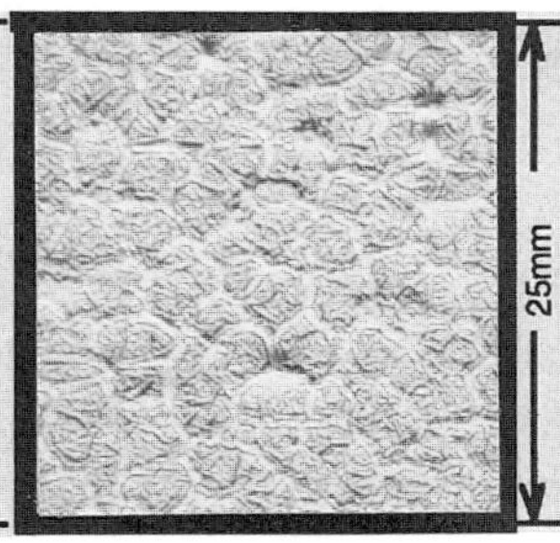

Fig. 21. Florida, 1 yr.
Thermoplastic Rubber

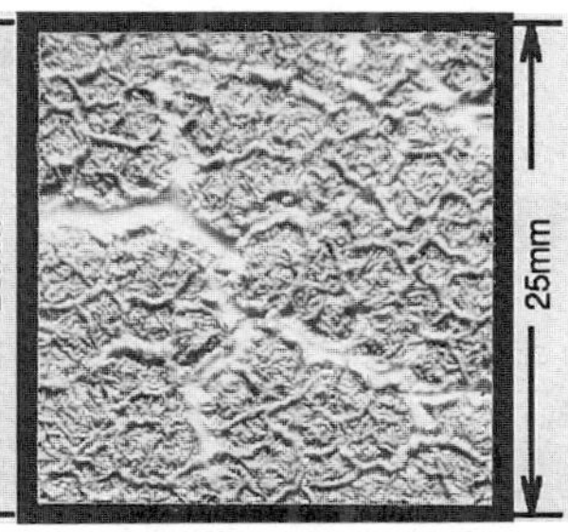

Fig. 22. Florida, 2 yrs.
Thermoplastic Rubber

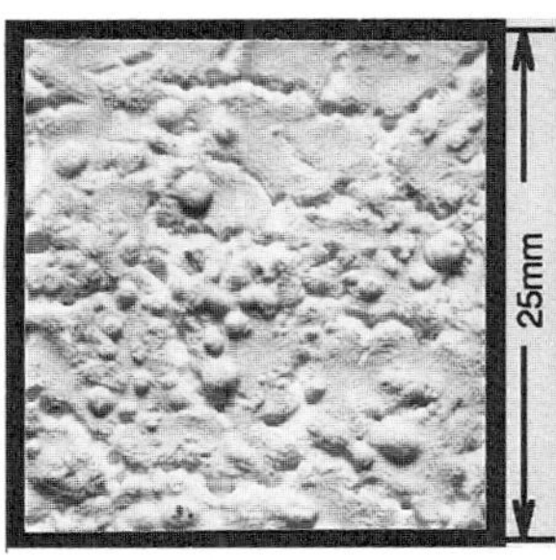

Fig. 23. UVA-340, 2000 hrs.
Thermoplastic Rubber

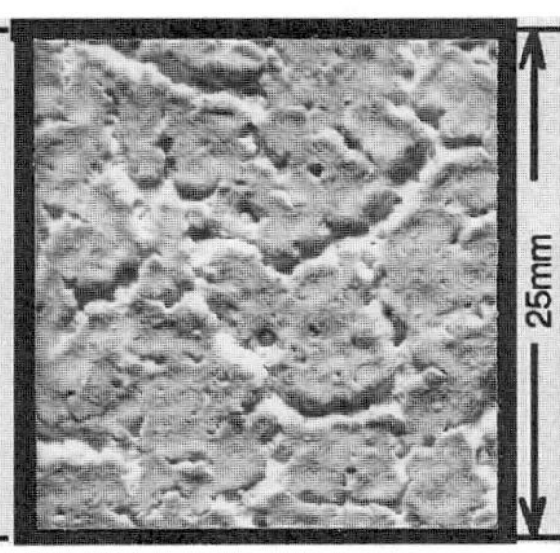

Fig. 24. UVB-313, 2000 hrs.
Thermoplastic Rubber

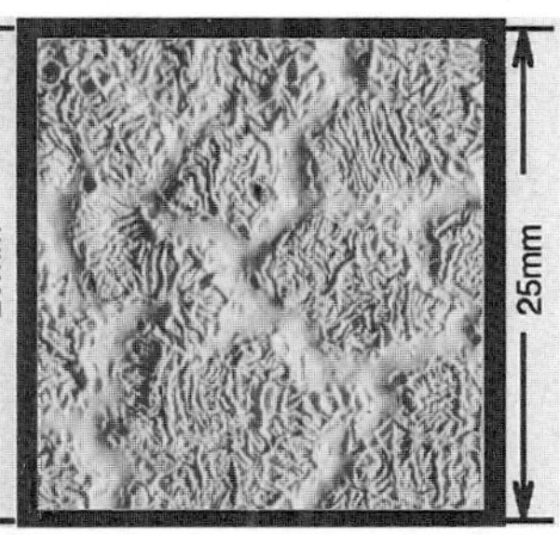

Fig. 25. Xenon, 2000 hrs.
Thermoplastic Rubber

Silicone (black)

Natural weathering: The only type of degradation apparent in any of the outdoor tests was a loss of gloss from the surface of the material. No other degradation of any kind was observed.

Accelerated weathering: The only type of degradation apparent in any of the accelerated tests was a loss of gloss from the surface of the material. No other degradation of any kind was observed.

Correlation: The only basis for correlating the accelerated and outdoor test results is gloss loss and that happened very quickly in both exposures. It was observed after 6 months outdoors and 500 hours in the accelerated devices. Longer exposure times are necessary if any other type of degradation is to be found.

Silicone (white)

Natural weathering: The only change in any of the outdoor test samples was that the surface became spotty from dirt and micro-organism growth which occurred in all of the tests with Florida being worse than Cleveland. No other degradation of any kind was observed. (ref. fig. 26)

Accelerated weathering: No degradation of any kind was observed.

Correlation: Since no degradation was observed in any of the test samples no correlation comparison can be made. Longer exposure times are necessary if any degradation is to be found.

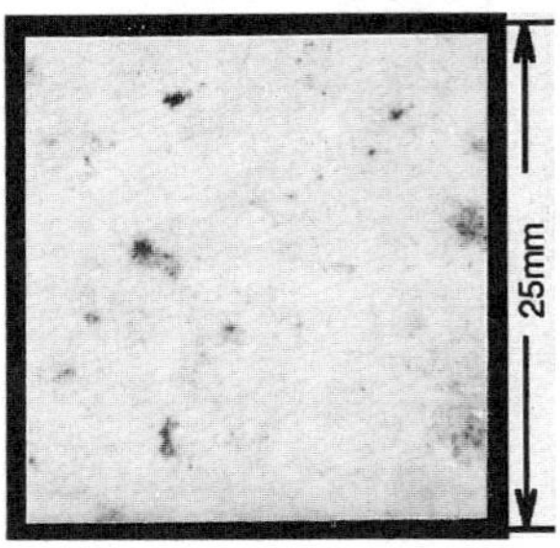

Fig. 26. Florida, 1 yr.
Silicone (white)

Acrylic - Solvent Type

Natural weathering: Micro-organism growth was present on all of the outdoor samples to an equal degree. The only other type of degradation found on any of the outdoor test samples was cracking on the 2 year insulated Florida samples. (ref. fig. 27)

Accelerated weathering: There was no micro-organism growth on any of the accelerated test samples. Blisters formed on all of the accelerated test samples. The UVA-340 and UVB-313 tests caused some relatively large blisters and the xenon test caused many small blisters. (ref. fig.'s 28, 29)

Correlation: The accelerated tests did not duplicate the micro-organism growth on the the outdoor samples. The blistering from the accelerated tests is not a natural type of degradation. It is probable that the accelerated test temperatures were excessively high.

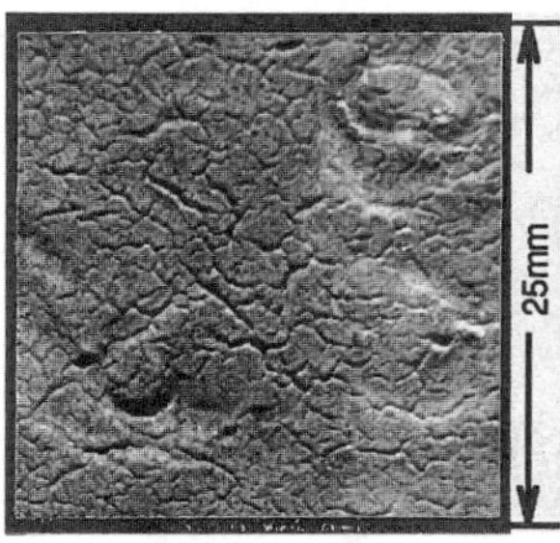

Fig. 27. Florida B.B., 2 yrs.
Acrylic - Solvent Type

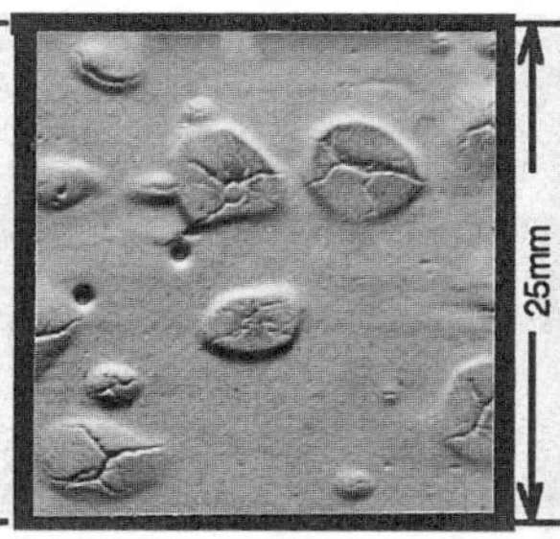

Fig. 28. UVA-340, 2000 hrs.
Acrylic - Solvent Type

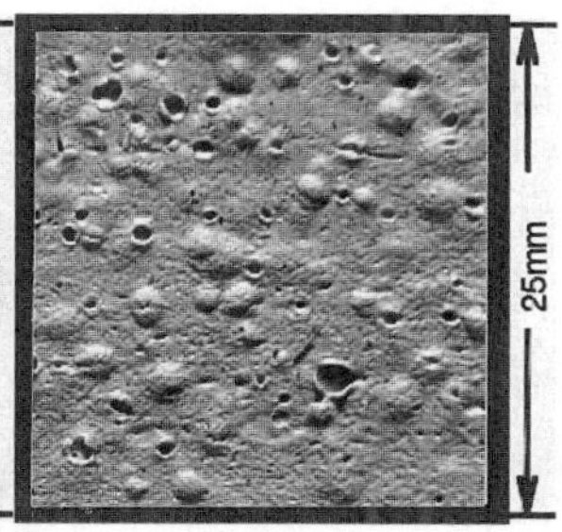

Fig. 29. Xenon, 2000 hrs.
Acrylic - Solvent Type

Acrylic - Latex

Natural weathering: Pitting occurred on the surface of all of the outdoor test samples and got worse with time. The severity after 1 year in Florida was equivalent to 2 years in Cleveland. The insulated samples had somewhat more severe pitting than the non-insulated samples. The test samples increased in hardness after 6 months with little increase after that. (ref. fig.'s 30, 31, 32)

Accelerated weathering: Pitting occurred on the surface of the UVA-340 test samples. Mild, shallow cracks formed on the UVB-313 and xenon test samples. The samples increased in hardness after 500 hours with little increase after that. (ref. fig.'s 32, 33, 34, 35)

Correlation: The pitting observed on the outdoor samples was duplicated only with the UVA-340 accelerated test. The severity at 2000 hours was the same as 2 years in Florida. The increase in hardness was duplicated with each of the accelerated tests. The cracking from the UVB-313 and xenon tests is not a natural type of degradation.

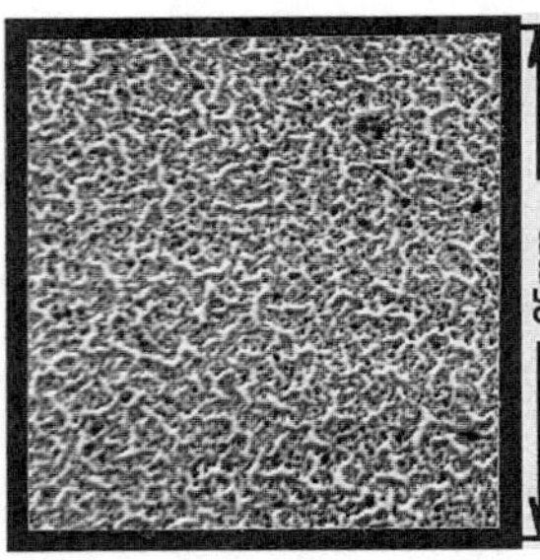

Fig. 30. Cleveland, 2 yrs.
Acrylic - Latex

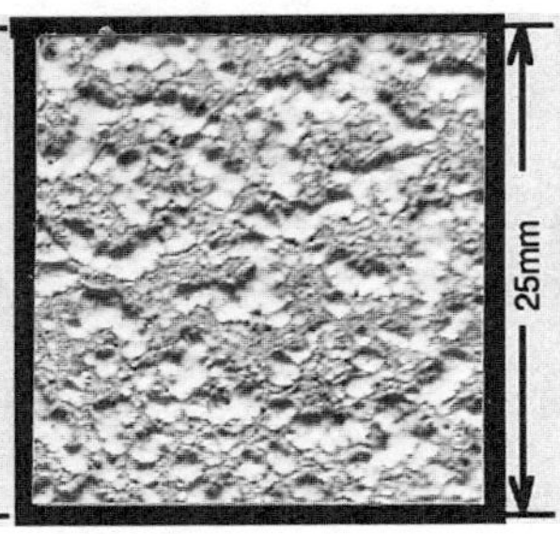

Fig. 31. Florida, 2 yrs.
Acrylic - Latex

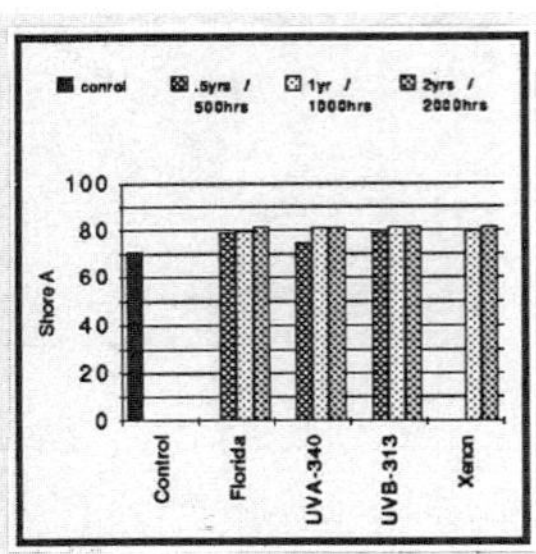

Fig. 32. Shore A Hardness.
Acrylic - Latex

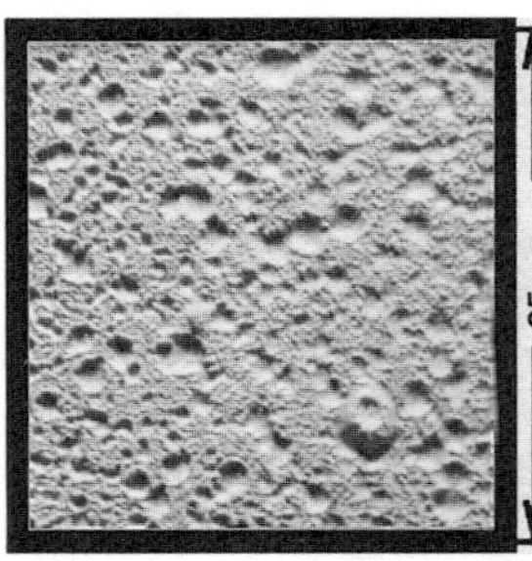

Fig. 33. UVA-340, 2000 hrs.
Acrylic - Latex

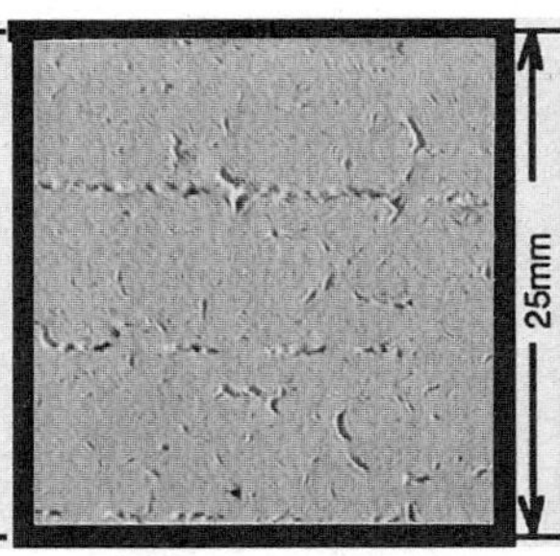

Fig. 34. UVB-313, 2000 hrs.
Acrylic - Latex

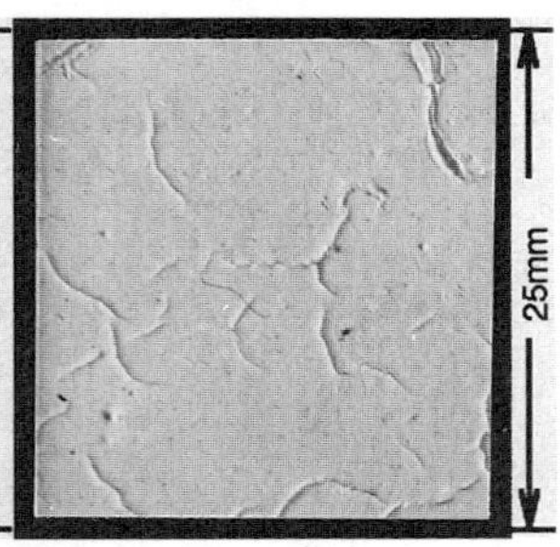

Fig. 35. Xenon, 2000 hrs.
Acrylic - Latex

Results Summary

degradation	natural exposures	accelerated exposures
cracking	3 materials experienced various types of cracking	same 3 materials experienced various types of cracking
pitting	1 out of 8 materials had surface pitting	the same material pitted but only from the UVA-340 test, the UVB-313 and xenon test did not produce pitting on any of the materials
blistering	1 material blistered but only from the black box exposure	2 materials blistered, 1 of which did not blister outdoors
color change	no color change on any of the 8 materials (4 of which were white)	no color change on any of the 8 materials except for the UVA-340 & UVB-313 tests which caused unnatural yellowing on 1 of the 8 materials
hardness	significant increase on 2 materials	significant increase on same 2 materials
micro-organisms, mildew, dirt, etc.	severe on 2 materials, mild on 4	none on any of the materials

CONCLUSIONS

The results of this study show that accelerated weathering test methods produce surface degradation typical of natural weathering in an accelerated manner and, therefore, can be useful in evaluating and developing better sealants. Each type of degradation produced by the three outdoor test environments was duplicated by one or more of the accelerated test methods with the exception of micro-organism growth, mildew, and dirt pick up. It should be noted, however, that since the scope of this study was limited to surface degradation and hardness, these conclusions do not apply to sealant performance factors such as adhesion to various substrates or cohesive failures under load.

The accelerated devices, on average, produced the equivalent of slightly less than two years of Florida degradation in 2000 hours. This acceleration rate applies only to the limited scope of this particular study. The acceleration rate will most likely be different when correlating other test cycles, materials, or properties such as movement capability. It is also possible that the acceleration rate will change as a function of exposure time. For instance, the degradation which occurs at exposure times longer than two years in Florida may continue at the same rate or may level off. The same is true for the degradation from the accelerated devices. As an extreme example, it may level off in Florida and continue in the accelerated devices. Therefore, a different acceleration rate is likely depending on what outdoor exposure time is used as the baseline. Nonetheless, for materials which are expected to last for years outdoors it would seem wise to test them for a minimum of 2000 hours in an accelerated device.

It is apparent that one accelerated test method is best for some materials and a different accelerated test method is better for other materials. This is because no one method can perfectly match all of the elements of natural weathering and because each material has its own unique weaknesses. The accelerated method which duplicates the elements to which a particular material is most susceptible (UV, moisture, or temperature) will produce the most realistic degradation. The results of this study show that for the particular formulas tested, the Q-U-V with UVA-340 lamps produced more realistic degradation than the xenon test method on the one component urethane and acrylic latex, but the xenon test produced more realistic degradation on the thermoplastic rubber. The Q-U-V with UVA-340 lamps produced more realistic degradation than the Q-U-V with UVB-313 lamps on the thermoplastic rubber and acrylic latex. In no case did the Q-U-V with UVB-313 lamps produce more realistic degradation than the Q-U-V with UVA-340 lamps.

The Q-U-V with UVA-340 lamps and the xenon arc device are therefore the perferred accelerated test methods of the three methods examined. Neither of these two test methods will guarantee that all types of degradation will always be discovered, but they did prove reliable for the materials tested in this study. For greater assurance that degradation from outdoor exposure is discovered before the material is in use, the material should be tested using both devices. It should be understood, however, that some unnatural degradation may occur from these accelerated tests. In other words, if a material is tested in one of these devices for a sufficient amount of time, there will usually not be any unpleasant surprises after the material has been in service for several years. There may, however, be pleasant surprises because certain materials may perform better than what the accelerated tests predicted.

The results also point out two areas where test procedures may be improved. The first is excessively high temperature in the accelerated test methods. Blistering of the two component urethane test samples in Florida was caused by insulating the samples and thereby increasing the temperature from 43°C to 48°C. Since blistering was caused by raising the temperature of this material, it is possible that the unnatural blistering of the solvent type acrylic material in the accelerated tests was caused by excessive temperature. The maximum temperature in the Florida black box for the solvent type acrylic was 53°C whereas in the accelerated tests it was 65°C. Therefore, a lower accelerated test temperature of, say, 55°C may be required to prevent the unnatural blistering of the solvent type acrylic.

The second improvement deals with accelerating the rate of cracking. All test samples in this study were free standing, meaning no stress was applied to the material. In many sealant applications stresses are applied to the material and the material moves as much as 25% or more. It has been reported by Sandberg [3] that applying tension to test samples during exposure greatly accelerates the rate of cracking in both outdoor and accelerated tests. Therefore if a material is intended for use in a moving joint it should be tested in a similar manner. This would preferably consist of alternating between an extended and compressed state, or as a minimum, continuous extension. Otherwise, the the test results may be less severe than what happens to the material undergoing cyclic movement in service. These improvements should be considered to further improve correlation and acceleration.

Lastly, two additional conclusions can be drawn concerning the outdoor exposures. The first is that the degradation from Florida exposures is approximately twice as fast as Cleveland exposures, and the second is, insulated exposures are slightly more severe than non-insulated exposures. The increased temperature from the insulated, or black box, test method accelerated the rate of degradation of three of the materials tested, and caused blistering on the two component urethane which wasn't caused otherwise. Since insulated applications of sealants are common in actual service, black box exposures should be considered for outdoor weathering tests.

REFERENCES

[1] Fedor, G. R., Brennan, P. J., "Correlation of Accelerated and Natural Weathering of Sealants", Journal of Adhesives Age, Vol. 33, No. 5, May 1990, pp. 22-27.

[2] DSET Laboratories, Inc., "Desert-ation, Climatological Data", Report, Vol. 32, No. 8, thru Vol. 34, No. 1.

[3] Sandberg, L. B., "Comparisons of Silicone and Urethane Sealant Durabilities", Journal of Materials in Civil Engineering, Vol. 3, No. 4, November 1991, pp. 278-291.

Joseph D. Tanzer and Larry S. Frankel

CLIMATE DRIVEN DURABILITY TESTER

REFERENCE: Tanzer, J. D., and Frankel, L. S., **"Climate Driven Durability Tester"**, Science and Technology of Building Seals, Sealants, Glazing, and Waterproofing: Second Volume, ASTM STP 1200, Jerome M. Klosowski, Ed., American Society for Testing and Materials, Philadelphia, 1992.

ABSTRACT: The "Climate Driven Durability Tester" (CDDT) tests channel movement capability of caulks and sealants under exposure conditions. A wide range of % movements, channel geometries, and substrates can be studied in the CDDT. Results are reported for a series of caulks exposed in aluminum and glass channels following 3, 7, and 16 months exposure. Adhesive failures noted in the present study do not completely correlate with ASTM C794 peel adhesion test results. The inability of ASTM C794 to distinguish between viscoelastic effects and adhesive bond strength may account for its unreliable prediction of adhesive performance in the CDDT.

KEY WORDS: exposure, caulk, sealant, acrylic, silicone, ASTM C794, channel movement, Climate Driven Durability Tester, artificial weathering

For most construction materials, laboratory experimental results provide a thorough understanding of how the materials will perform in an application. For example, mechanical property testing of a group of metals will determine the one capable of supporting the greatest load in high rise construction. Caulks and sealants present a far more complicated problem, as they are composed of viscoelastic polymers. Laboratory experiments must account for time-temperature responses to accurately predict performance in an exterior application. Consider a sealant used to patch a crack in the sidewalk. The crack will move according to temperature fluctuations at a variable rate, depending on the rate of temperature change. One day the temperature may drop slowly, causing the crack to open at 0.0001 m/h and the next day it may drop swiftly resulting in rapid crack movement at a rate of 0.01 m/h. The same sealant may in one case perform flawlessly, while in the other fail adhesively or cohesively. To predict the sealant's functional durability in the crack, one could study channel movement capability as a function

Dr. Tanzer is a research scientist and Dr. Frankel a Section Manager, both at Rohm and Haas Research Laboratories, 727 Norristown Road, Spring House, PA 19477.

of deformation rate and temperature. Considering the slow rates and low temperatures experienced in caulking applications, any laboratory testing would inherently involve tedious, time consuming measurements. For example, the most widely accepted laboratory test of caulk performance in the U.S. is ASTM C719 [1], otherwise known as the "Aymar Durability Test" or "Hockman Cycle". It attempts to measure performance at +/- 25% or +/- 12.5% joint movement at a slow deformation rate (.003 m/h) and specific combination of low and moderate temperatures. The test requires two months to complete at the stated conditions which do not account for varying deformation rates or temperature histories. While C719 represents a reasonable first approximation of a sealant's durability, its two month turn around time is not conducive to studies as a function of deformation rate and temperature.

Adhesive properties of sealants are usually judged by their performance in ASTM C794 [2], which tests 180 degree peel adhesion at a 3 m/h cross-head speed separation at standard laboratory conditions (23 ^{o}C, 50% humidity). While C794 does afford some idea of whether or not a material adheres to a given substrate, the present study will demonstrate the danger of using C794 to determine relative caulk adhesive performance. The test does not adequately account for testing rate, temperature, and viscoelastic nature of the sealant. However, C794 still remains the industry standard and there are no other good alternative tests available at this time.

The present paper does not suggest an alternative to C794 or C719, but instead offers a simple exterior exposure experiment that can be performed with the control of a laboratory type experiment. The Climate-Driven-Durability-Tester (CDDT), described in the Experimental Section and shown in Figures 1-4, enables mounting of a sealant channel over a building's expansion joint. The sealant must expand and contract in concert with the movement of the expansion joint as temperature and moisture conditions are controlled by the environment. Consequently, the CDDT affords a realistic test of sealant performance. The experimenter can vary substrate material, joint design (width to depth ratio), and +/- % joint movement by controlling the initial channel width. Of course, CDDT results depend on the specific weathering conditions encountered during exposure, preventing exact replication at a later date or in a different geographical location. In addition, the channel movement rate and amplitude depend on the weather, and as such cannot be preset. For these reasons, the CDDT is best used to determine relative sealant performance during a given exposure period.

Karpati [3-5] previously developed a strain cycling rack which bears similarity to the CDDT. The strain cycling rack is exposed to exterior conditions, and enables channel movement through temperature controlled expansion of thermally conductive metal bars. Like the CDDT, the specific amplitudes and rates of movement are determined by weathering conditions. The CDDT is considerably simpler in design, and as such much cheaper to construct. Karpati [6] has since developed an economical version of the rack, which relies on manual operation to effect channel movement. Manual control facilitates experiments as a function of precise channel movements, but doesn't simulate exact application conditions where slower, temperature induced movements result.

FIGURE 1 -- CDDT exposure racks mounted on exterior wall

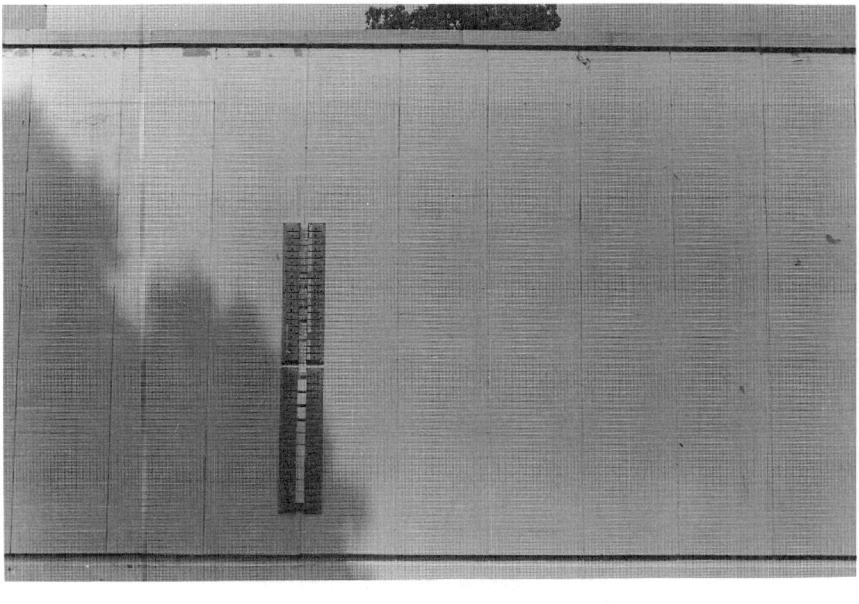

FIGURE 2 -- Closer view of CDDT exposure racks

EXPERIMENTAL

Climate-Driven-Durability-Tester

The CDDT consists of aluminum clamps mounted on aluminum panels that are positioned on either side of a building's expansion joint (Figures 1-4). After sample preparation, the channels were mounted with the 25.4 mm length parallel to the vertical expansion joint. Channel ends were clamped down firmly to the aluminum panels mounted to either side of the expansion joint. In the present study, all of the aluminum samples were mounted at the same time at an air temperature of 20 ^{o}C. Since the movement amplitude may be a function of sample position, three replicates of each caulk were spaced along the series. One replicate of each caulk was placed in the top, middle, and bottom third sections (Figure 5). The glass samples were mounted 8 months later at an air

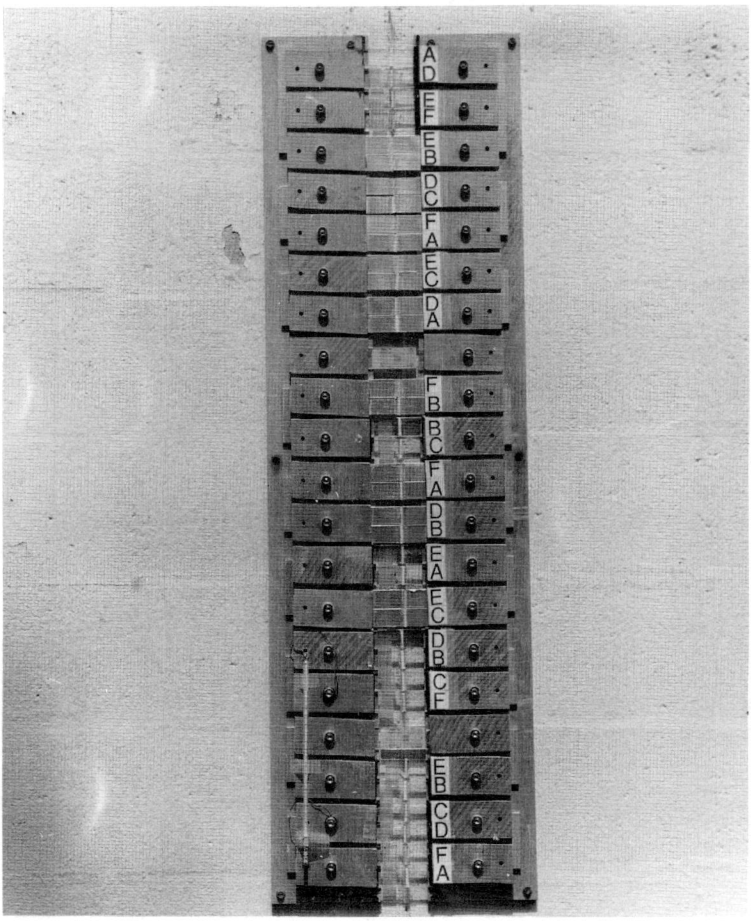

FIGURE 3 -- Current caulk exposures on glass and aluminum.

FIGURE 4 -- Glass channels mounted in CDDT.

temperature of 10 °C. Triplicate samples were again carefully arranged along the expansion joint (Figure 3). The concrete wall on which the CDDT is mounted faces southwest, and catches direct afternoon sun. The climate at the exposure site normally experiences lows in the -20 to -12 °C range during winter months, and highs around 38 °C in the summer. At 3 °C, a 17 °C drop from the initial exposure temperature, a gauge mounted in the center of the aluminum series indicated 30% channel movement (based on the initial 3.2 mm channel width). The measured expansion is less than the predicted value of 40% based on [7]

$$dC/dT = aW$$

where dC is the expansion joint movement effected by a change in temperature (dT), $a = 1.0 \times 10^{-5}\ {}^{o}C^{-1}$ [8] is the linear coefficient of expansion for concrete, and W = 7.6 m is the width of the concrete wall sections between expansion joints. At temperature extremes of -12 °C and 38 °C, the predicted values would be 77% expansion/43% compression. Similarly for the glass series exposed at 10 °C, the predicted range of movement would be 67% expansion/53% compression. These predicted values exceed the observed values of approximately 50% expansion/20% compression for the aluminum series and 40% expansion/35% compression for the glass series.

The aluminum series, exposed in April 1990, has experienced all

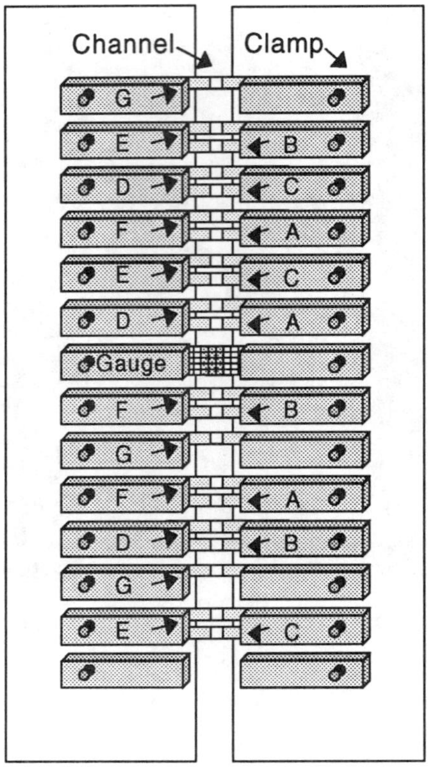

FIGURE 5 -- Aluminum channels mounted in CDDT

seasons at least once. The glass series, exposed in January 1991, has experienced a partial winter, full spring, and full summer of weathering.

Sample Preparation for Climate-Driven-Durability-Tester

In the first series, each sample was prepared as a sandwich between 2 pieces of anodized aluminum. The aluminum surfaces were cleaned with Methyl Ethyl Ketone and permitted to dry thoroughly before layup. Each channel was constructed by placing two 25.4 x 76.2 x 6.4 mm thick aluminum plates (identical to ASTM C719 substrate) next to each other on end to form a 6.4 mm deep by 25.4 mm long channel. In the present study, the plates were separated by 3.2 mm +/- 0.1 mm gap to form the sample width. The exit orifices of the caulk cartridges were modified with very narrow extensions to facilitate the flow of caulk into the narrow channels. Samples were cured 1 week at 23 °C and 50% relative humidity and carefully inspected for defects before mounting in the CDDT.

The second series was similar to the first, with glass substituted for aluminum as the substrate. Again, a Methyl Ethyl Ketone cleaning preceded layup in the 3.2 mm wide channels. Samples were cured 1 week at 23 °C and 50% relative humidity, inspected for defects, and mounted in the CDDT.

ASTM C794 Peel Adhesion Test

Testing was performed according to the C794 specification. Cure conditions were 1 week at 23 °C and 50% relative humidity, 1 week at 50 °C, and 1 week at 23 °C and 50% relative humidity. Wet peel adhesion was performed after a 1 week water soak in deionized water.

Fish Tanks - Wet Adhesion to Glass

Fish tanks were constructed out of 6.4 mm thick glass plates to the following dimensions: 0.15 x 0.3 m bottom, two 0.15 x 0.3 m sides, two 0.3 x 0.3 m sides. To construct each tank, the side plates were placed against the bottom plate and caulked in place from the inside with clamps used to support the glass during cure. The tanks were cured for 7 or 14 days at 23 °C and 50% relative humidity. After curing, the clamps were removed and the tanks filled to a height of 0.13 m with deionized water. The tanks were observed daily for any sign of leakage.

RESULTS AND DISCUSSION

A range of do-it-yourself caulks was tested including two RTV silicones (A,B), a silicone latex (C), an experimental acrylic (D) possessing elastomeric properties similar to silicones, a top-of-the-line acrylic (E) passing +/- 25% channel movement according to ASTM C920, an intermediate performing acrylic (F) passing +/- 12.5% per C920, and a low performance acrylic (G) These caulks were exposed on the CDDT to determine their relative performance in an exterior application venue.

CDDT Aluminum Series:

The aluminum substrate series was placed out in April and read after 8 months in December, and 16 months in August the following year. The series experienced maximum temperature drops of about 4 °C and 16 °C after 8 and 16 months respectively, resulting in maximum extensions of greater than 30%. Results reported in Table 1 identify the 25% acrylic (E) and the experimental acrylic (D) as the only survivors of 16 month exposure without failure. As expected, the low end acrylic G failed adhesively after 8 months, and the 12.5% acrylic developed center-line cracks in 2 of the 3 channels after 16 months. Both RTV silicones failed adhesively after 8 months, which is not surprising since most RTV silicones require a primer for use over aluminum. The silicone latex failed adhesively in middle and bottom channels.

TABLE 1 -- CDDT results on aluminum substrate

CAULK	POSITION	8 MONTH FAILURE[a,b]	16 MONTH FAILURE[a,b]
A (SILICONE)	TOP	100% A	100% A
	MIDDLE	100% A	100% A
	BOTTOM	100% A	100% A
B (SILICONE)	TOP	100% A	100% A
	MIDDLE	90% A, 10% C	90% A, 10% C
	BOTTOM	60% A	90% A, 10% C
C (SILICONE LATEX)	TOP	NONE	NONE
	MIDDLE	100% A	100% A
	BOTTOM	100% A	100% A
D (EXPT'L ACRYLIC)	TOP	NONE	NONE
	MIDDLE	NONE	NONE
	BOTTOM	NONE	NONE
E (25% ACRYLIC)	TOP	NONE	NONE
	MIDDLE	NONE	NONE
	BOTTOM	NONE	NONE
F (12.5% ACRYLIC)	TOP	NONE	NONE
	MIDDLE	NONE	30% CR
	BOTTOM	NONE	100% CR
G (ACRYLIC)	TOP	100% A	100% A
	MIDDLE	100% A	100% A
	BOTTOM	60% A	NOT TESTED

[a]% of channel length failing noted

[b]Failure key: A (adhesive), C (cohesive), CR (surface crack).

CDDT Glass Series:

The glass series, placed out in January, experienced a minimum temperature of about -12 °C for both the 3 and 7 month readings reported in Table 2. Only caulk D performed flawlessly over the 7 month period. Most of the other caulks failed in their top positions while surviving the middle and lower exposures without failure.. Based on these results, it is likely that the top of the exposure rack experiences the greatest movements. A lack of space prevented continuous placement of the glass samples along the exposure rack. As shown in Figure 3, large gaps within the top third and between the top and middle thirds of the series exist, which may contribute to the results as a function of position. Adhesive failures of the RTV silicones, the 12.5% acrylic, and 25% acrylic were unexpected since all are reputed to have excellent glass adhesion, and all pass the glass adhesion requirement of ASTM C920.

ASTM C794 Adhesion:

Aluminum dry and wet peel adhesion results are reported in Table 3. Dry peel results are best for caulks A, D, and E and relatively poor for

B. Wet peels are quite different, as sample G nearly floats off the substrate while A and C have relatively low values. Caulks B, D, E, and F all have comparably good wet adhesion values, and would pass the adhesion requirement of ASTM C920 (Federal Specification TTS230C) to aluminum (90 kg/m, cohesive failure).

Dry and wet peel adhesion results on glass are reported in Table 4. One of the RTV silicones has excellent dry and wet peel values, while the other RTV silicone has relatively low wet and dry peel ratings. The experimental acrylic (D) rates at the bottom for wet peel and at the top for dry peel. Caulks A, C, E, and F all have sufficient adhesion to pass the adhesion requirement of ASTM C920 to glass (90 kg/m, cohesive failure).

Comparison of Test Results:

The CDDT is representative of a true application test stressing adhesion and cohesion of caulks. ASTM C794 is a well accepted industry wide test of adhesion, and a critical requirement of ASTM C920. The CDDT results reported here do not completely correlate with the C794 results. With its vertical exposure the CDDT does not experience ponded water conditions as per wet peel testing described in C794, and should probably best correlate with dry peel adhesion results. Tables 5 and 6 compare dry peel adhesion with CDDT results. On aluminum, caulk A recorded the

TABLE 2 -- CDDT results on glass substrate

CAULK	POSITION	3 MONTH FAILURE[a,b]	7 MONTH FAILURE[a,b]
A (SILICONE)	TOP	50% A	70% A
	MIDDLE	NONE	10% C
	BOTTOM	50% A	70% A
B (SILICONE)	TOP	10% A, 90% C	10% A, 90% C
	MIDDLE	NONE	NONE
	BOTTOM	NONE	NONE
C (SILICONE LATEX)	TOP	NONE	30% A, 30% C
	MIDDLE	NONE	NONE
	BOTTOM	NONE	NONE
D (EXPT'L ACRYLIC)	TOP	NONE	NONE
	MIDDLE	NONE	NONE
	BOTTOM	NONE	NONE
E (25% ACRYLIC)	TOP	50% A, 50% C	50% A, 50% C
	MIDDLE	NONE	NONE
	BOTTOM	NONE	NONE
F (12.5% ACRYLIC)	TOP	20% A	40% A
	MIDDLE	NONE	NONE
	BOTTOM	NONE	NONE

[a]% of channel length failing noted

[b]Failure key: A (adhesive), C (cohesive), CR (surface crack).

TABLE 3 -- Adhesion to aluminum per ASTM C794

CAULK	DRY PEEL ADHESION[a,b] (kg/m)		WET PEEL ADHESION[a,b] (kg/m)	
A	378-432	CP	18	A
B	40	C	90-108	C
C	144-180	A	18	A
D	270-342	C	144	CP
E	342	CP	72-90	CP
F	198-234	CP	108-126	C
G	144-288	CP	2	A

[a]Range reported represents high and low values measured

[b]Failure modes: A (adhesive), C (cohesive), CP (cohesive peak)

TABLE 4 -- Adhesion to glass per ASTM C794

CAULK	DRY PEEL ADHESION[a,b] (kg/m)		WET PEEL ADHESION[a,b] (kg/m)	
A	455-464	C	225-261	CP
B	94-106	C	45-90	C
C	261-266	CP	99-106	F
D	454-464	CP	61-76	A
E	331-360	CP	81-101	CP
F	130-140	LC	113-121	CP

[a]Range reported represents high and low values measured

[b]Failure modes: A (adhesive), C (cohesive), CP (cohesive peak), F (fabric delamination)

highest dry peel value, yet failed adhesively in the CDDT. On glass caulks A and D had comparable C794 adhesion, but caulk A failed adhesively while D experienced no failure. Similarly, wet peel results do not completely correlate with the CDDT (Tables 7 and 8). Sample B, exposed on aluminum, fails the CDDT while its C794 value is comparable to two other samples that pass. Perhaps the best illustration is sample D, which records the lowest wet peel to glass value but was the only 7 month survivor on the CDDT.

To understand the apparent contradiction, consider the factors at work in each of the experiments. In the CDDT channel geometry, weather conditions, caulk modulus as a function of temperature, caulk shrinkage, caulk adhesive bond strength to substrate, and caulk viscoelasticity all contribute to performance.

Weather conditions: Control the number of deformation cycles, rate of deformation, amplitude of deformation, and other intangibles like moisture content in the sample during deformation, polymer degradation from UV exposure, etc.

Channel geometry: Affects stress distribution when the channel experiences expansions and contractions. For example, a rectangular butt

TABLE 5 -- Aluminum adhesion results comparison

CAULK	DRY PEEL ADHESION (kg/m)		CDDT ADHESIVE FAILURE
A	378-432	CP	YES
E	342	CP	NO
D	270-342	C	NO
F	198-234	CP	NO
C	144-180	A	YES
B	40	C	YES

TABLE 6 -- Glass adhesion results comparison

CAULK	DRY PEEL ADHESION (kg/m)		CDDT ADHESIVE FAILURE
A	455-464	C	YES
D	454-464	CP	NO
E	331-360	CP	YES
C	261-266	CP	YES
F	130-140	C	YES
B	94-106	C	YES

joint whose width/depth is 2/1 will produce much smaller stresses than one with width/depth = 1/2.[9] Additionally, the joint will encounter end effects if the length to cross sectional area (L/A) is too small.[10] In the present study L/A = 1.3 mm^{-1}, which may produce some end effects. By comparison, these end effects should not be as significant as those encountered in ASTM C719, where L/A = 0.3 mm^{-1}.

Caulk modulus: Determines the forces transmitted to the caulk-substrate boundary during deformation. A high modulus caulk transmits high forces to the boundary, and as such requires a better adhesive bond than a low modulus version. At -18 °C for example, Young's modulus for RTV silicones can be as much as 50 times smaller than a low performance acrylic. Based on modulus arguments alone, the low end acrylic will have a 50X greater adhesion requirement than the silicone.

Shrinkage (concave profile): Influences stress distribution during deformation. Consider again the low end acrylic and RTV silicone. The silicone does not shrink leaving a constant thickness along the deformation (width) axis . Acrylics shrink so that the center becomes the thinnest section. A caulk will accomplish deformation at its weakest (thinnest) point. For the acrylic, the greatest amount of width axis deformation occurs in a volume element at the center of the channel, and the smallest width deformations occur near the boundary. The initial silicone deformation is essentially homogeneous, with all volume elements deforming equally. As such the silicone undergoes greater deformation near the substrate boundary than the acrylic. The essentially incompressible nature of caulks and sealants requires that a deformation in the width dimension be accompanied by deformations in the thickness and/or longitudinal dimensions. At the boundary these deformations are restricted, resulting in the build up of high stresses. Shrinkage arguments alone would therefore subject the silicone to greater boundary

TABLE 7 -- Aluminum adhesion results comparison

CAULK	WET PEEL ADHESION (kg/m)	CDDT ADHESIVE FAILURE
D	144 CP	NO
F	108-126 CP	NO
B	90-108 C	YES
E	72-90 CP	NO
A	18 A	YES
C	18 A	YES

TABLE 8 -- Glass adhesion results comparison

CAULK	% SOLIDS (WEIGHT)	WET PEEL ADHESION (kg/m)	CDDT ADHESIVE FAILURE
A	94	225-261 CP	YES
F	81	113-121 CP	YES
C	81	99-106 F	YES
E	85	81-101 CP	YES
B	96	45-90 C	YES
D	60	61-76 A	NO

stresses and poorer adhesive performance than the acrylic. This argument is consistent with simulated stress calculations performed by Meyers [9] and Catsiff et al [11], from which a concave profile is shown to produce lower boundary stresses than its rectangular counterpart. It is also noteworthy that curved backer rods are now recommended for use with non-shrink sealants, presumably to impose a concave surface and reduce stresses.[12]

Caulk adhesive bond strength: Adhesive failure occurs when the bond strength is exceeded.

Caulk viscoelasticity: The viscoelastic response to an applied stress varies for different caulks. At a given temperature, a caulk with a strong viscous response will have an easier time relieving stress than one with a strong elastic response. However, the strong elastic response is favorable for avoiding fatigue failure from repetitive stress cycles.

In the present study, sample geometry and weather conditions were constant for all caulks tested, leaving shrinkage, adhesive bond strength, modulus, and viscoelasticity as possible factors contributing to the results. The adhesive bond strength could not be determined in our laboratories. As noted below, peel adhesion testing (ASTM C794) failed to sort out viscoelastic and modulus effects from the adhesive bond contribution, and therefore did not accurately reflect relative adhesive strengths. Lack of shrinkage acted as a negative influence for silicones, while the relatively low modulus of silicones at low temperatures was a positive influence. The elastomeric nature of silicones prevented viscous stress relief during deformation, which increased internal stresses. Similarly, the elastomeric structure of the experimental acrylic (D) prevented stress relief through viscous flow. Sample D did benefit from its substantial shrinkage, reducing the stresses at the adhesive boundary by concentrating the deformation into the center of the channel. The elastomeric nature of D enabled it to withstand repeated strain cycling without failure while preserving microscopic structure. As such, D was well suited to withstand cohesive fatigue at the center of the channel. Samples E, F, and G enjoyed the shrinkage benefit but had relatively high modulii that tended to increase upon aging as plasticizer was leached out of the system. Unlike A, B, and D, these samples relieved stress through viscoelastic flow, but were more susceptible to cohesive failure from fatigue.

ASTM C794 actually measures a combination of adhesive bond strength, viscoelasticity, and modulus. The schematic in Figure 6 illustrates the deformation of a caulk in the C794 test. Initially, as the strip is pulled the full length, from the top of the strip to the substrate, is under stress. What happens next depends on the viscoelastic nature of the caulk. A caulk displaying a strong viscous component will tend to flow in the region between the bent portion of the strip and the substrate, preventing stress build up at the adhesive boundary. The test then becomes a tensile test with cohesive failure the likely outcome. Does the result provide any information about the adhesive bond strength? No, because we cannot sort out the adhesive bond contribution to the measured force. If the caulk fails cohesively at 90 kg/m can we at least conclude that the adhesive bond strength is greater than 90 kg/m? The answer is again no. The reported units for the force are kg/m, requiring the caulk to fail adhesively or cohesively at the instant the strip begins to pull. If the caulk failure lags behind, which is typically the case, the force will reflect contributions from more than the linear profile contacting the deformation region. Remember, the force recorded

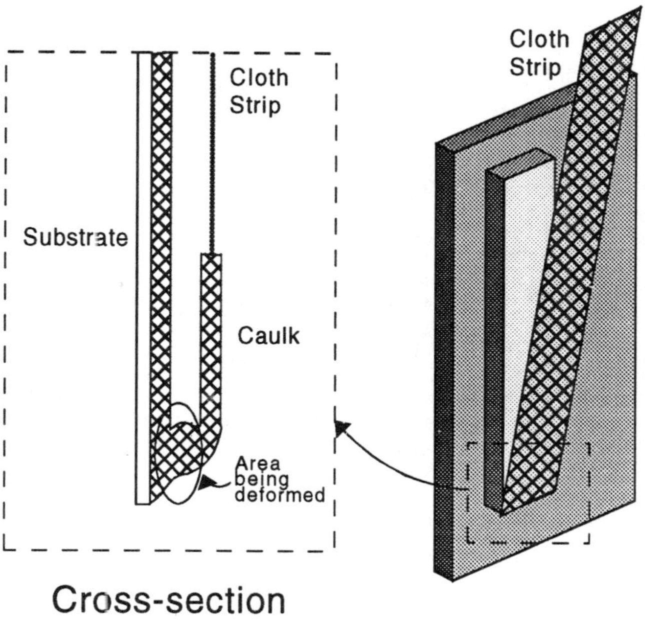

FIGURE 6 -- ASTM C794 peel adhesion mechanics

TABLE 9 -- Wet peel adhesion to glass vs. fish tank results

CAULK	CURE	FISH TANK FAILURE[a] 1 DAY	FISH TANK FAILURE[a] 2 DAY	FISH TANK FAILURE[a] 4 MONTH	WET PEEL ADHESION (kg/m)	
A	7 DAY	NONE	NONE	NONE	225-261	CP
F	7 DAY	FAIL	FAIL	FAIL	113-121	CP
	14 DAY	NONE	FAIL	FAIL		
C	7 DAY	NONE	FAIL	FAIL	99-106	F
E	7 DAY	NONE	FAIL	FAIL	81-101	CP
	14 DAY	NONE	FAIL	FAIL		
B	7 DAY	NONE	NONE	NONE	45-90	C
D	7 DAY	NONE	NONE	NONE	61-76	A

[a]Fish tank failure indicates complete loss of water

in a tensile test is proportional to the cross-sectional area and makes sense only if reported in stress units (kg/m^2). The measured force will be proportional to the cross-sectional area being deformed, which will differ from one material to another depending on their particular viscoelastic characteristics. The measured force will also be influenced by the modulus of the caulk. To determine the adhesive bond strength, one would have to sort out all of these factors, a task well beyond the scope of this paper.

Fish Tanks:

The fish tanks were constructed to further test the validity of ASTM C794 as a measure of adhesive bond strength. The water soak environment in the tanks enabled direct comparison with wet peel adhesion to glass results (Table 9). The fish tank results indicate no failure for the RTV silicones (A,B) and the experimental acrylic (D). The silicone latex (C) and other acrylics all failed adhesively along the bottom of the tank within 2 days. To achieve adhesive failure, the adhesive bond strengths must have been exceeded. Viscoelastic effects may have influenced the results, since the caulks were under stress from the weight of the glass side panels. However, viscoelastic contributions to adhesive failure would be strongest for RTV silicones and the experimental acrylic, the only caulks to avoid failure.

In conclusion, ASTM C794 is not an accurate measure of relative adhesive performance. Sample D was the only acrylic survivor in the fish tanks even though it could not pass the wet peel adhesion to glass requirements of C920, greater than 90 kg/m per C794 testing. Samples E and F both passed the adhesion requirement of C920 to glass, yet both failed in the fish tanks within 2 days.

CONCLUSIONS:

The Climate-Driven-Durability-Tester provides a means for distinguishing relative channel movement abilities within a series of caulks. By controlling the initial channel width, a wide range of %

movements is possible. In the present series, differences within the acrylic family and between acrylics and silicones were demonstrated.

Peel adhesion testing per ASTM C794 cannot determine adhesive bond strength because other contributing factors such as viscoelasticity and deformation mode confound the results. For this reason, C794 does not completely correlate with results from the CDDT and fish tank applications.

Finally, according to CDDT test results, the experimental acrylic outperforms silicones and other acrylics with respect to channel movement capability on glass and aluminum. According to fish tank results, the new acrylic outperforms other acrylics as a seal to glass in a water soak environment which may correlate with performance in a tub-n-tile application. To date the experimental acrylic has equalled RTV silicone performance in the fish tanks.

ACKNOWLEDGMENTS

The authors are grateful to Dr. John Van Curen for computer generation of Figures 5 and 6, and Mr. Cary Moore and Mr. Jeff Panera for the photographs in Figures 1-4.

REFERENCES

[1] American Society for Testing and Materials, "Standard Test Method for Adhesion and Cohesion of Elastomeric Sealants Under Cyclic Movement," 1991 Annual Book of ASTM Standards, Vol. 04.07, Section 4, S. P. Canning et al, editors, 1991, pp 58-62.

[2] American Society for Testing and Materials, "Standard Test Method for Adhesion-in-Peel of Elastomeric Joint Sealants," 1991 Annual Book of ASTM Standards, Vol. 04.07, Section 4, S. P. Canning et al, editors, 1991, pp 95-97.

[3] Karpati, K. K., Solvason, K. R., and Sereda, P. J., "Weathering Rack for Sealants," Journal of Coatings Technology, Vol. 49, No 626, March 1977, pp 44-47.

[4] Karpati, K. K., "Quick Weathering Test for Screening Silicone Sealants," Journal of Coatings Technology, Vol. 56, No 710, March 1984, pp 29-32.

[5] Karpati, K. K., "Exposure Evaluation of Sealants with Low Movement Capability," Adhesives Age, May 1988, pp 20-23.

[6] Karpati, K. K., "Device for Weathering Sealants Undergoing Cyclic Movements," Journal of Coatings Technology, Vol. 50, No 641, June 1978, pp 27-30.

[7] Elfring, W. H. Jr. and Rosano, W. J., "Climate and Performance of Elastomeric Wall Coatings," Science and Technology of Building Seals, Sealants, Glazing, and Waterproofing: Second Volume, ASTM STP 1142, Jerome M. Klosowski, editor, Philadelphia, 1992.

[8] Shortley, G. and Dudley, W., Elements of Physics, Englewood Cliffs Prentice Hall Co., 1971, p 71.

[9] Myers, J. C., "Behavior of Fillet Sealant Joints," Building Sealants: Material, Properties and Performance, ASTM STP 1069, Thomas F. O'Connor, Ed., American Society for Testing and Materials, Philadelphia, 1990.

[10] Catsiff, E. H., Hoffman, R. F., and Kowalewski, R. T., "Predicting Joint Sealant Performance of Elastomers by Computer Simulation. II. Results in Simple Extension and Compression," Journal of Applied Polymer Science, Vol. 14, 1970, pp 1159-1178.

[11] Catsiff, E. H., "Predicting Joint Sealant Performance of Elastomers by Computer Simulation. III. Simulation of Single- and Multi-Step Extension of a Stress-Relaxing Material," Journal of Applied Polymer Science, Vol. 15, 1971, pp 1021-1028.

[12] American Society for Testing and Materials, "Standard Guide for Use of Elastomeric Joint Sealants," 1991 Annual Book of ASTM Standards, Vol. 04.07, Section 4, S. P. Canning et al, editors, 1991, p 158.

Todd J. Bridgewater[1] and Lawrence D. Carbary[1]

ACCELERATED WEATHERING AND HEAT STABILITY OF VARIOUS PERIMETER SEALANTS

REFERENCE: Bridgewater, T. J., and Carbary, L. D., **"Accelerated Weathering and Heat Stability of Various Perimeter Sealants"**, Science and Technology of Building Seals, Sealants, Glazing, and Waterproofing: Second Volume, ASTM STP 1200, Jerome M. Klosowski, Ed., American Society for Testing and Materials, Philadelphia, 1992.

ABSTRACT: All perimeter sealants do not react the same when exposed to long term UV and extreme heat caused by the environment. The resistance to weathering is achieved by the inherent nature of the sealant polymer system. Some manufacturers achieve some degree of resistance to weathering by the use of fillers which shade the polymer from the UV and others use chemical additives to absorb the UV radiation and protect the polymer system. This paper will look at combinations of deteriorating mechanisms and give some indication as to the key factors causing the deterioration or sealant changes with the sealants studied. Some urethanes were shown to revert back into their original liquid state during UV exposure where the silicones did not. By the use of accelerated weathering lab equipment and lab ovens, the detrimental physical effects were demonstrated on various two-part urethane and one-part silicone perimeter sealants currently used in construction. Tensile adhesion data, durometer and visual observations are reviewed following a specific exposure time. This testing evaluation showed a faster sealant degradation with heat and ultraviolet light when compared to cyclic UV and condensation cycles with UVB313 bulbs. No significant degradation occurred with samples exposed to 100% humidity at 30°C (86°F) or 50°C (122°F) with the exception of loss of adhesion to the precast concrete. This study provides the framework of a test method to screen perimeter sealants for durability and reversion resistance.

KEYWORDS: Reversion, urethane, silicone, QUV, UV, weathering, durometer, humidity

[1]Construction Sealants Technical Service and Development Representative, Technical Service and Development Specialist, respectively, Dow Corning Corporation, 2200 W. Salzburg Road, Midland, MI 48686.

INTRODUCTION

Perimeter sealants face many extreme conditions depending upon which building face they are placed on, type of substrate and geographic region of the country. Therefore various forms and degrees of temperature, UV exposure, humidity are addressed in this paper to assist in understanding the effect the environment has on perimeter sealant materials in working joints. This study involved seven different perimeter sealants on fourteen time specific tests and ultimately yielded over 580 specimens.

EXPERIMENTAL METHOD

Seven perimeter sealants currently available in the marketplace and used extensively were prepared per ASTM C 1135 on pre-cast concrete samples prepared per Federal Specs SS-S-200D both with and without the manufacturers recommended primer for pre-cast concrete. See Table #1 for specific perimeter sealants. The pre-cast samples were 25.4 by 25.4 by 76.2 mm (1.0 by 1.0 by 3.0 in). The sealant shape between the two pre-cast concrete substrates was 12.7 by 12.7 by 50.8 ± 0.25 mm (0.5 by 0.5 by 2.0 ± 0.01 in), See Figure #1.

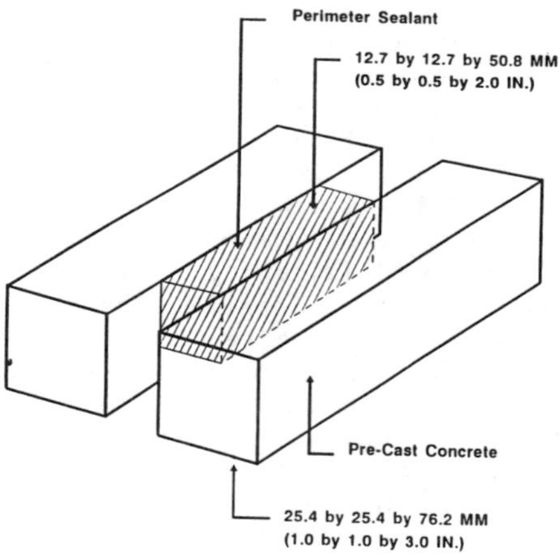

FIGURE 1 -- Sealant Test Specimen

The samples were allowed to cure for 21 days and then subject to a series of tests over specific time periods. Control values for all the perimeter sealants were taken following the 21 day cure time. Every data point represents the average of three actual samples run through each test condition.

ACCELERATED TESTS

Accelerated Tests	Test Duration (Hours)
1. QUV (ASTM G 53)	(1) Control, 1000, 2500, 4000
2. UV (ASTM C 1087)	Control, 1000, 2500, 4000
3. 30°C (86°F) @ 100% Humidity	Control, 1000, 2500, 4000
4. 50°C (122°F) @ 100% Humidity	Control, 1000, 2500, 4000

(1) Control value was run at 21 days room temperature cure

TABLE 1 -- Perimeter Sealant Description

Perimeter Sealant	Type
LM Silicone	One component, low modulus silicone sealant
MM Silicone	One component, medium modulus silicone sealant
Urethane #1	Two part, polyurethane sealant
Urethane #2	Three part expoxidized polyurethane terpolymer sealant
Urethane #3	Two part, low modulus/high elongation polyurethane sealant
Urethane #4	Two part, non priming polyurethane sealant
Urethane #5	Two part, non sag, low modulus polyurethane sealant

QUV

The standard test used for QUV was in accordance with ASTM G53, operating light and water-exposure apparatus (Fluorescent UV - condensation type) for exposure of nonmetallic materials. The accelerated QUV weathering tester was manufactured by the Q-Panel Company, Model # QUV. This procedure consisted of a continuous cycle of 4 h UV exposure at 60°C (140°F)(UVB 313 Bulbs); and 4 h condensation exposure at 40°C (104°F). The various perimeter sealants were exposed to this weathering for the above prescribed timeframe and then evaluated for tensile adhesion and durometer.

UV

The standard test used for UV was in accordance with ASTM C 1087 with the exception of the substrate which was pre-cast concrete, and the samples were continually exposed to the UV light. The type of UV light was incandescent tanning bulbs and the average reading was 2000 $\mu w/cm^2$ at 320 nm and an average temperature of 52°C (125 °F).

30°C AND 50°C AT 100% HUMIDITY

These two tests consisted of subjecting the test specimens to 30°C (86°F) and 50°C (122°F) at 100% humidity controlled condition. These conditions were held constant throughout the duration of the tests. The test specimens were allowed to return to room temperature conditions prior to tensile adhesion testing.

PHYSICAL TESTING

The tensile joint samples were measured for durometer before tensile testing using a hand held Shore A durometer instrument. The joints were all pulled in tension per ASTM C 1135 at 50.80 mm/ minute (2.0 in/minute) and the stress/strain curves were constructed for each individual sample, (see Fig. 2). The stress at 25%, 50% and ultimate elongation was recorded. The primer values were notably higher at all three points, 25%, 50% and ultimate elongation whereas the durometer results were the same for primed vs. unprimed. The primed substrates gave slightly higher and more consistent results for 25%, 50% and ultimate elongation because of better adhesion to the precast concrete. Therefore, all reported values are primed. After further review of the data no other correlation could be made on primed vs. unprimed. If the material degraded with primer it also degraded without.

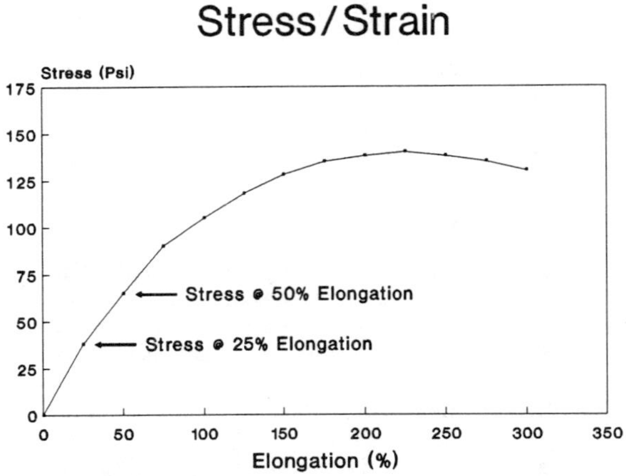

Fig. 2 Mechanical Properties Recorded during Tension Tests. (1 psi = 6.89 kpa)

RESULTS

The data is reported in the attached charts showing durometer, 25% modulus, and 50% modulus vs. control (21 day cure) and the various exposures and times. The ultimate elongation results are not presented in this study because the maximum movement of all the perimeter sealants evaluated was 50% therefore the authors felt this data point was not representative of the design criteria for joint movement. The silicones were separated from the urethanes for clarity. The samples were exposed to four different deteriorating mechanisms, two temperatures (30 and 50°C) with 100% humidity, QUV (UV and 100% Humidity) and UV.

DUROMETER

The testing results following the 30°C and 50°C at 100% humidity showed that Urethane #3 exhibited a slight increase in durometer from a control value of 35 to 50 Shore A points after 4000 hours. All other urethane and silicone samples showed only slight variances in durometer within 10 Shore A points. The durometer following the QUV (UV and 100% humidity) showed much different results. Urethane #1 showed an increase in durometer from 26 to 44 Shore A points and Urethanes #2 and #4 showed drastic decreases in durometer from 32 and 26 down to 6 and 4 Shore A points, respectively. Urethanes #3 and #5 and both silicones showed only slight variances in durometer within 10 Shore A points. The durometer following UV exposure showed the most severe durometer swings. Urethanes #1, #2, #4 and #5 showed decreases in durometer from 26, 32, 26 and 32 down to 5, 6, 5 and 3 Shore A points, respectively. Urethane #3 showed only a slight decrease in durometer within 10 Shore A points the two silicones showed a increase in variance of 12 Shore A points. Based on the standard ASTM tests run during this study QUV and UV alone both showed deteriorating mechanism to some of the urethanes. The 30°C and 50°C at 100% humidity showed no deteriorating mechanism to any of the sealants tested.

STRESS AT 25% AND 50% ELONGATION

The testing results were similar to the durometer results. The stress values for the 50% elongation were consistently higher than the stress at 25% elongation therefore the discussion will address only the stress at 25% elongation to eliminate redundancy. During the 30°C and 50°C at 100% humidity exposure Urethane #2 and #3 showed a substantial increase from 28 and 32 up to 53 and 79 psi, respectively. All other urethanes and silicones showed variances within 10 psi. The stress at 25% elongation following QUV again showed some urethanes loosing their physical properties. Urethane #3 showed an increase from 32 to 73 psi and Urethanes #2 and #4 showed decreases from 26 and 26 down to 5 and 4 psi, respectively. All other urethanes and silicones showed variances within 10 psi. The stress at 25% elongation following UV once again showed the most severe results. Urethanes #1, #2, #4, #5 all showed decreases from 26, 26, 24 and 34 down to 2, 3, 3, 4 psi, respectively. Urethane #3 again increased from 32 up to 60 psi. Both silicones showed increases within a 12 psi variance. In terms of stress at 25% elongation as compared to durometer the results of this study clearly show that the two temperatures at 100% humidity only caused Urethane #3 to increase in strength (harden). During the 30°C at 100% humidity stress at 50% elongation testing Urethane #2 lost adhesion to the precast concrete which is shown on the graph as a loss in tensile strength. Once again during the QUV and UV exposures both showed deteriorating mechanisms to some of the urethanes. In this test the 30°C and 50°C at 100% humidity clearly causes Urethane #3 to increase in stress (harden) all other sealants in the test showed variances less than 10 psi.

SUMMARY

Based on all the testing in this paper the 30°C and 50°C at 100% humidity exposures showed no deteriorating mechanism to the sealants tested with the exception of Urethane #3, which showed an increase in stress (hardened) at both 25% and 50% elongation. It was very apparent that the UV testing showed the most severe deteriorating mechanism on all urethanes, one increased in durometer and stress at 25% and 50% elongation which was Urethane #3 and the rest of the urethanes decreased in durometer and stress at 25% and 50% elongation. The QUV testing showed similar results to that of UV with the exception of Urethane #5 not showing a severe loss in physical properties. The reasons for the differences in weathering are due to the QUV weathering exposed the samples to UV 50% of the time and humidity the other 50%. Therefore the samples see only 1/2 the UV exposure. Also, it is possible that some of the short wave radiation given off during the 100% UV weathering could further accelerate the degradation of the sealants. But the difference in the QUV vs. UV weathering was the rate of degradation and not the nature of weathering. The results of this paper are consistent with that reported by L. Bogue Sandberg[1], which showed deteriorating stress values at 25% elongation after exposure to QUV for 5000 hours. Sealant D, (Low Modulus, two part urethane) in L. Bogue Sandberg's paper is the same as (Urethane #2) in this study. It was also apparent that the five 2-part urethanes tested in this study are different in terms of the failure mode and weathering characteristics. Urethane #3 always increased in durometer and stress values verses the other four urethane values decreased. In all cases the silicones were not affected by the accelerated testing in this study which is consistent with L. Bogue Sandberg's and Gregory Fedor and Patrick Brennan.[2] All of the conditions tested within the scope of this test can be found on exterior vertical walls where high performance perimeter sealants must perform. All of the sealants tested in this study are well known currently available high performance perimeter sealants and are many times specified as equals. This set of tests which use the constant UV and QUV cycles provide a framework for physical properties testing within a reasonable time frame with reasonable temperatures. This testing should be considered to be in the ASTM C 920 specification.

ACKNOWLEDGEMENT

The authors would like to acknowledge Richard Badour and Ronita Snyder for preparing, testing and coordinating the data for this study. Without their work this study would not have been possible.

REFERENCES

[1] L. Bogue Sandberg, "Comparisons of Silicone and Urethane Sealant Durabilities," Journal of Materials in Civil Engineering, Vol. 3, No. 4, November, 1991.

[2] Fedor, Gregory and Brennan, Patrick, "Correlation of Accelerated and Natural Weathering of Sealants," The Applicator, Sealant, Weatherproofing & Restoration Institute, Fall 1990.

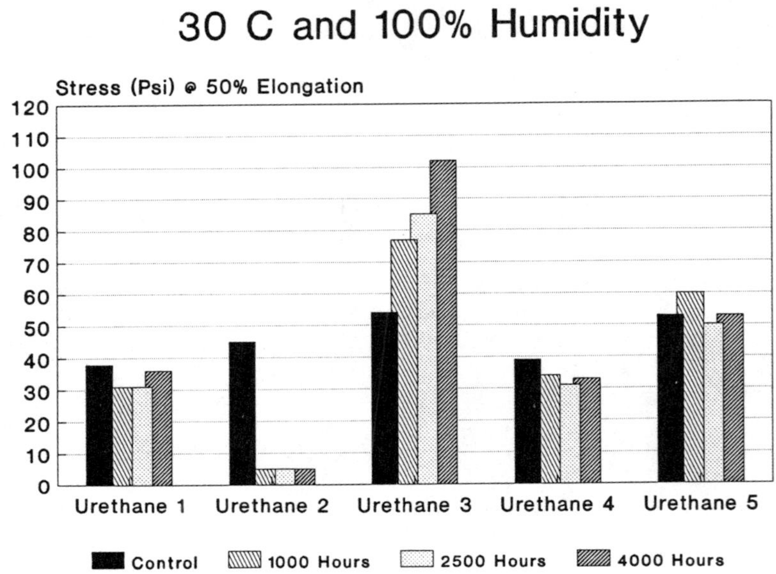

Note: 1 psi = 6.89 kpa

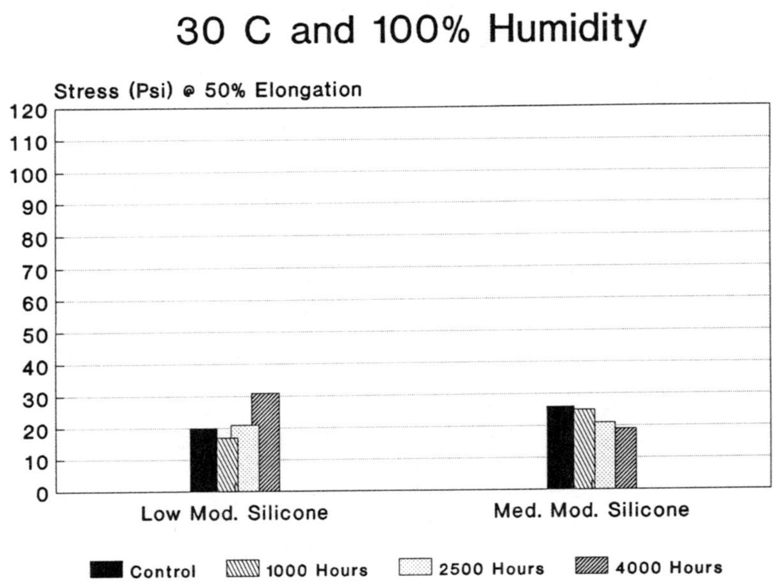

Note: 1 psi = 6.89 kpa

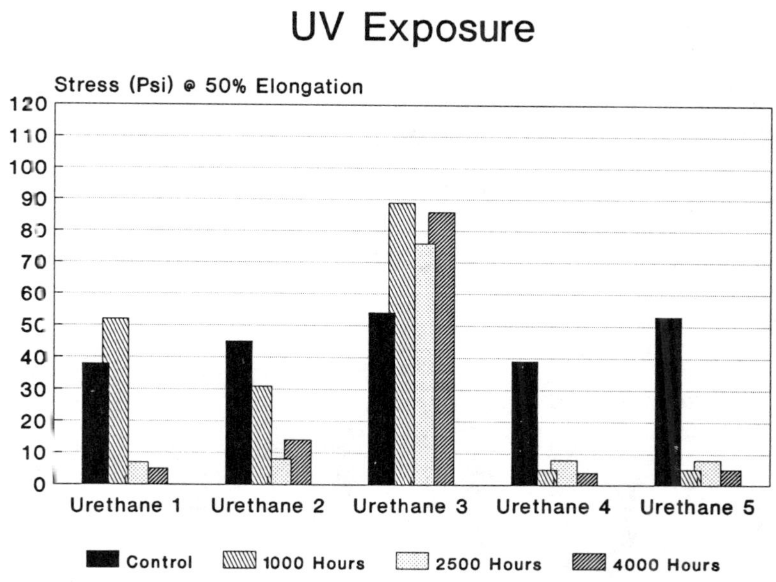

Note: 1 psi = 6.89 kpa

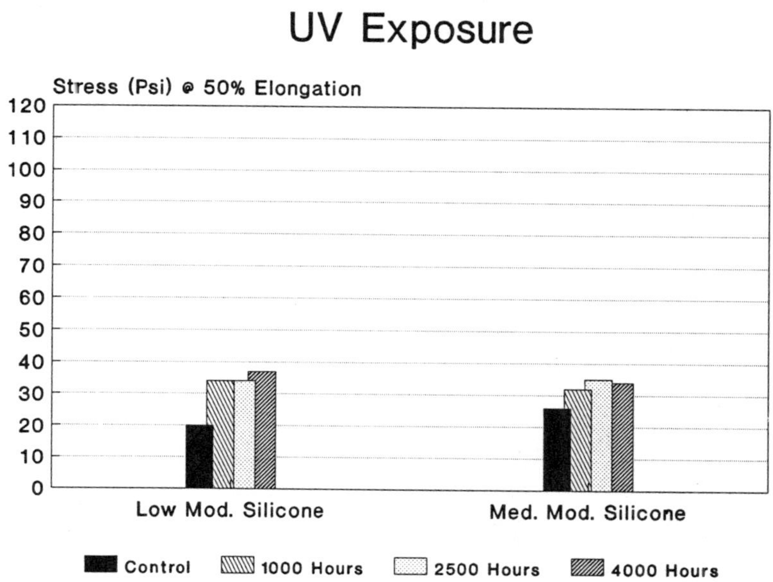

Note: 1 psi = 6.89 kpa

50 C and 100% Humidity

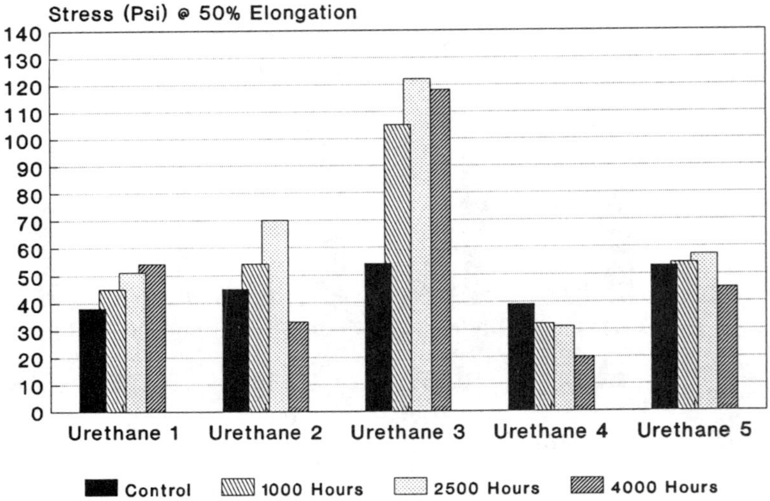

Note: 1 psi = 6.89 kpa

50 C and 100% Humidity

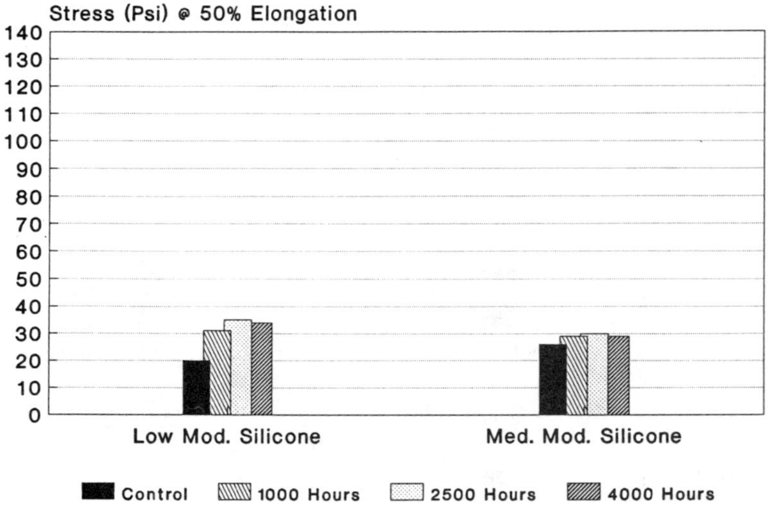

Note: 1 psi = 6.89 kpa

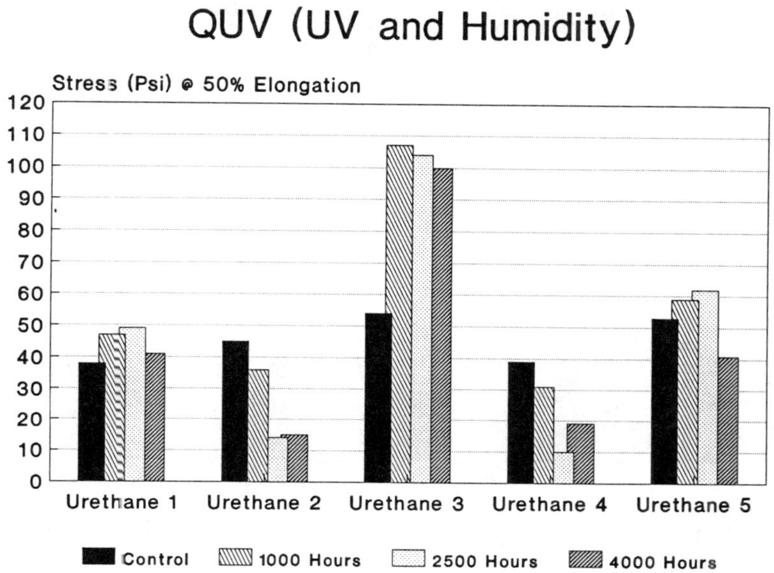

Note: 1 psi = 6.89 kpa

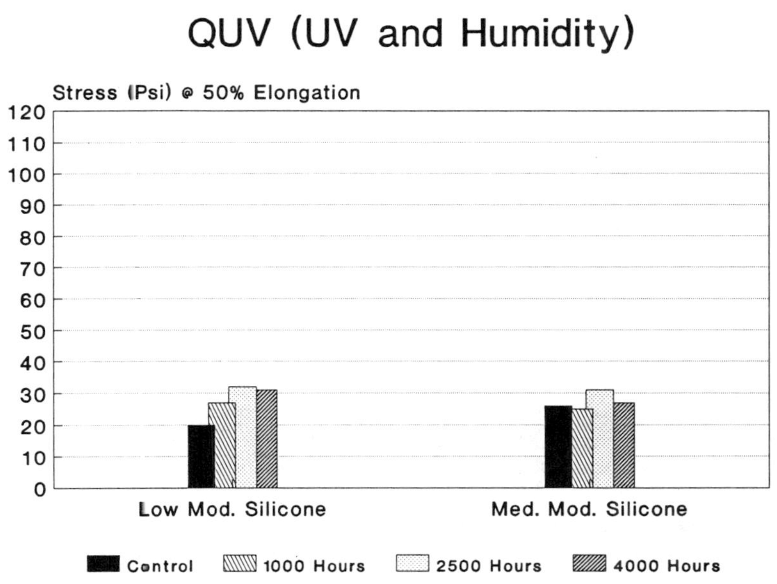

Note: 1 psi = 6.89 kpa

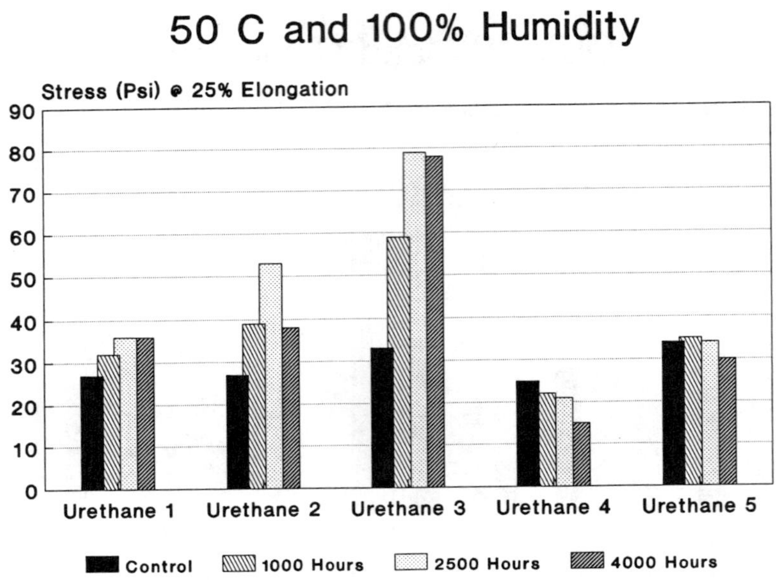

Note: 1 psi = 6.89 kpa

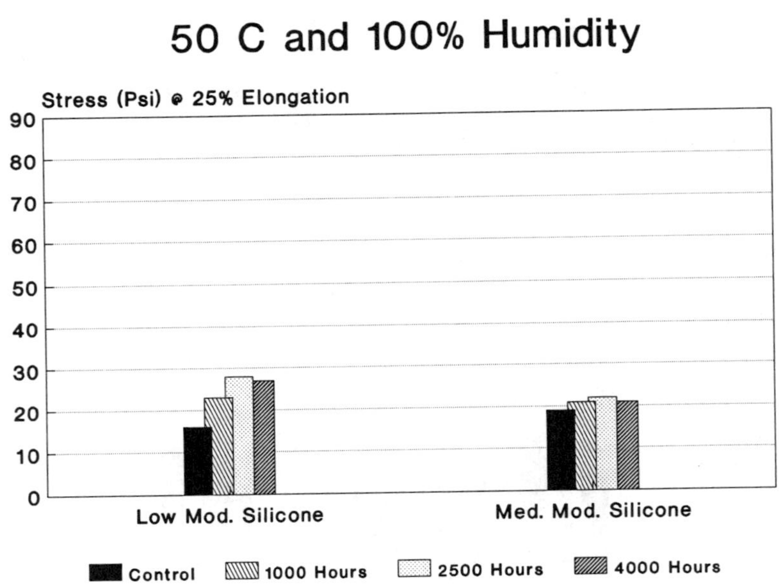

Note: 1 psi = 6.89 kpa

30 C and 100% Humidity

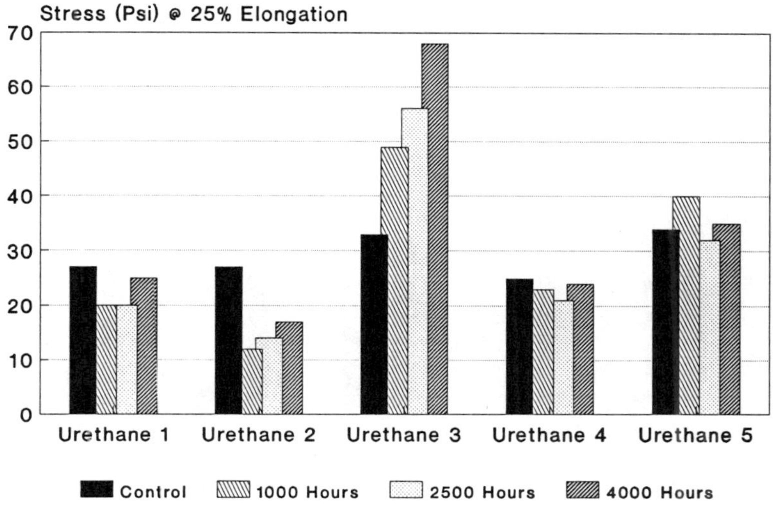

Note: 1 psi = 6.89 kpa

30 C and 100% Humidity

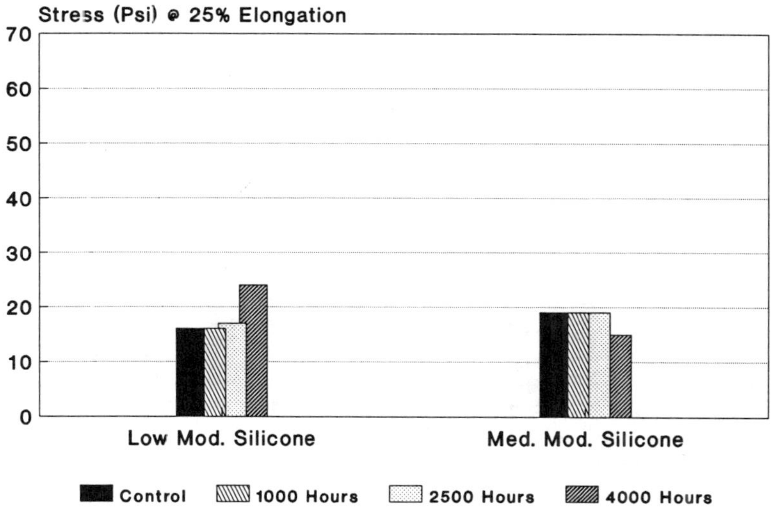

Note: 1 psi = 6.89 kpa

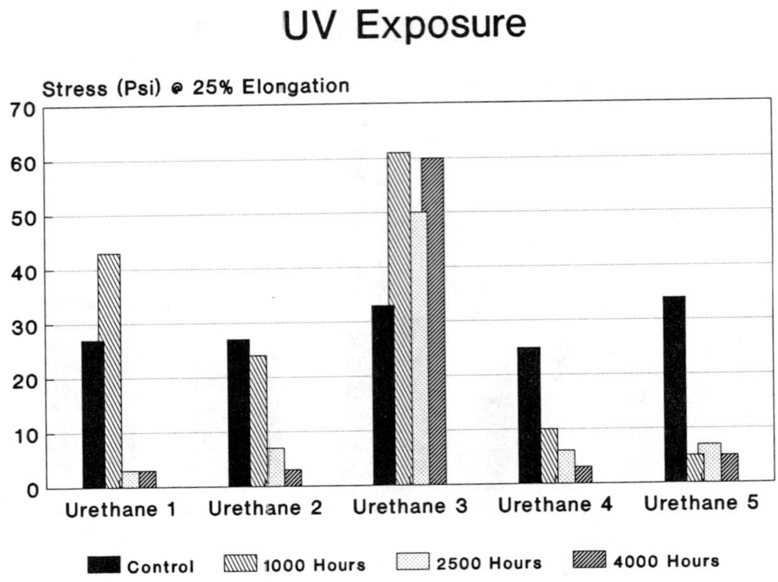

Note: 1 psi = 6.89 kpa

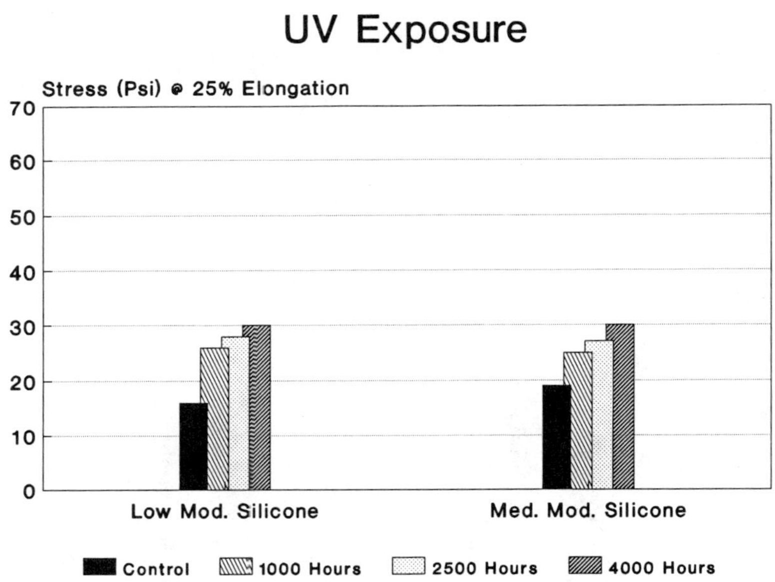

Note: 1 psi = 6.89 kpa

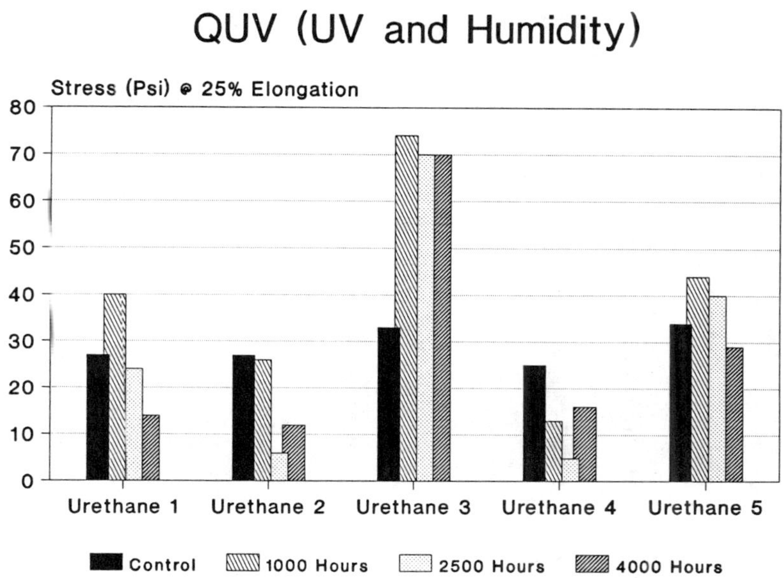

Note: 1 psi = 6.89 kpa

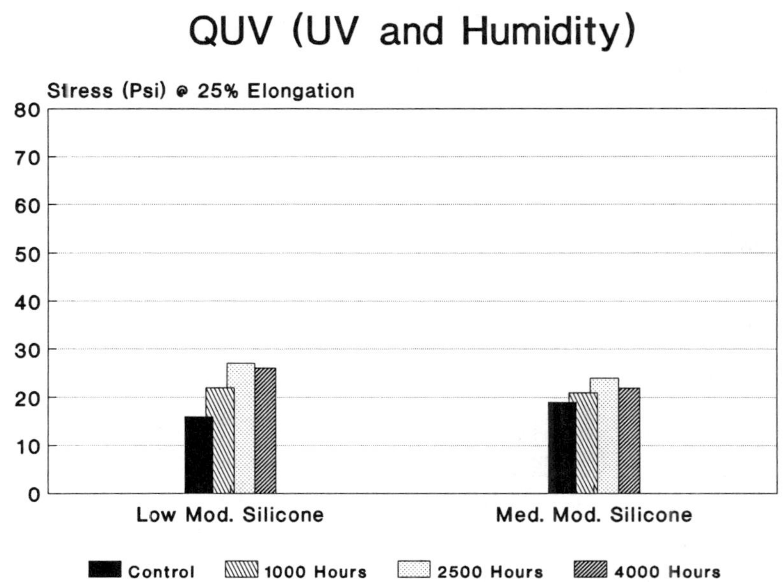

Note: 1 psi = 6.89 kpa

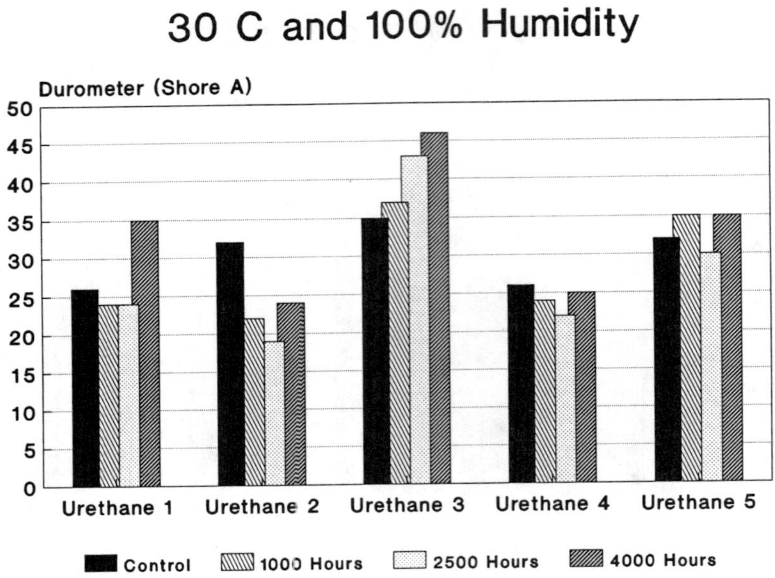
30 C and 100% Humidity
Durometer (Shore A)
50
45
40
35
30
25
20
15
10
5
0
Urethane 1
Urethane 2
Urethane 3
Urethane 4
Urethane 5
Control
1000 Hours
2500 Hours
4000 Hours

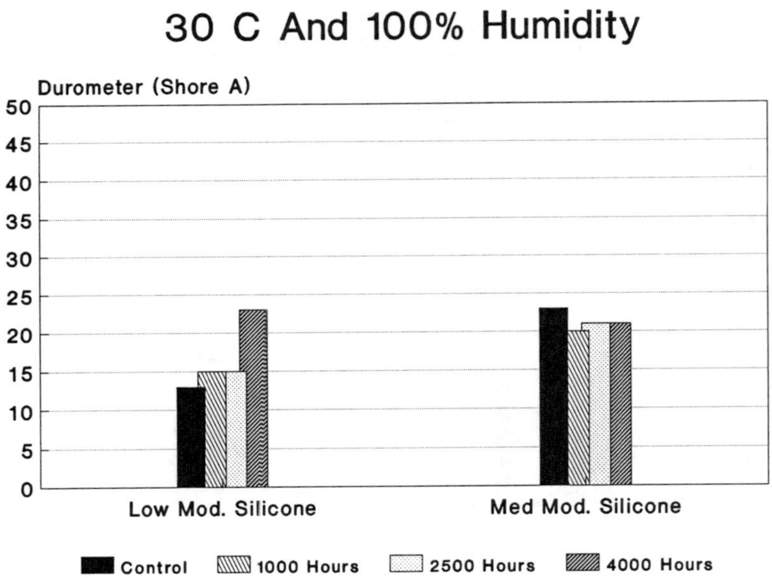
30 C And 100% Humidity
Durometer (Shore A)
50
45
40
35
30
25
20
15
10
5
0
Low Mod. Silicone
Med Mod. Silicone
Control
1000 Hours
2500 Hours
4000 Hours

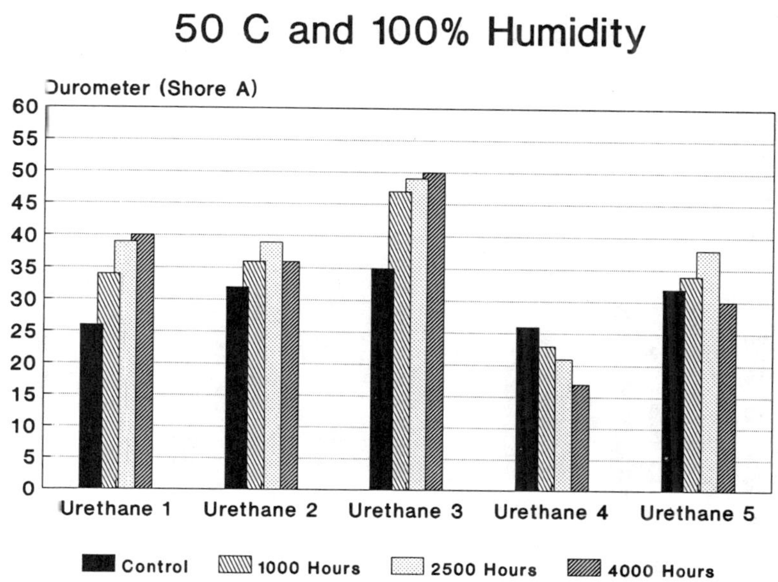
50 C and 100% Humidity
Durometer (Shore A)
60
55
50
45
40
35
30
25
20
15
10
5
0
Urethane 1
Urethane 2
Urethane 3
Urethane 4
Urethane 5
Control
1000 Hours
2500 Hours
4000 Hours

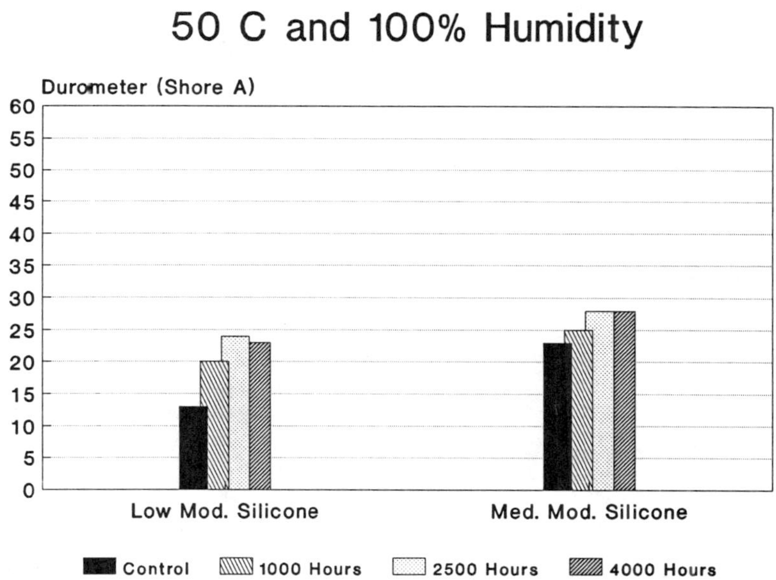
50 C and 100% Humidity
Durometer (Shore A)
60
55
50
45
40
35
30
25
20
15
10
5
0
Low Mod. Silicone
Med. Mod. Silicone
Control
1000 Hours
2500 Hours
4000 Hours

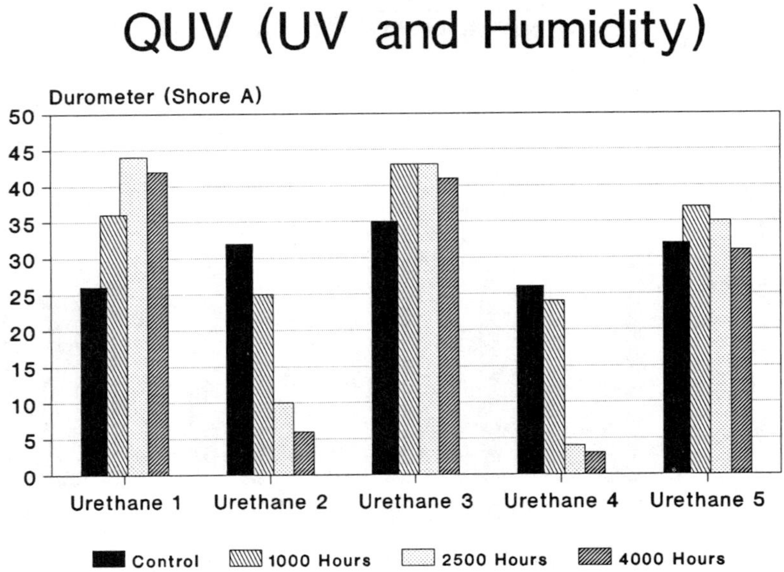
QUV (UV and Humidity)
Durometer (Shore A)
50
45
40
35
30
25
20
15
10
5
0
Urethane 1
Urethane 2
Urethane 3
Urethane 4
Urethane 5
Control
1000 Hours
2500 Hours
4000 Hours

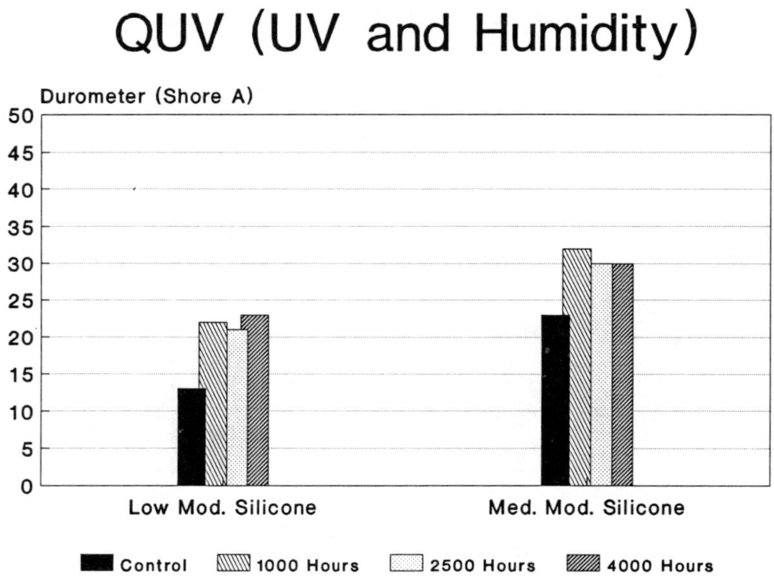
QUV (UV and Humidity)
Durometer (Shore A)
50
45
40
35
30
25
20
15
10
5
0
Low Mod. Silicone
Med. Mod. Silicone
Control
1000 Hours
2500 Hours
4000 Hours

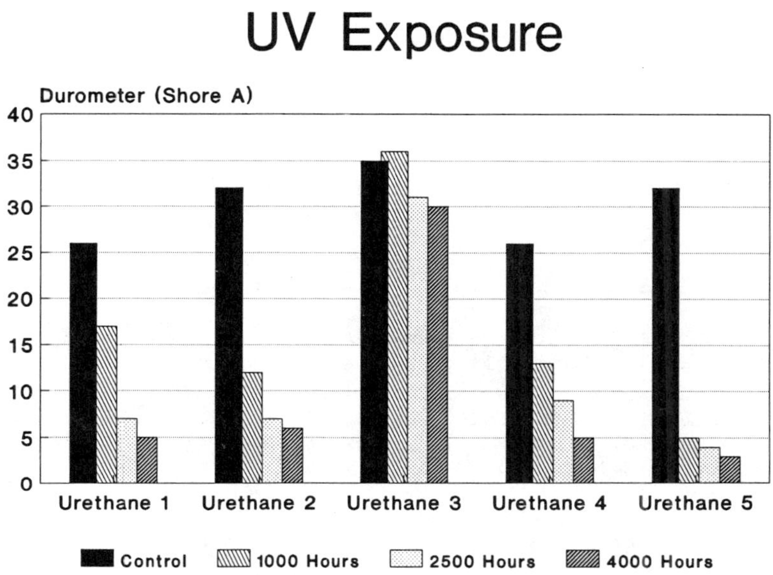
UV Exposure
Durometer (Shore A)
40
35
30
25
20
15
10
5
0
Urethane 1
Urethane 2
Urethane 3
Urethane 4
Urethane 5
Control
1000 Hours
2500 Hours
4000 Hours

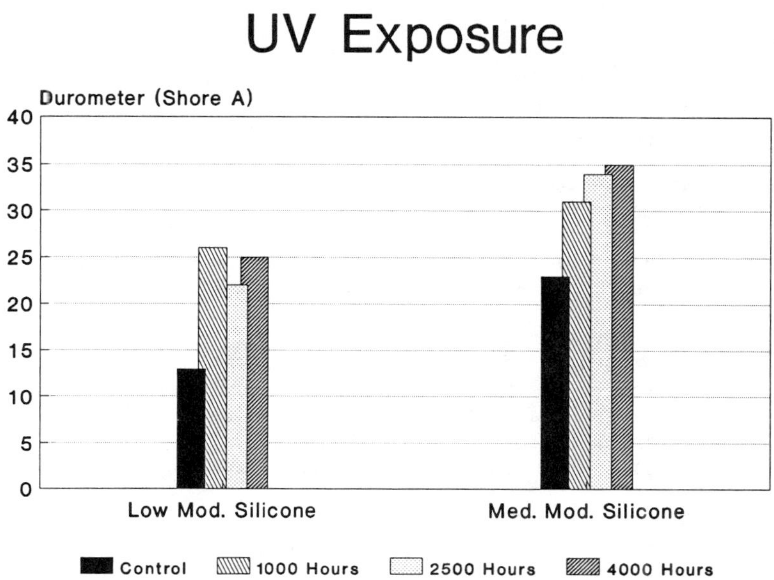
UV Exposure
Durometer (Shore A)
40
35
30
25
20
15
10
5
0
Low Mod. Silicone
Med. Mod. Silicone
Control
1000 Hours
2500 Hours
4000 Hours

John Beech and John Beasley

EVALUATION OF CURE AND DURABILITY ASPECTS OF BUILDING SEALANTS

REFERENCE: Beech, J. and Beasley, J. L., **"Evaluation of Cure and Durability Aspects of Buiding Sealants"**, Science and Technology of Building Seals, Sealants, Glazing, and Waterproofing: Second Volume, ASTM STP 1200, Jerome M. Klosowski, Ed., American Society for Testing and Materials, Philadelphia, 1992.

ABSTRACT: Durability characteristics of sealants have been assessed by measuring the changes in certain key properties ("performance indices") after exposure of specimens to natural and to artificial weathering. A major problem in such assessments is the long duration of the cure processes of sealants after specimens are made up: in some cases these extend to 12 months or longer, during which periods continuing changes in sealant properties are observed.

This paper reports the results of a controlled study which attempts to distinguish between those changes in properties due to the continuing cure process, and those caused by changes in the sealant due to exposure to various environments which simulate the effects of weather. The latter included laboratory regimes in which the sealant samples were exposed to elevated temperature (70°C), both with and without UV(B) radiation; to artificial weathering including UV(B) radiation and water condensation; and exposure to weather in the UK, for up to 12 months.

The study involves two phases, in the first of which, reported here, freshly-prepared sealant specimens were exposed to ageing or to weather before tensile testing. In the second phase the same procedures were followed using specimens which had first been stored under standard conditions in the laboratory for 12 months.

Some results of the first phase of this study of cure and durability characteristics are given in this paper, and discussed in relation to the development of test methods for the assessment of the durability of sealants. Modulus values measured on sealant samples after exposure to artificial weathering and to heat ageing in the laboratory show highly-significant correlations with data from naturally-weathered samples.

KEYWORDS: sealants, durability, cure, weather, accelerated ageing, modulus changes.

INTRODUCTION

The scientific challenge of predicting the durability of building materials centres upon the correct simulation of the conditions of service in the laboratory: this may involve numerous individual service factors and their combinations. In investigating the durability of building sealants two additional factors must be taken into account. The first is the slow cure of many of these products, so that a period of months may elapse before stable properties are attained. Clearly, unless a stable datum is achieved for each type of sealant under consideration, before subjecting samples to accelerated ageing or to weather, it may be impossible to distinguish between cure effects and those caused by the ageing or weather factors.

Secondly, it is necessary to identify what we have previously defined[1] as "performance indices": properties of the sealant which can readily be measured (or derived from measurements) in the laboratory, and which change in service, or in simulated weathering or ageing, in ways which influence the performance by the sealant of its essential functions in the building joint.

In an investigation of the water resistance of sealants, Beech and Mansfield[1] proposed that the secant modulus (measured in that research at 25 per cent extension in a tensile test) and the extensibility (taken as the extension of the specimen at maximum load in the tensile test), were suitable performance indices when taken in conjunction after a suitable exposure to aqueous media. In an earlier study[2] of sealant durability other properties were derived from tensile test data in addition to the modulus at 25 per cent extension: the elastic recovery and stress relaxation rate after the same extension.

One of the objectives of the present study of cure and durability aspects of sealants is to assess a number of potential performance indices derived from tensile test data, in terms of their utility in correlating test data from naturally-weathered sealant specimens with those obtained from samples exposed to various laboratory conditions. The latter simulate, to some degree, the factors which are known to cause degradation of sealants in the service environment.

The second objective is to evaluate the cure characteristics, in the short and longer terms, of the various types of sealant on the market. The change in tensile modulus is the primary index of cure, and this has been measured with exposure of specimens to weather at Garston, England; to various artificial weather and ageing regimes; and using control samples stored in the laboratory at standard conditions.

Finally, we wish to identify the separate respective effects of continuing cure of the sealants, and of their exposure to weather and ageing factors and to natural weather. Two sets of specimens were prepared at the outset of the experimental programme. The first were used in the various test exposures after an initial period of cure (Phase I). The second set was stored under standard conditions for 12 months, before repeating the exposures and tensile measurements (Phase II). A comparison of the two sets of experimental data will eventually enable cure effects to be distinguished from the effects of artificial and natural weathering.

However at this stage only data from Phase I are available, and some of these will be presented and discussed in this paper. The results for modulus changes have mainly been selected for this purpose, through considerations of space.

EXPERIMENTAL

Materials

Nine commercial elastomeric sealants were used in the tests. The coding below is used in the paper.

P13 Moisture-curing one-part polysulphide to BS5215:1983[3]

2P10)
2P15) Two-part chemically-cured polysulphides to BS4254:1986[4]

S6 Silicone sealant to BS5889:1989[5] low modulus

S12 Silicone sealant to BS5889:1989 low modulus

S13 High-modulus silicone sealant to BS5889:1989

U1)
U6) One-part polyurethane sealants

U3 Multi-part epoxy-polyurethane sealant.

Preparation of Specimens

The test specimens each comprised a 12 x 12 x 50 mm bead of sealant between the large surfaces of cement mortar and an aluminium adherends. The cement mortar pieces were 12 x 50 x 8 mm in dimension and were prepared by the method described in BS4254[(4)]. Aluminium test pieces were 12 x 50 x 10 mm, with a milled finish.

Five replicate specimens were prepared of each sealant for each of the test exposures, using the manufacturers' recommended primers for the two substrates. The specimens were stored for 28 days at 23°C and 50% r.h. (at 27°C and 95% r.h. in the case of P13) to allow initial cure to take place, drawing the spacer bars back from the sealant beads after a few days in the case of the moisture-curing materials.

Test Exposures

The specimens were allocated randomly to the test exposures listed below, for each set of specimens made using one of the nine products. [As mentioned previously a second set for each product was set aside for 12 months storage at standard conditions, for use in the second phase of the study].

The test exposures used were:

Control(C): stored at standard conditions of 23°C ± 2°C and 50 ± 5% r.h.

H70: storage in an oven at 70°C ± 2°C

U70: continuous exposure to UV(B) radiation in a QUV apparatus, with temperature maintained at 70°C

U/cond: artificial weathering in a QUV apparatus with cycles of 4 hours of UV(B) radiation at 70°C followed by 4 hours condensation at 40°C

Weather(W): samples were exposed facing south on racks inclined at 45°, on the weathering site at Garston (a moderate climate in southern England).

The coding above is used throughout the paper.

Tensile Testing

Five specimens of each sealant were removed from each of the five test environments after the following durations of exposure: 2, 4, 8, 16, 28, 40 and 56 weeks. They were conditioned for 24 hours at standard conditions before being placed in a tensile test machine and pulled to destruction at 5 mm/min. Force and elongation data were recorded at 0.1 mm increments of strain and stored on disc for subsequent analysis.

A set of five control samples of each sealant was also tested in this manner after the initial storage period of 28 days under standard conditions.

From the tensile test data the following parameters were derived for each specimen: force (N) at 15 mm joint width, which was taken as equivalent to the secant modulus at 25 per cent extension; the maximum force on the force/extension curve (N); and the extension (mm) at which maximum force was attained, and hence the % elongation at this force.

RESULTS AND DISCUSSION

The modulus changes observed during the 12 month period of the first stage of the study are presented in figures 1 to 9, as the force in Newtons at 3 mm extension (25 per cent of joint width). Each data point is normally the mean of 5 force readings, though in a few cases only 4 data were available. For each of the nine sealants the control data represent the cure characteristics over the period from 28 days to 52 weeks after the specimens were prepared. In the cases of exposure to natural weather and the three laboratory simulations of weather factors, however, the modulus changes observed were due to both continuing cure of the sealant, and to the effects of UV, water, heat (or combinations of these, as appropriate). Clearly the relative magnitudes of the two influences will differ between the various sealants, with their different cure mechanisms and responses to the weather factors.

Cure Effects

To estimate the magnitude of the cure effects and the degree of cure of each sealant at the start and at the conclusion of the period of exposure two sets of parameters have been derived from the data shown in figs 1 to 9. In Table 1 the rate of change of modulus (as force in Newtons per day) initially (0-2 weeks) and during the latter half of the 52 week period has been calculated for the control specimens as a percentage of their initial mean values. Table 1 also shows the initial modulus as a fraction of the "final" value attained at 52 weeks in standard conditions.

TABLE 1 -- Modulus changes in control specimens

Mean Force (N) at 25% Extn.	Polysulphide Sealants			Silicone Sealants			Polyurethane Sealants		
	P13	2P10	2P15	S6	S12	S13	U1	U3	U6
Initial	8.3	60.4	57.6	27.4	25.2	143.2	54.4	36.7	36.5
At 2 weeks	46.0	63.0	63.7	35.4	59.0	140.6	67.0	47.8	43.1
At 28 weeks	111.9	64.9	74.2	52.7	102.5	149.1	109.8	61.4	67.2
At 52 weeks	124.2	73.0	81.8	79.3	105.0	161.5	130.2	67.4	72.0
Initial modulus as fraction of M52	0.06	0.83	0.70	0.35	0.24	0.89	0.42	0.54	0.51
Rates of modulus change (% per day) (a)	32.2	0.3	0.8	2.1	9.6	-0.1	1.7	2.2	1.3
(b)	0.88	0.08	0.08	0.58	0.06	0.05	0.22	0.10	0.08

(a) initial rate; (b) final rate (see text)

It is apparent from the data in Table 1 that there is considerable variation among the sealants in the degree of cure at the outset of the tests. Two of the moisture-curing silicones gave modulus values of less than a third of the final value while the one-part polysulphide was barely cured (though sufficiently rubbery to be handled and tested. Two of these materials (P13 and S6) were also still showing marked changes in modulus values [(b) in Table 1] towards the end of the 12 month period. For the majority of materials, showing final rates of modulus change of less than 0.1% per day in Table 1, stable properties had been attained at that stage.

Further data relating to the longer-term cure characteristics will be obtained from the second phase of this study.

TABLE 2 -- Percentage differences from Controls at 52 weeks

Sealants	Exposures			
	Natural Weather	Heat 70°C	UV(B) 70°C	UV + Condensation
Polysulphides				
P13	-29	-17	-41	-50
2P10	-19	+31	+23	+6
2P15	-7	+39	+11	+17
Silicones				
S6	-25	+10	+5	-17
S12	+2	+27	+12	+11
S13	-10	+36	+36	-16
Polyurethanes				
U1	+4	+113	+46	-20
U3	-28	+23	+23	-59
U6	+4	-29	+41	+2

Durability Effects

The changes in modulus (F_{25}) at 52 weeks are shown as percentage changes from the control values at this time in Table 2. There is no apparent relationship between the changes caused by weather and those due to artificial weathering. However, 52 weeks is not a realistic period for the simulation of relatively short-term weather effects for a moderate climate, using these laboratory conditions, and a more detailed comparison of the data for shorter periods of laboratory exposure (2, 4, 8 and 16 weeks) was made.

The mean values of the actual modulus (F_{25}) values were compared, and the degree of correlation was computed between the values measured for naturally-weathered specimens, and for those obtained from specimens exposed to the artificial weather regimes. The results are summarised in Table 3. In general the QUV weathering cycles of UV radiation alternating with condensation give generally the best correlations with the data for 12 months English weather. The exposure for 8 weeks yielded a linear correlation which was significant at the 0.1% level; this is illustrated in figure 10.

This correlation takes into account both changes due to prolonged cure (where these are of significance) and also those caused by exposure to the weather elements. In view of the wide disparity of sealants formulations involved, the degree of correlation observed is encouraging and suggests that the QUV weathering apparatus has potential for predicting the longer term durability of sealants exposed to temperate climates.

Another aspect of modulus changes was also investigated. The degree of correlation between data for natural and artificial weathering was determined using the difference between modulus values for exposed and control samples. The results are summarised in Table 4, in which it will be noted that a number of durations of exposure to the three laboratory regimes gave linear correlations, at the 1% significance level, with the data from naturally-weathered samples after 1 year's exposure. However the best correlations, at the 0.1% level of significance were obtained with 2 weeks exposure to heat at 70°C (both with and without UV(B) radiation). Since it is unlikely that such a short duration of UV radiation has any but superficial effects on the sealants specimens, it seems likely that heat is the effective agency of modulus change in both these exposures. The correlations are illustrated in figures 11 and 12.

The modulus changes measured in these correlations relate apparently only to the influence of the weather factors involved. It will be noted that both positive and negative changes were observed relative to the "same-age" controls, in both weather and laboratory exposures. It may be doubted therefore whether the simple assumption is correct, that modulus changes with continuing cure, and those caused by exposure to heat, UV or moisture (and by combinations of these factors) are independent.

TABLE 3 -- Correlation Coefficients for linear regression: Modulus. 12 months Natural Weather with laboratory tests

Exposures	Durations (weeks)			
	2	4	8	16
Heat 70°C	0.726 *	0.706 *	0.713 *	0.607 NS
Heat/UV 70°C	0.735 *	0.725 *	0.708 *	0.724 *
UV + Condensation cycles	0.845 **	0.170 NS	0.915 ***	0.831 **

TABLE 4 -- Correlation Coefficients for modulus differences from Controls. 12 months weather and laboratory tests

Exposures	Durations (weeks)			
	2	4	8	16
Heat 70°C	0.879 ***	0.838 **	0.639 *	0.502 NS
Heat/UV 70°C	0.875 ***	0.807 **	0.804 **	0.696 *
UV + Condensation cycles	0.775 **	0.814 **	0.595 NS	0.773 **

Key to significance levels in Tables 3 and 4:

*** at 0.1% level
** at 1% level
* at %% level
NS Not significant

It is believed that an examination of the changes in one parameter in isolation, such as modulus, could give a misleading impression of the nature of the changes in the sealants due to weather exposures. Earlier work to study the water resistance of sealants (1) suggested that only by studying changes in two or more relevant performance indices can a true evaluation be made. It is therefore intended to complete the first stage of this study by a careful examination of the data pertaining to changes in other measured properties, particularly the extensibility.

At present, data for this property have been analysed for only two of the sealants tested. These data are given in figures 13 and 14 for illustrative purposes. The sealants are polysulphide materials from the same manufacturer: a single-part product (P13) and a two-part (2P10). Even without comparing these data with those for the other seven sealants, some interesting differences between the two materials can be identified.

A general observation is that both the laboratory exposures involving UV radiation produce reductions in extensibility (E) compared to controls or to naturally-weathered samples, even after the briefest exposures. This effect is probably not significant in the case of the two-part polysulphide sealant, though it is very marked for the one part sealant. It is believed that this phenomenon is caused by the formation of a well cured skin over the exposed surfaces of the samples of P13, which have very high resistance to the permeation of water vapour, and therefore impede the further cure of material deeper within the sealant bead.

A comprehensive review of the extensibility changes must await availability of further data. An interim conclusion appears to be that the magnitude of changes due to weather and to laboratory exposures is much related to the state of cure of the sealants at the onset of the exposures. Data from the second phase of work, using well-cured samples, will clarify this effect.

CONCLUSIONS

1. The data for cure characteristics reveal marked differences in the rates of cure for different sealants formulations , which may cause problems in obtaining representative data for performance related properties of the slow curing materials using standard test methods. It is recommended that these properties should be measured after using two cure procedures: the "nominal cure" stipulated in the appropriate test method; and after a supplementary cure schedule, involving exposure to (e.g.) heat and water, such as those given in ASTM and ISO standard test methods. Two weeks exposure to a QUV weather cycle was found to achieve cure of these materials to similar extent to that found after 12 months exposure to weather.

2. Data showing the changes in modulus values for a range of generic types of sealants after exposure to different weather factors indicate that no general conclusions can be drawn on the basis of generic type. Within each class of sealant (e.g. Polyurethanes) differences in both the magnitude and direction of modulus change may be found.

3. Modulus values of specimens after exposure to weather for 12 months have been compared with those measured on samples exposed to various artificial weathering and ageing regimes in the laboratory. It was found that 8 weeks exposure in a QUV weathering apparatus gave a highly significant correlation when mean modulus values were compared with those of naturally-weathered sealants (i.e. an "acceleration factor" of 6.5 for this QUV cycle).

4. The differences of modulus values from "same-age" controls have been examined as putative performance indices of the effects of various weather factors. It appears that heating samples for two weeks at 70°C can give results having a highly significant correlation with corresponding mean values obtained by testing specimens exposed to weather.

5. On the basis solely of the data for modulus changes, reported herein, artificial weathering including UV, heat and moisture, and simple heat ageing at 70°C, both appear to have potential for the prediction of sealant durability for service in temperate climates. It will be necessary to complete the analysis of data for other performance indices (e.g. extensibility) before a firm recommendation regarding predictive test procedures can be made.

REFERENCES

(1) Beech, J.C. and Mansfield, C., "The water resistance of sealants for construction", Building Sealants: Materials Properties and Performance, ASTM STP.1069, Thomas F. O'Connor, ed. American Society for Testing and Materials, Philadelphia, 1990.

(2) Beech, J.C. and Mansfield, C., "Prediction of the durability of building sealants", Proceedings of the Fourth International Conferences on Durability of Building Materials and Components, Singapore, RILEM., Pergamon Press, Oxford, 1987.

(3) British Standard Specification BS5215:1983, One-part gun grade polysulphide-based sealants, BSI, London, 1983.

(4) British Standard Specification BS4254:1986, Two-part polysulphide-based sealants, BSI, London, 1986.

(5) British Standard Specification BS5889:1989, One-part gun-grade silicone-based sealants, BSI, London, 1989.

William R. O'Brien and Daniel T. Berg

A SCREENING EVALUATION OF SURFACE CLEANING FLUIDS AND THEIR EFFECTS UPON THE ADHESION OF VARIOUS WEATHERSEALING SEALANTS

REFERENCE: O'Brien, W. R. and Berg, D. T., **"A Screening Evaluation of Surface Cleaning Fluids and Their Effects Upon the Adhesion of Various Weathersealing Sealants"**, Science and Technology of Building Seals, Sealants, Glazing, and Waterproofing: Second Volume, ASTM STP 1200, Jerome M. Klosowski, Ed., American Society for Testing and Materials, Philadelphia, 1992.

ABSTRACT: A low volatile silicone cleaning/priming solution was evaluated in comparison with a variety of organic cleaning solvents and one organic solvent based primer. The evaluation also reviewed a wide variety of substrates, sealants, and surface contaminations. In virtually all situations the use of the silicone cleaner/primer alternative to organic solvents actually helped promote sealant adhesion better than all other solvents evaluated.

KEYWORDS: sealants, surface preparation cleaning solvents, adhesion.

The long term trouble free performance of weatherseals and other sealant joints are critical to the success of the building envelope. Sealant joints are required to prevent water infiltration, to absorb building movements and in some cases to actually hold sections of the building together as seen in silicone structural glazing.

Assuring the success of the sealant joints requires careful attention to joint preparation as well as careful selection of the sealants. One vital area relating to the joint preparation is the cleaning of the substrate surface with a solvent cleaner. Frequently cleaning the surface is followed by a primer step to chemically activate the surface or to form a film coating over the surface.

Increasingly the world is becoming concerned with the environmental and health issues relating to the use of the organic solvents used frequently to clean joint surfaces prior to sealant application. The purpose of this report is to review the effect of a variety of cleaning solvents on the ability of sealants to adhere to a wide range of building substrates.

Mr. O'Brien is a Technical Service Representative and Mr. Berg is a Developmental Chemist at the Dow Corning Corporation, 2200 W. Salzburg Rd., Auburn, MI 48611.

EXPERIMENTAL METHOD

This study contains a review in three areas:

1. A review of several different types of solvent with several types of sealants to determine the broad effect of the type of cleaning solution on sealant adhesion.

2. A detailed review of the ability of a medium modulus silicone sealant to adhere to a wide variety of surfaces utilizing different cleaning techniques.

3. A review of the ability of the solvents to remove specific contaminants and promote sealant adhesion.

The test procedure utilized in sections one and two was a modified ASTM C–794 Peel Adhesion Test. The standard test procedure was modified by replacing the 1" wide cloth strip specified with a 1/2" wide screen. The surface preparation techniques and cure conditions are as documented below.

In all cases the sealants adhesion was measured based upon the sealant percent cohesive failure to the substrate. This measurement technique can be considered a reasonable screening comparison of a single sealant to different substrates, but is not intended as a method of comparing between different sealants.

In all studies performed the use of a low volatile silicone cleaning/priming solution achieved the best most consistent sealant adhesion over the more volatile organic solvent alternatives.

General Sealant Evaluation

The purpose of this study is to review the ability of a variety of cleaners to promote adhesion to a variety of surfaces with a variety of sealants.

The following types of sealants were evaluated:

1. Low modulus neutral cure silicone.
2. Medium modulus neutral cure silicone.
3. Acetoxy curing silicone.
4. One–part polyurethane.
5. Acrylic latex.

The sealants were evaluated for their adhesion to the following substrates:

1. Precast concrete.
2. Polycarbonate sheet.
3. Bronze anodized aluminum.
4. Orential Grey Duranar.
5. Mill finish aluminum.

Each substrate was cleaned by a one step procedure utilizing the following cleaning techniques:

1. Xylene solvent.
2. Isopropyl alcohol solvent
3. Citrus based solvent.
4. Silicone cleaner/primer (silicone polymer containing active compounds).

Each substrate was cleaned and primed by a two–step procedure utilizing the following materials:

1. Xylene cleaner plus organic solvent based primer.

Table 1 is a listing of the adhesion test results generated utilizing a modified ASTM C–794 peel adhesion test. The cure conditions were 7 day room temperature plus 21 days at 40°C and 7 days water immersion. In each case a standard two cloth cleaning technique was utilized for surface preparation.

TABLE 1
Broad Sealant Adhesion Evaluation
Peel Adhesion and % Cohesive Failure
After 7 Days Room Temperature +
7 Days Water Immersion
(% Cohesive Failure)

	One Step Cleaning Technique				Two Step Cleaning & Priming
Sealant	Xylene Solvent	IPA Solvent	Alternative Organic Solvent	Silicone Cleaner/ Primer	Xylene Solvent Based Primer
Substrate – Orential Grey Duranar					
Low Modulus Silicone	100	100	100	100	100
Medium Modulus Silicone	100	100	100	100	100
Acetoxy Silicone	0	10	10	100	100
One Part Polyurethane	70	70	0	100	100
Acrylic Latex	50	0	50	100	100
Average	64	56	52	100	100
Substrate – Mill Finish Aluminum					
Low Modulus Silicone	10	100	100	100	100
Medium Modulus Silicone	70	100	100	90	100

TABLE 1 – continued
Broad Sealant Adhesion Evaluation
Peel Adhesion and % Cohesive Failure
After 7 Days Room Temperature +
7 Days Water Immersion
(% Cohesive Failure)

	One Step Cleaning Technique				Two Step Cleaning & Priming
Sealant	Xylene Solvent	IPA Solvent	Alternative Organic Solvent	Silicone Cleaner/ Primer	Xylene Solvent Based Primer
Substrate – Mill Finish Aluminum					
Acetoxy Silicone	0	20	20	0	100
One Part Polyurethane	0	70	70	100	100
Acrylic Latex	100	0	0	10	0
Average	36	58	58	60	80
Substrate – Bronze Anodized – Aluminum					
Low Modulus Silicone	100	100	100	100	100
Medium Modulus Silicone	100	100	100	100	100
Acetoxy Silicone	50	50	0	0	40
One Part Polyurethane	90	50	100	100	90
Acrylic Latex	0	0	50	0	0
Average	68	60	70	60	66
Substrate – Precast Concrete					
Low Modulus Silicone	80	100	90	80	100
Medium Modulus Silicone	10	0	0	0	0
Acetoxy Silicone	0	0	0	0	0
One Part Polyurethane	100	100	100	100	100
Acrylic Latex	0	0	0	0	0
Average	38	40	38	36	40

TABLE 1 – continued
Broad Sealant Adhesion Evaluation
Peel Adhesion and % Cohesive Failure
After 7 Days Room Temperature +
7 Days Water Immersion
(% Cohesive Failure)

	One Step Cleaning Technique				Two Step Cleaning & Priming
Sealant	Xylene Solvent	IPA Solvent	Alternative Organic Solvent	Silicone Cleaner/ Primer	Xylene Solvent Based Primer
Substrate – Polycarbonate Sheet					
Low Modulus Silicone	0	0	0	0	0
Medium Modulus Silicone	0	0	100	100	0
Acetoxy Silicone	20	0	0	0	0
One Part Polyurethane	0	0	0	0	0
Acrylic Latex	20	0	0	90	0
Average	8	0	20	38	0
Overall Average % Cohesive Failure					
	42.8	42.8	47.6	58.8	51.2

As illustrated, the use of the three organic solvent alternatives were approximately equivalent. However, the use of the one step silicone based cleaner/primer actually improved the ability of the sealants to adhere. This improvement is most pronounced with the silicone sealants and the common non–porous construction substrates (polycarbonate sheet, grey duranar, bronze anodized aluminum, mill finish). In most cases the adhesion was actually equivalent to the adhesion obtained utilizing a two step clean then prime technique.

The following section focuses on silicone sealants and a wide variety of construction substrates.

Construction Substrate Evaluation

In the previous section the use of a silicone based cleaner/primer was shown to improve the ability of sealants, particularily silicone to adhere to some construction substrates. The purpose of this evaluation is to focus on a medium modulus silicone sealant and compare its adhesion to a wide variety of construction substrates when cleaned utilizing different cleaning techniques.

Results

Table 2 documents the test results obtained utilizing a modified ASTM C–794 peel adhesion test. Cure conditions consisted of 7 day room temperature plus 7 days water immersion. The table compares cleaning the sample with xylene, to cleaning the sample with the silicone cleaner/primer solution and finally to cleaning the sample with xylene followed by a second step when the sample is activated by a solvent based primer solution.

TABLE 2
A Review Of The Adhesion Of A Medium Modulus Silicone Sealant To A Variety Of Substrates Cleaned With Different Cleaning Solutions

Substrate	Xylene	Silicone Cleaner/ Primer	Xylene Cleaned + Solvent Based Primer
Alodine Aluminum			
Alodine Finish Alum. Ext.	85	100	100
Alodine Plated Alum. Ext.	100	100	90
Alodine Plated Alum. Ext.	100	100	100
Anodized Alum. Pressure Plate	100	0	100
Anodized Aluminum			
Black Anodized Aluminum	100	100	100
Black Anodized Aluminum	100	100	100
Black Anodized Panel	99	100	100
Bronze Anodized	100	100	100
Class I Black Anodized Alum. Ext.	100	80	100
Clear Anodized Aluminum	10	90	100
Clear Anodized Alum. Ext.	100	100	100
Clear Anodized Alum. Ext.	100	100	100
Clear Anodized Alum. Ext.	100	100	100
Clear Anodized Alum. Ext.	100	100	100
Clear Anodized Alum. Ext.	100	100	100
Clear Anodized Alum. Ext.	100	100	100
Clear Anodized Aluminum Mullion	100	100	100
Clear Anodized Alum. Silicone Glazing Wedge	100	100	100
Clear Anodized Alum. Slider Piece	100	100	100

TABLE 2 – continued
A Review Of The Adhesion Of A Medium Modulus Silicone Sealant To A Variety Of Substrates Cleaned With Different Cleaning Solutions

Substrate	Xylene	Silicone Cleaner/ Primer	Xylene Cleaned + Solvent Based Primer
Anodized Aluminum			
Clear Anodized Alum. Extrusion	50	100	15
Clear Anodized Brakenshapes	100	100	0
Clear Anodized Panel	15	50	100
Clear Anodized Panel	100	100	100
Clear Anodized Plated Alum. Ext.	90	100	100
Clear Anodized Plated Alum. Ext.	100	100	100
Dark Bronze Anodized Plated Alum. Ext.	100	100	100
Dark Bronze Class II Anodized Panel	100	100	100
Painted Aluminum			
Beige Kynar Painted Alum. Ext.	100	100	100
Beige Silicone Polyester Painted Alum. Ext.	100	100	100
Black Kynar Painted (Exotic Texture) Alum. Ext.	0	90	100
Black Kynar Painted Alum. Ext.	100	97	100
Black Kynar Painted Panel	0	100	100
Black Silicone Polyester Painted Alum. Ext.	100	100	100
Blue Kynar Painted Alum. Ext.	0	100	100
Bone White Duranar Painted Alum. Ext.	100	100	100
Bone White Fluopon Painted Alum. Ext.	40	100	100
Bronze Duranar Painted Alum. Ext.	20	100	100
Bronze Kynar Painted Alum. Ext	80	100	100
Bronze Kynar Painted Panel	0	100	100
Bronze Sheet with Aliphatic Ctg.	15	100	100
Bronze Silicone Polyester Painted Alum. Ext.	100	100	100

TABLE 2 – continued
A Review Of The Adhesion Of A Medium Modulus Silicone Sealant To A Variety Of Substrates Cleaned With Different Cleaning Solutions

Substrate	Xylene	Silicone Cleaner/ Primer	Xylene Cleaned + Solvent Based Primer
Painted Aluminum			
Bronze Sheet with Clear Powder Coat	**100**	**100**	**100**
Burgandy Kynar Painted Alum. Ext.	**0**	**100**	**25**
Burgandy Silicone Polyester Painted Alum. Ext.	**100**	**100**	**100**
Fiord Blue Fluroceram Painted Alum. Ext.	**15**	**100**	**100**
Gloss White Painted Panel	**100**	**100**	**100**
Green Kynar Finish Face Cap	**80**	**100**	**100**
Green Kynar Painted Alum. Ext.	**50**	**100**	**100**
Green Kynar 500 Painted Alum. Ext.	**40**	**100**	**100**
Green Kynar 500 Painted Alum. Ext.	**10**	**100**	**100**
Green Polyester Finish Sill Member	**100**	**100**	**100**
Green Silicone Polyester Paint Chip	**25**	**100**	**40**
Green Silicone Polyester Painted Alum. Ext.	**100**	**100**	**95**
Grey Duranar XL Painted Alum. Ext.	**25**	**30**	**100**
Grey Silicone Polyester Painted Alum. Ext.	**100**	**100**	**100**
Hartford Green Fluropon Painted Alum. Ext.	**5**	**100**	**100**
Jaymont Silver Baked Enamel Finished Alum. Ext.	**100**	**100**	**100**
Jaymont Silver Silicone Polyester Painted Alum. Ext.	**100**	**100**	**100**
Lafaenza Green Duranar Painted Alum. Ext.	**100**	**100**	**100**
Lapis Blue Silicone Polyester Painted Steel Sheet Metal	**100**	**100**	**100**

TABLE 2 – continued
A Review Of The Adhesion Of A Medium Modulus Silicone Sealant To A Variety Of Substrates Cleaned With Different Cleaning Solutions

Substrate	Xylene	Silicone Cleaner/ Primer	Xylene Cleaned + Solvent Based Primer
Painted Aluminum			
Light Grey Fluoropon Painted Alum. Ext.	100	100	100
Maroon Kynar Painted Alum. Ext.	100	100	100
Maroon Kynar Painted Alum. Ext. – Overspray	70	100	100
Medium Bronze Painted Panel	80	100	100
Monterey Pine Fluropon Painted Alum. Ext.	80	100	0
Off White Silicone Polyester Painted Alum. Ext.	100	100	100
White Painted Panel	100	100	100
Pewter Duranar Painted Sample Panel	65	100	100
Pewter Kynar Painted Alum. Ext. (L Angle)	100	100	100
Polar Bear Fluoroceram Painted Alum. Ext.	100	100	100
Polar Bear Fluoroceram Painted Alum. Ext.	100	100	100
Rome Green Fluoroceram Painted Alum. Ext.	80	100	100
Sandstone Kynar 500 Paint Chip	100	100	100
Silver Kynar Painted Alum. Sample	80	100	100
Silver Duranar Painted Alum. Ext.	40	100	95
Silver Kynar Painted (Exotic Texture) Alum. Ext.	10	90	100
Silver Kynar Painted Alum. Ext. (L Angle)	90	100	100
Smoke Black Fluoropon Painted Alum. Ext.	100	100	100
Spruce Green (DG-112-01) Fluoroceram Painted alum. Ext.	100	100	100
Thistle II Green Kynar Painted Alum. Ext.	100	100	100

TABLE 2 – continued
A Review Of The Adhesion Of A Medium Modulus Silicone Sealant To A Variety Of Substrates Cleaned With Different Cleaning Solutions

Substrate	Xylene	Silicone Cleaner/ Primer	Xylene Cleaned + Solvent Based Primer
Painted Aluminum			
UC54434 Aged Copper Duranar Painted Alum. Ext.	75	100	100
White Diamond Pearl Kynar Painted Steel Panel	100	100	100
White Kynar Painted Alum. Ext.	90	100	100
White Kynar Painted Alum. Ext.	100	100	100
White Kynar Painted Alum. Ext.	50	100	100
White Kynar 500 Painted Alum. Ext.	100	100	100
White Painted Alum. Ext. (L Angle)	100	100	100
White Polycron Painted Alum. Ext.	0	100	100
White Silicone Polyester Painted Alum. Ext.	100	100	100
Winter White Kynar Paint Chip	100	100	100
Glass			
CP–14 on clear Base, 1/4" Monolithic Glass Sample	100	100	100
Green Spandrelite Ceramic Frit Spandrel Glass Sample	100	100	100
Green Reflective Glass – Edge	100	100	100
HC 380 DW65 Solarflex	100	100	100
HC 380 Neutral Solarflex Spandrel	100	100	100
Insulated Glass Unit	100	100	100
Laminated Glass	100	100	100
Solargray Vision Glass	80	100	100
Reflective Glass	80	100	100
Cool Lite SS 508	100	100	100
SL 105 Neutral Solarflex Laminated Glass with Edge Sealer – Edges	100	100	100
SPI Glass S2–14	97	100	100

TABLE 2 – continued
A Review Of The Adhesion Of A Medium Modulus Silicone Sealant To A Variety Of Substrates Cleaned With Different Cleaning Solutions

Substrate	Xylene	Silicone Cleaner/ Primer	Xylene Cleaned + Solvent Based Primer
Glass			
SPI P2–13 Reflective Glass	100	100	100
SS–14 on Green Base, 1/4" Monolithic Sample	100	100	100
SS–14 Laminated Glass	100	100	100
1/4" VA3–18 Reflective Vision Glass	100	100	100
VT 5–20 Spandrel Glass with Opacifier Cut Back	100	100	100
VT 5–20 Vision Glass	100	100	100
1/4" Clear Glass	100	100	100
1/4" Blue Glass	100	100	100
1/4" Green Tint Glass Sample	100	100	100
1" Insulating Glass Unit	100	100	100
12 x 12 x 3/8 Vision Glass	100	100	100
3/4" Clear Tempered Glass	100	100	100
9/16 Laminated Glass Sample	90	90	95
Concrete			
Precast Concrete	0	0	0
Granite			
3cm Carnation Rose Granite – Block 3714 Backside	90	85	99
3cm Carnation Rose Granite – Block 3714 Polished Face	100	98	100
Other			
Acrylic Sheet	100	0	0
Copper Sheet Metal	100	100	100
Mill Finish Alum. Ext.	0	100	100
Mill Finish Alum. Ext.	90	100	100
Plexiglass	100	100	100
PVC	0	100	100
Sanded Plywood	0	0	0
Unsealed Plywood	0	0	0

TABLE 2 – continued
A Review Of The Adhesion Of A Medium Modulus Silicone Sealant To A Variety Of Substrates Cleaned With Different Cleaning Solutions

Substrate	Xylene	Silicone Cleaner/ Primer	Xylene Cleaned + Solvent Based Primer
Other			
3/4" A/A Plywood Treated with Sealant Lacquer	3	25	8
Red Primer on Steel	100	100	100
AVERAGE % COHESIVE FAILURE	79.62	94.83	92.95

This study indicates that to a wide variety of substrates the use of the low volatile cleaner/primer actually improves adhesion beyond that obtained through the use of a common organic cleaning solvent even to the point where adhesion is as good as cleaning followed by a second priming step.

Cleaning Review

Since the performance of a cleaning solvent is by definition a function of how well that solvent removes a variety of surface contaminations this section is intended as a review of solvents ability to remove various contaminations.

Experimental Method

The investigation was performed by purposely contaminating surfaces, cleaning the contaminated surfaces with a variety of cleaners and determining if an acetoxy silicone sealant would adhere to the newly cleaned surface. In this test a simple tab adhesion test was performed to determine if the sealant would fail adhesively or cohesively.

Results

The results of this investigation are documented in Table 3. As illustrated in this table the adhesion of the sealant onto the surface which was cleaned and the silicone based cleaner/primer was superior than the adhesion obtained utilizing any other cleaning technique.

TABLE 3
Tab Adhesion of Acetoxy Silicone Onto Bronze Anodized Aluminum Surfaces Contaminated Prior To Cleaning

Contamination*	Tab Adhesion (% Cohesive Failure 7 Day Cure) Cleaning Techniques		
	Xylene	**IPA**	**Silicone Cleaner/Primer**
None	0	0	60
Pencil Lead	0	30	100
Household Oil	0	0	30
Average (% C.F.)	**0**	**10**	**63**

* In each case the bronze anodized aluminum was completely covered with the referenced contaminant.

Conclusions

As regulations prohibiting the use of materials containing a high percentage of volatile organic compounds continue to tighten, some manufacturers are beginning to introduce alternative solutions for cleaning and priming construction joints. This study compared one product, a silicone based cleaner/primer recommended for use with non–porous substrates and one organic solvent which is reported to be more environmentally safe to conventional cleaning and priming techniques.

Surprisingly the use of the low volatile silicone based cleaner/primer not only out performed the organic solvents it also promoted silicone sealant adhesion in one step as well as the traditional two step cleaning/priming procedures.

Voytek S. Gutowski[1], Lee Russell[1] and Anthony Cerra[1]

NEW TESTS FOR ADHESION OF SILICONE SEALANTS

REFERENCE: Gutowski, V. S., Russell, L., and Cerra, A., **"New Tests for Adhesion of Silicone Sealants"**, Science and Technology of Building Seals, Sealants, Glazing, and Waterproofing: Second Volume, ASTM STP 1200, Jerome M. Klosowski, Ed., American Society for Testing and Materials, Philadelphia, 1992.

ABSTRACT: It has been widely accepted that the ASTM C–794 peel test is suitable for determining adhesion between the sealants and substrates used in curtain wall construction. However, it is shown in this paper that this test has certain deficiencies which may allow the acceptance of some materials which in fact have poor initial adhesion and long-term durability.

The influence of the specimen's geometry, sealant type, sealant thickness, cross-head speed (test rate), etc, on the resultant failure mode and quantitative results are discussed in detail.

Finally, two novel adhesion tests are proposed and discussed, i.e.:

(a) 'thin' peel specimen test
(b) 'thin' lap-shear specimen test.

Comparative results, obtained using a range of common commercially available sealants, show that the proposed tests have significant advantages over the accepted ASTM C–794 test as well as the DIN 52455 and ISO/DIS 9047 tests.

KEY WORDS: silicone sealant, adhesion, surface energy, polarity, acid-base interactions, stress concentration, peel tests, tensile tests, lap-shear tests.

[1]Dr Gutowski is a Principal Research Scientist; Mr Cerra and Ms Russell are Experimental Scientists at the CSIRO Division of Building, Construction and Engineering, Melbourne, Australia. Postal address: PO Box 56, Highett, Victoria 3190, Australia

INTRODUCTION

Structural glazing is becoming increasingly popular as a means of finishing building facades. In this method, glass, aluminium, stone or other materials are bonded to a building structure by silicone sealants. In the USA about 30% of new glass curtain walls are structurally glazed. A similar trend can be observed in Australia and the South Pacific regions.

Silicone 'sealants' are presently the only materials that possess the necessary properties for this application: they can bond reliably to various materials, e.g. glass, anodised or coated aluminium, stone, etc. even when the cured sealant is wet. They have sufficient elastic properties to accommodate the thermal and structural movements of the building components. Both the adhesive and cohesive properties of silicone sealants are little affected by ultra-violet radiation and other environmental factors.

Most glazing systems still use either, entirely mechanical fastening, e.g. preformed gaskets, or a combination of mechanical connectors and silicone sealant to fasten a glass panel to a frame. However, there is a trend amongst developers and architects to fix all four sides of a panel in place without any mechanical support. In the latter case, the structural integrity, and thus safety of the system depends on the quality of adhesion between the sealant, external panel and building frame. However, the emergence of various finishes applied to the surface of structural members (e.g. polymeric coatings on aluminium, metallic and ceramic coatings on glass and various finishes on anodised aluminium), results in significant changes to the surface properties of the substrate which can adversely affect sealant adhesion.

Although the outstanding bonding capacity of silicone sealants is already recognised, the mechanism of adhesion is not yet fully understood, and this may lead to inadvertent design problems in the facade. There is also a great deal of controversy regarding the available techniques for assessing bond quality e.g. ASTM C–794, DIN 52455, ISO 9047 and how the results of these tests relate to the real in-service situation.

It must be emphasised here that the extensive experience of the sealant manufacturers and construction industry provides a high level of confidence in the optimisation of a sealant/substrate system. However, it is desirable to extend the scientific understanding of the interaction between the various components in the system.

This paper presents the results of an investigation that examines present test methods, investigates the effect of test parameters, e.g. test rate and sealant thickness on failure mode for a typical range of facade materials and gives a theoretical explanation of the mechanism of the peel test. Finally, as part of this work a novel technique for adhesion testing was developed as a result of the deficiencies identified in existing methods. The scope of this paper does not allow for a detailed discussion of results on all silicone sealants investigated. However, the examples shown and analysed are representative for all phenomena and trends observed within the whole range of materials considered in this work.

EXPERIMENTAL

1. Materials

A wide range of curtain wall materials were used in the experiments, i.e. anodised aluminium including various colouring techniques such as electrolytical-cobalt salts; electrolytical-tin salts, organic dye, integral colour; organic coatings, i.e. powder polyester, wet polyester, PVF2; reflective glass; clear float glass; stone; fibrous cement. Also a range of engineering plastics such as polymethyl methacrylate (PMMA), acrylonitrile-butadiene-styrene terpolymer (ABS), nylon 6-6, high impact polystyrene (PS), vinyl copolymer, acetal, polypropylene – natural (PP) and filled (PP Filled), polyethylene – low density (LDPE), high density filled (HDPE), and ultra-high molecular weight (UHMW-PE), were selected to provide a broad range of surface properties.

A number of typical silicone sealants were used which included 1 part alcoxy, 1 part oxime, 1 part acetoxy and 2 part RTV type curing systems.

2. Test Methods

Peel strength: This was evaluated using ASTM C–794 except that the specimen size was reduced from 152 x 76 x 6.3 to 100 x 25 x 4 mm. All tests were done in duplicate at the standard strain rate of 50 mm/min as well as 0.05, 1.0, 25.0, 50.0 and 1000 mm/min. The sealant thickness was controlled at 1.6 mm as per the standard (Figure 1a). A 'thin peel' specimen was also tested with a sealant thickness of 0.2 mm (see Appendix).

Shear strength: This was determined by single lap-shear specimens as illustrated in Figure 1b. The tests were done in triplicate at strain rates of 0.05, 5, 10, 50, 250 and 1000 mm/min using the following sealant thicknesses 0.2, 0.5, 1.0 and 3.0 mm. The preparation of specimens with thin sealant layers is described in the Appendix.

Tensile strength: This was determined using DIN 52455 and ISO/DIS 9047 (12 x 12 x 50 mm) and DIN 52455 (10 x 6 x 50 mm) specimens illustrated in Figure 1c. Tests were done in triplicate at strain rates of 0.05; 10 and 50 mm/min.

3. Surface Preparation

All substrates were thoroughly cleaned in accordance with the recommendations of the sealant manufacturers. Typically, the following cleaning liquids were used: ethyl alcohol, iso-propyl alcohol, MEK (methyl-ethyl ketone) and white spirit. All substrates were initially wiped with ethyl alcohol (3 times) in order to remove surface contaminants, prior to final cleaning with one of the above liquids, e.g. glass substrates: MEK; anodised aluminium: MEK; organic coatings: ethyl alcohol, white spirit or isopropyl alcohol – depending on recommendations.

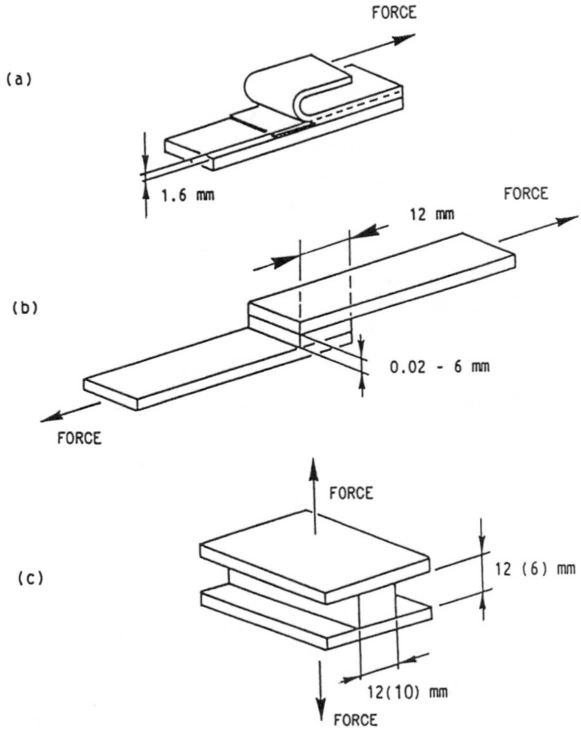

Figure 1. Geometry of test specimens: (a) ASTM C–794 peel specimen; (b) typical single lap-shear specimen; (c) typical tensile specimens standardised by DIN 52455 and ISO/DIS 9047.

All cleaned specimens were bonded on the same day typically within 2 hours after cleaning.

4. Cure of Specimens

All bonded assemblies were allowed to cure for four weeks as follows prior to testing: 2 weeks at 23°C/50% RH, 1 week at 38°C/98% RH, 1 week at 23°C/50% RH.

RESULTS AND DISCUSSION

Deficiency of Existing Techniques for Sealant Adhesion

The scope of this paper does not allow for a detailed discussion on the mechanism of adhesion between silicone sealants and organic or inorganic substrates as used in facade construction. The authors refer the reader to earlier publications on this topic [1,2].

It is shown in earlier publications [1, 2] that the strength of adhesion between any silicone sealant and any organic or inorganic substrate depends on the energy of acid-base interactions, W_A^{ab}. The typical relationship: strength *v.* W_A^{ab} is illustrated in Figure 2.

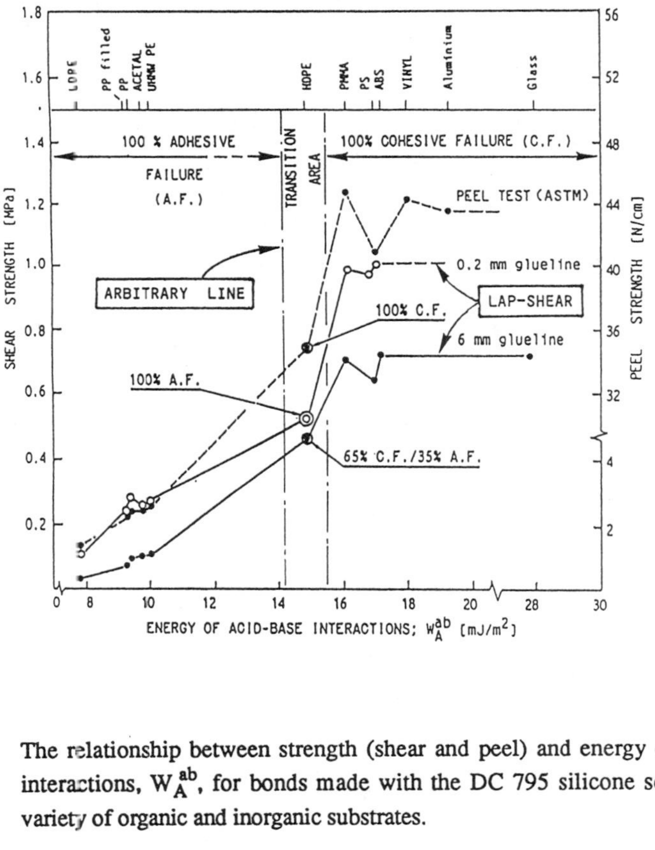

Figure 2. **The relationship between strength (shear and peel) and energy of acid-base interactions, W_A^{ab}, for bonds made with the DC 795 silicone sealant and a variety of organic and inorganic substrates.**

It is seen from Figure 2 that the bond strength increases monotonically with increasing acid-base interaction energy, up to the point where the strength is determined by the cohesive strength of a sealant.

As expected, the lowest strength and 100% adhesive failure (delamination at the sealant/substrate interface) occurs with the substrates exhibiting the lowest polar component of the total surface energy, e.g. LDPE with γ_1^p = 2.65 mJ/m2 (see Table 1). As the value of γ_1^p increases for a series of substrate materials, the resultant bond strength also increases monotonically for such plastics as PP, Acetal and UHMW PE.

However, with HDPE, a material which should be inherently difficult to bond due to the low γ_1^p the following observations have been made:

(a) ASTM C–794 peel specimens (Figure 1a) exhibit 100% cohesive failure within the sealant,

(b) the 6 mm thick sealant bead exhibited a mixed failure mode (i.e. 35% adhesive/65% cohesive failure) when tested using a lap-shear specimen (see Figure 1b),

(c) the 6 mm thick sealant bead in tensile specimens, as recommended by the DIN 52455 (results not shown in Figure 1) exhibited 100% cohesive failure within the sealant,

(d) a CSIRO-modified 'thin peel specimen' employing a 0.2 mm thick sealant bead exhibited 100% adhesive failure at the sealant/substrate interface.

Table 1. Surface properties of substrates used in bonding with sealant DC 795 and the relevant energy of acid-base interaction (values of W_A^{ab} for bonds with sealant DC 795 [Ref. 1]).

Material	γ_1 (mJ/m^2)	W_A^{ab} (mJ/m^2)	γ_1^p (mJ/m^2)
Glass	59.0	31.57	33.6
Aluminium	38.5	21.38	15.4
Nylon 6-6	39.44	21.86	16.11
ABS	39.97	19.21	12.68
Vinyl copolymer	43.36	20.36	13.99
PS	41.00	19.41	12.44
PMMA	37.84	18.25	11.15
HDPE	37.26	16.79	9.58
UHMW–PE	28.47	11.29	4.33
Acetal	29.68	11.06	4.12
PP	24.36	10.70	3.82
PP filled	24.16	10.59	3.73
LDPE	26.34	8.85	2.65

W_A^{ab} calculated using the following sealant properties: γ_2 = 18.09 mJ/m^2; γ_2^p = 7.42 mJ/m^2 (all sealant properties refer to its uncured state).

The results described in (a) to (c) above are somewhat disturbing for a substrate such as HDPE and indicate that the DIN 52455 tensile test as well as the ASTM C–794 peel test are less severe under certain conditions than perhaps they should be. The observed transition between adhesive and cohesive modes of failure occurs at the point where the energy of acid-base interactions, W_A^{ab}, equals about 15 mJ/m^2.

All other systems with W_A^{ab} values greater than 16 mJ/m^2 exhibit 100% cohesive failure within the sealant, for both organic and inorganic substrates, within the scope of the experiment.

From the results obtained with HDPE it is perceived by the authors that such factors as (a) the thickness of the sealant layer between the substrates, and (b) the test rate (i.e. the cross-head speed), are important in governing failure mode. The importance of these factors is now examined.

Figure 3 shows photographs of two sets of peel specimens prepared using DC 795 silicone sealant and high density polyethylene (HDPE). The specimens were as follows:

(a) the ASTM C-794 peel specimen with the recommended 1.6 mm sealant thickness, and

(b) the CSIRO 'thin peel specimen' prepared with a 0.2 mm sealant thickness. Both types of specimens were tested using the following peel rates: 0.05; 0.5; 2.5 and 5 cm/min.

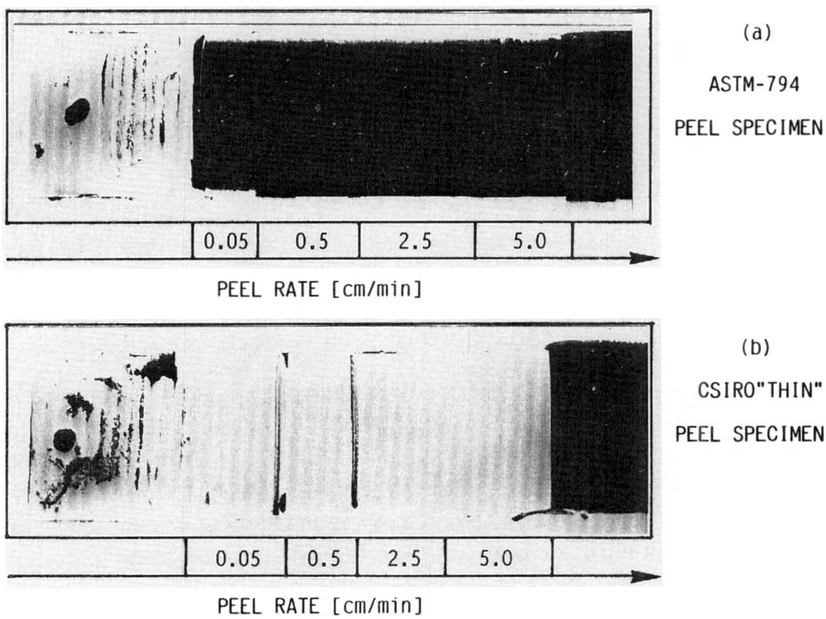

Figure 3. Appearance of the fracture surfaces of peel specimens prepared using DC 795 sealant and High Density Polyethylene (HDPE) substrate: (a) ASTM C-794 peel specimen, and (b) CSIRO 'thin bead' peel specimen. Both specimens were tested at the following peel rates: 0.05; 0.5; 2.5 and 5.0 cm/min.

It is evident from the photos that the CSIRO 'thin peel' specimens exhibit 100% AF independently of the peel rate, thus detecting the expected inherent adhesion problems associated with this particular substrate. Also the ASTM C–794 specimen exhibits 100% cohesive failure when tested at 50 mm/min – the standard rate. Significantly lower test speeds, i.e. 0.05 mm/min and 0.5 mm/min also result in 100% CF.

The results for the lap-shear tests using HDPE substrates and DC 795 silicone sealant are shown in Table 2. Once again for the CSIRO thin specimens (0.2 mm) adhesive failure occurs for all test speeds with the extent of cohesive failure increasing

Table 2. The influence of bondline thickness and test rate on the failure modes of lap shear specimens made with HDPE substrates bonded using 795 Dow Corning Silicone Sealant

Bondline Thickness (mm)	Cross-head Speed (cm/min)				
	0.05	0.5	5.0	25	100
0.2	AF	AF	AF	AF	AF
0.5	AF	AF	C (95%)	C (85%)	AF (50%) CF (50%)
1.0	C (98%)	C	C	C	C
3.0	C	C	C	C	C

as the sealant thickness increases. For certain thicknesses the cross-head speed appears to be the controlling factor in determining the failure mode with cohesive failure occurring at higher speeds. It may thus be possible to increase the level of adhesive failure in the thicker specimens by reducing the test speed below 0.05 cm/min.

Another set of results proving our conclusions regarding the influence of test parameters such as sealant bead thickness and the test rate is shown in Table 3. In this set of experiments the substrates were typical for facade construction. They consisted of anodised aluminium specially prepared with two anodised layer thicknesses, three seal methods and a range of etch levels. The oxime cured, RTV1 sealant used was known to exhibit inherent adhesion problems. The results confirm the above observation that reducing the testing speed tends to increase the severity of the tests as indicated by the increased level of adhesive failures for the CSIRO 'Thin Peel' tests at 0.05 mm/min compared to 50 mm/min. Similar results were obtained with several other silicone sealants investigated in this work. On the other hand, those sealants that exhibit inherently good adhesion, e.g. DC 795, revealed 100% cohesive failure within the sealant bead whatever the sealant bead thickness and the test rate.

Table 3. A comparison of peel strength and failure mode (% of adhesive failure) for anodised aluminium/silicone sealant (oxime cure) peel samples tested using ASTM C-794 and the new CSIRO type of specimens (dry specimens tested after 4 weeks cure)

	ASTM C-794 (1.6 mm glueline) test rate: 50 mm/min		CSIRO 'THIN PEEL' (0.2 mm glueline) test rate: 50 mm/min		test rate: 0.05mm/min	
SUBSTRATE	Strength [N/cm]	Adh. failure [%]	Strength [N/cm]	Adh. failure [%]	Strength [N/cm]	Adh. failure [%]
Black Anodised						
25 µm/HS	30.4	0	10.0	95	2.4	100
25 µm/CS	16.0	60	5.2	100	1.6	100
25 µm/AS	24.8	0	9.2	60	3.6	100
15 µm/HS	32.4	0	8.0	95	1.8	100
15 µm/CS	30.0	5	8.8	95	1.8	100
15 µm/AS	32.0	0	10.4	95	2.0	100
Clear Anodised						
25 µm/HS	31.0	0	9.6	100	2.2	100
25 µm/CS	32.0	0	10.8	75	2.0	100
25 µm/AS	31.0	0	11.2	85	2.0	100
15 µm/HS	28.8	10	18.4	75	5.8	75
15 µm/CS	30.0	0	7.4	100	2.0	100
15 µm/AS	32.8	0	7.2	100	2.2	100
Bronze Anodised						
25 µm/HS	29.6	5	7.6	100	2.0	100
25 µm/CS	29.6	5	7.6	100	2.0	100
25 µm/AS	30.4	5	11.2	70	3.6	90
15 µm/HS	29.6	5	13.6	60	2.0	100
15 µm/CS	31.2	10	8.8	95	2.2	100
15 µm/AS	28.8	20	6.8	100	1.8	100

Symbols: 15 µm/25 µm – anodised layer thickness; HS – hot sealing, AS – accelerated sealing; CS – cold sealing. Base alloy type: 6063. Colouring: electrolytic/cobalt salt colour. All specimens etched at 100 g/m^2.

The results again confirm our earlier results and show that adhesive failure is promoted in sealant/substrate systems with inherent adhesion problems, by reducing the thickness of the sealant layer between the substrates to about 0.2 mm as for the CSIRO Peel Test. This is evident by a comparison of results for C–794 and the CSIRO test at 50 mm/min.

An Analysis of the Observed Experimental Results

The reasons for the effect of sealant thickness, as observed above, can be explained if we examine in more detail one of the tests used such as the peel test (see Figure 1a), whose mechanics is discussed in detail in References [3–8]. Figure 4 illustrates the distribution of the tensile (or cleavage) stress σ_{11} at the distance from the peel front, given by the following relationship originally developed by Kaelble [5–7].

$$\sigma_{11} = \sigma_0(\cos\beta_p x + \kappa \sin\beta_p x) \exp(\beta_p x), \tag{1}$$

where:

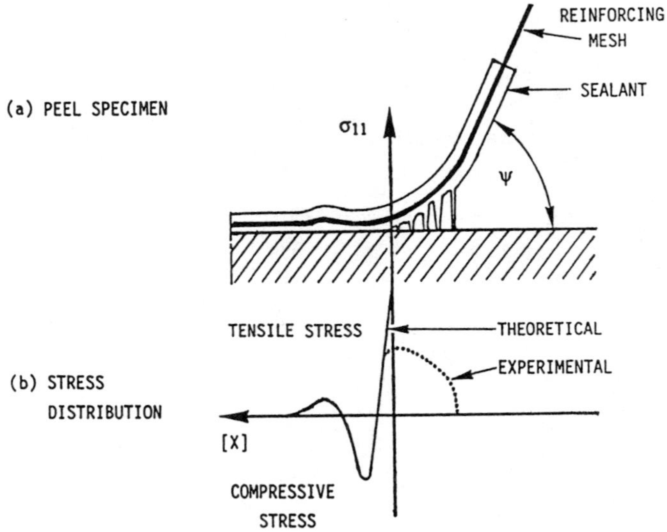

Figure 4. **Distribution of the normal stresses σ_{11} at the sealant/substrate interface ahead of the advancing peel front [6].**

$$\beta_p = \left(\frac{E_a b}{4EIh_a}\right)^{1/4} \tag{2}$$

$$\kappa = \frac{\beta_p m}{\beta_p m + \sin\psi} \quad \text{, and} \tag{3}$$

σ_0 is the boundary cleavage stress at x = 0; E_a is the elastic tensile modulus of the sealant; E is the elastic tensile modulus of the backing material (reinforcing mesh or continuous strip), h_a is the thickness of the sealant layer between the rigid substrate and backing material; I is the moment of inertia of the peeling strip cross-section; b is the bond width; m is the moment of the peel force and ψ is the peel angle.

Eqn (1) implies that the peel stress is represented by a highly damped harmonic function in which alternating compressive/tensile stress zones exist along the peel direction. Theoretically (see Figure 4) the peel stress falls from maximum to zero at the peel front but in practical terms some tensile stress exists in the nominally detached zone due to the presence of some unbroken fibrils of sealant.

According to Kaelble's theory [5–7] the peel force F per width of specimen is given by

$$\frac{F}{b} = \frac{h_a \kappa^2 \sigma_o^2}{2E_a\left(1 - \cos \psi\right)}. \tag{4}$$

The quantity 'κ' is shown [see Eqns (3) and (4)] to be a function of the parameter β, the moment of the peel force 'm' and the peel angle ψ. Thus, Eqn (4) implicitly considers the elastic properties of the flexible backing material, e.g. reinforcing mesh or tape.

One drawback of the Kaelble's equation (4) is that it does not consider the influence of the peel rate (cross-head speed) on the resultant peel force. The authors of this paper carried out extensive tests to determine the influence of peel rate on the peel strength of specimens with the controlled sealant thickness. The results of these tests are illustrated in Figure 5 and it can be seen that because of the effect of peel rate, the

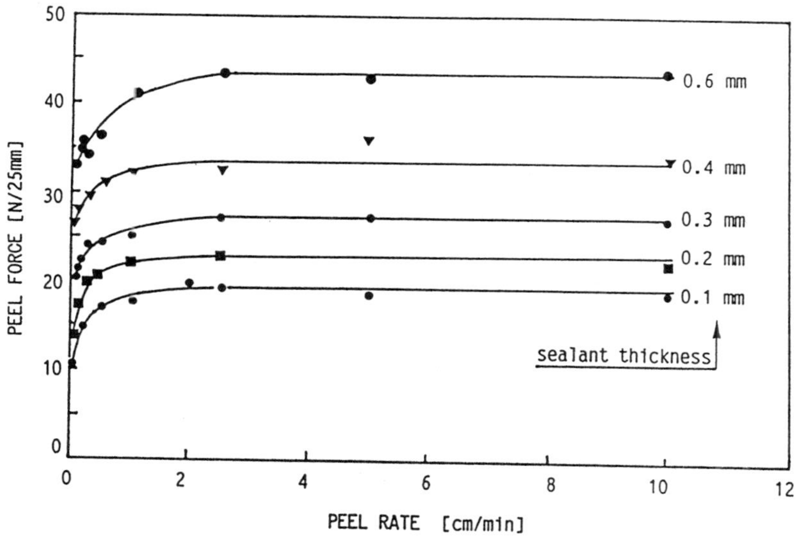

Figure 5. The relationship between peel strength and peel rate for specimens prepared using DC 795 silicone sealant and PVC substrate. Specimens were prepared using different sealant bead thicknesses as illustrated in the graph. (All specimens exhibited 100% cohesive failure within the sealant, independent of peel rate.)

Kaelble equation (4) should be corrected as follows:

$$\frac{F}{b} = \frac{h_a \kappa^2 \sigma_o^2}{2E_a(1 - \cos\psi)} \cdot f(\dot{a}) \tag{5}$$

where $\dot{a}$ is the peel rate. This will be analysed in greater detail in a future paper.

For the most commonly used peel techniques, e.g. ASTM C–794, we have a peel angle $\psi = 180°$ (see Figure 1a). It follows from Eqns (3) and (5) that

$$\kappa = 1.0, \text{ and} \tag{6}$$

$$\frac{F}{b} \cong \frac{h_a \sigma_o^2}{4E_a} \cdot f(\dot{a}) \tag{7}$$

The authors of this paper use the approximation symbol in Eqn (7) because it can be perceived that although $\kappa = 1$ for $\psi = 180°$, there is some influence due to the backing material.

It can now be deduced from Eqn (7) that the stress at the peel front for a peel angle $\psi = 180°$ is

$$\sigma_o \cong 2\left[\frac{F E_a}{bh_a} \cdot \frac{1}{f(\dot{a})}\right]^{1/2} \tag{8}$$

In order to fully analyse equation (8) one has to know the relationship between the peel force (F) and sealant thickness as well as that between the elastic tensile modulus (E_a) and the sealant thickness.

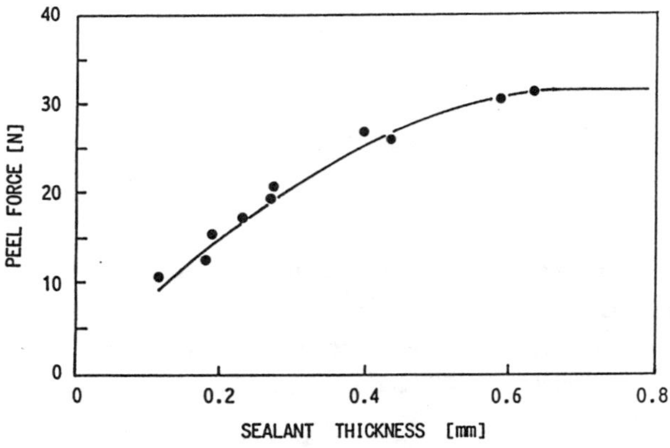

Figure 6. **Influence of sealant thickness on the peel force (DC 795 sealant and PVC substrate. Test rate: 0.05 cm/min).**

With regard to the peel force it is shown in Figure 6, that an increase in the elastomer's thickness results initially in the proportional increase in the peel force. At a certain threshold value, however, the peel force becomes constant and independent of any further increase in the sealant's thickness. A similar relationship for another type of elastomer has been obtained by Gent and Hamed [8].

In another work of the authors [9] it is shown that the elastic tensile modulus (Young's modulus) of the silicone sealant is strongly dependent on the thickness of the sealant bead. An example of this relationship is illustrated in Figure 7.

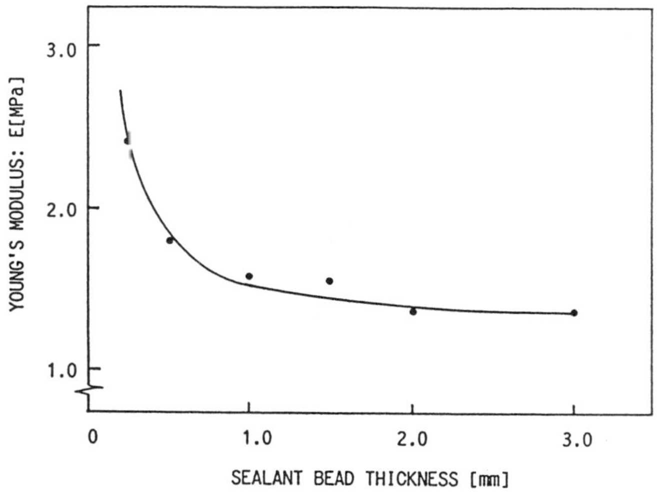

Figure 7. **Influence of sealant thickness on the dynamic elastic tensile modulus (Young's modulus) of silicone sealant.**

It is apparent from this Figure that for thin sealant beads the modulus is relatively high. An increase in the sealant thickness from 0.25 mm to 2.0 mm results in a gradual decrease of modulus of elasticity. Any further increase of sealant thickness above this threshold value (about 2 mm) does not change the value of the dynamic modulus of elasticity of the silicone sealant which then becomes constant and independent of the elastomer's thickness.

Knowing the relationship between peel force and sealant thickness and that between Young's modulus and sealant thickness we can now estimate from Eqn (8) that the stress at the peel front is dependent on the sealant bead thickness as shown by the continuous line in Figure 8. Therefore, it can be concluded that the stress at the peel front may be increased by reducing the thickness of the sealant bead.

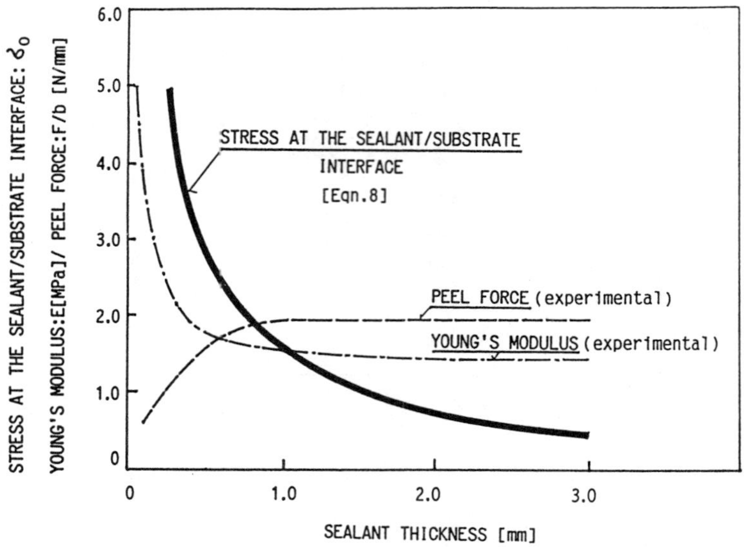

Figure 8. **Stress at the sealant/substrate interface at the peel front, as a function of the parameters shown in Eqn (8): dynamic modulus of elasticity [E]; sealant layer thickness [h_g]; peel rate [ȧ] and the apparent peel force [F/b].**

It can be seen from Figure 8 that a reduction of sealant thickness from 1.6 mm (typical for ASTM C–794 test) to 0.2 mm as suggested in the CSIRO test procedure results in a 4–fold increase of stress at the sealant/substrate interface. This may lead, as shown earlier in Table 3, to the detection of any inherent adhesion problems for a given substrate/sealant combination. Furthermore, it was shown (see Table 2) that a decrease in the test rate may further increase the severity of the test, thereby promoting adhesive failure. These two factors, i.e. sealant thickness and peel rate, can be now used to detect unsuitable materials, and to establish more reliable safety factors.

CONCLUSIONS

1. Existing standardised testing procedures, i.e ISO/DIN 9047, DIN 52455 and ASTM C–794, .may fail to identify sealant/substrate combinations with marginal adhesion due to inappropriate test and specimen parameters. It has been shown that usually the sealant thickness and test rate are too great and new test methods are proposed, designed to promote adhesive failure.

2. It is possible to provide a theoretical explanation for the relationship between failure mode and the critical test parameters such as sealant thickness and testing speed in the peel specimens used for testing the quality of adhesion at the sealant/substrate interface.

3. The new procedures for testing adhesion of silicone sealants were developed based on theory and experimental evidence presented in this paper.
These procedures employ the following specimens:

   (a) a modified 'thin' peel specimen with a thin (0.2 mm) sealant layer between the rigid substrate and backing tape.

   (b) 'thin' lap-shear specimen with a thin (0.2 mm) sealant bead between the substrates.

   It has been shown that the severity of the test may further be increased by reducing the test rate to about 0.05–0.1 mm/min.

ACKNOWLEDGMENTS

The authors wish to thank the Consortium of sponsors involved in the 'Structural Glazing Project' for the financial support of this work.

REFERENCES

[1] Gutowski, W.S., Adhesive Properties of Silicone Sealants, in Building Seals – Materials, Properties and Performance, (ed. F. O'Connor) ASTM STP 1069, Philadelphia, 1990, p. 174.

[2] Gutowski, W.S., Cerra, A. and Russell, L., Adhesion of Silicone Sealants, in Durability of Building Materials and Components (ed. J. M. Baber *et al.*), Chapman and Hall, London, New York, Tokyo, Melbourne, Madras, 1990, p. 671.

[3] Spies, G.J., Aircraft Engineering, **25**, 1953, p. 64.

[4] Bikerman, J.J., Theory of Peeling Through Hookean Solid, Journal of Applied Physics, **28**, 1957, p. 1484.

[5] Kaelbe, D.H., Theory and analysis of Peel Adhesion: Bond Stress and Distributions, Transactions of Society of Rheology, **3**, 1959, p. 161.

[6] Kaelbe, D.H., Theory and analysis of Peel Adhesion: Bond Stress and Distributions,Transactions of Society of Rheology, **4**, 1960, p. 45.

[7] Kaelbe, D.H., Peel Adhesion: Micro-fracture Mechanism of Interfacial in Bonding of Polymers, Transactions of Society of Rheology, **9**, 1965, p. 135.

[8] Gent, A.N. and Hamed, G.R., Peel Mechanics of an Elastic-Plastic Adherent, Journal of Applied Polymer Science, **21**, 1977, p. 2817.

[9] Gutowski, W.S. and Russell, L., Rheological Properties of Silicone Sealants, CSIRO, Australia, 1991, (unpublished report).

[10] Kendal, K., Peel Adhesion of Solid Films – The Surface and Bulk Effects, Journal of Adhesion, **5**, 1973, p. 179.

APPENDIX

1. PREPARATION OF THE 'THIN' PEEL SPECIMEN

1.1. Materials required:

(a) one piece of substrate material: 100 mm long, 25 mm wide thickness optional

(b) PET film (about 0.1 to 0.15 mm thick) or polyester or nylon 'fly screen' mesh 150 mm long, 25 mm wide

(c) glass rod, spatula or stencils (see Figure 9b) for controlling sealant thickness.

1.2. Specimen preparation:

(a) substrate must be cleaned to the manufacturer's specification;

(b) PET film, or polyester or nylon 'fly screen' mesh must be thoroughly cleaned by immersion in MEK or ethyl alcohol;

(c) very thin layer of sealant should be evenly distributed over the length of the substrate;

(d) the sealant thickness can be controlled as shown in Figure 9b if a continuous film e.g. PET is used as the flexible substrate, by placing a few single glass beads in the sealant layer over the length of the substrate. It is suggested that the glass beads are placed at each corner of the specimen as well as at 1/3 and 2/3 of the specimen length along its both sides.

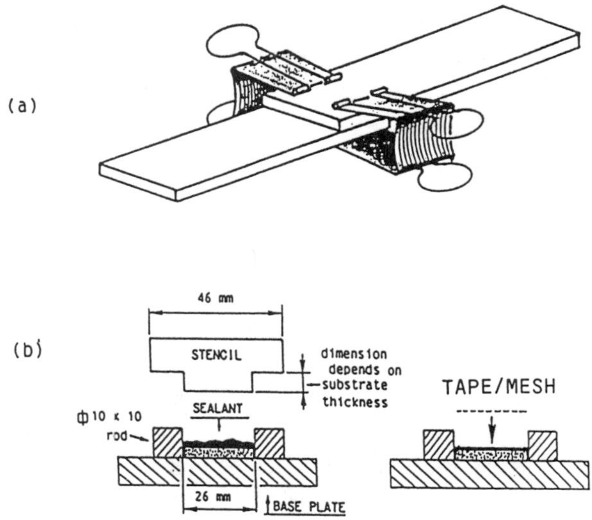

Figure 9. **CSIRO specimens for testing adhesion of silicone sealants: (a) a typical assembly of the 'thin' lap-shear specimen; (b) a method for controlling sealant bead thickness for a 'thin peel specimen'.**

2. PREPARATION OF 'THIN' LAP SHEAR SPECIMEN

2.1. Material required:

(a) two pieces of the substrate material: 100 mm long, 25 mm wide, optional thickness within the range 0.5 to 10 mm, depending on substrate strength

(b) glass beads about 0.2 mm diameter for control of the sealant thickness

(c) two fold-back stationary clips

(d) glass rod or spatula

2.2. Specimen preparation:

(a) substrates should be cleaned to the manufacturer's specification, e.g. by wiping with ethyl alcohol, MEK, white spirit or other suggested solvent;

(b) after the appropriate drying time some quantity of the sealant must be placed at one end of the substrate. Sealant should then be evenly distributed over the overlap area (12.5 to 25 mm) by the use of a glass rod or spatula. Sufficient sealant should be applied to ensure that the joint will be completely filled, i.e. excess sealant will be squeezed from the joint when the specimen is assembled;

(c) single glass beads (ϕ0.2 mm) should be placed on the sealant layer at each corner of the overlap area.

(d) the second piece of substrate material should be placed firmly on top of the sealant layer to obtain the overlap required (i.e. 12.5 to 25 mm);

(e) the assembly should be clipped using fold-back clips as illustrated in Figure 9a.

(f) a 'very thin' lap shear specimen can be prepared by using a procedure as above, without the use of glass beads [i.e. exclusion of step 2.2(c)].

Specimen cure

(1) thin 'lap shear' specimen: 4 weeks at 50% RH and 23 ± 2°C or 2 weeks at 50%/23°C, 1 week at 38 ± 2°C/98% RH and 1 week at 50% RH/23°C.

(2) 'thin' peel specimen: as above, but this may be shortened to the following: 1 week at 50% RH/23°C followed by 1 week at 98% RH/38°C followed by 2 days conditioning at 50% RH/23°C prior to testing.

Specimen testing

Both 'thin' lap shear specimens and 'thin' peel specimens can be tested in a standard testing machine with controlled testing rate, at *any* rate between 0.1 cm/min to 5 cm/min. Recommended test rates are **1 cm/min** or 5 cm/min

Jerome M. Klosowski, Stanton J. Dent

USE OF THE CONSTRAINED BLISTER TEST TO STUDY SEALANT ADHESION

REFERENCE: Klosowski, J. M., and Dent S. J., **"Use of the Constrained Blister Test to Study Sealant Adhesion"**, Science and Technology of Building Seals, Sealants, Glazing, and Waterproofing: Second Volume, ASTM STP 1200, Jerome M. Klosowski, Ed., American Society for Testing and Materials, Philadelphia, 1992.

ABSTRACT: An attempt was made to modify the constrained blister test as designed and reported by Dr. David A. Dillard of Virginia Polytech. (VPI) and independently by Dr. A. (Sam) Moet of Case–Western Reserve Univ. to apply to sealants. The test device is detailed and modifications of the device to make it applicable to mullions and samples collected from the field are explained. Variables such as sealant thickness, distance between sealant and the constraining plate and the rate of pressure were also studied. The data from the constrained blister is compared to the peel test and the lap shear test. Lastly the validity of the test modifications as indicated are discussed.

KEYWORDS: constrained blister, lap shear, adhesion, adhesive failure, peel test

The constrained blister test has been reported by Moet of Case–Western Research and Dillard of VPI of having the ability to determine adhesion strength in excess of the cohesive strengths of the bulk material. Further, the above authors have indicated that this type of test is applicable to sealants. Both the above researchers have performed their test at a constant pressure and studied time to adhesive failure – a creep rupture type study. This study stressed the adhesive bond of the sealant by increasing the pressure at a constant rate. The thought was to have a faster test, one that approached the ASTM C–1135 Tensile Adhesion Test. The constained blister test has been studied as an alternative to the tests presently available. It is an improvement on some but it is not the perfect test for all sealant adhesion information. All of the present tests have some advantages and disadvantages. The constrained blister test also has some advantages which will be discussed. All through this paper the comparison will be made to the peel test. While the authors feel this test has much more to offer than the peel test, they feel that the constrained blister test does not belong in a general building sealant specification, any more than a peel test belongs in a construction sealant specification. Lastly the constrained blister test as performed by

Mr. Klosowski is a research scientist and Mr. Dent an associate project chemist at the Dow Corning Corporation, 2200 Salzburg Rd., Auburn, MI, 48686–0994.

Dillard and Moet were versions of a creep rupture test in that they held the test piece at a constant pressure and studied time to failure while the modifications presented here used increasing pressure looking for the start of blister growth.

Constrained Blister Experimental Set–up

The constrained blister test apparatus we used to evaluate the adhesive properties of sealants is similar in set–up to the apparatus used by others to evaluate pressure sensitive adhesives. The basic set–up as depicted in Fig. 1 consists of a substrate on which is cured a sealant. The typical cure scheme was room temperature cured at ambient conditions for 3 to 4 weeks. The constraining plate was a clear 1/2 inch polycarbonate sheet which was held over the top of the 1/4 inch sealant using polyethylene spacers. The whole arrangement is clamped together at the two sides. The pressurizing gas used was air, which was supplied by an industrial cylinder. The pressure of the gas was controlled by using a computer controlled flow valve and a down stream throttling valve to give relatively linear pressure increases in the range 10 psig to 160 psig. A commercial computer program (Strawberry Tree, v1.1) was used to control and monitor the flow controller. The gas pressure was typically ramped at 5 psi/min. The pressure and growth of the constrained blister was monitored by a large calibrated pressure gauge and a video camera. The video camera was utilized to play back the blister growth and accurately measure blister growth as a function of pressure. The pressure gauge, flow controller and clamping mechanism were all mounted on an aluminum test stand.

Figure 1. shows the sealant adhered to the substrate prior to being subjected to the pressurizing gas and subsequent adhesive failure. It was sufficiently demonstrated that the sealant was failing in a completely adhesive manner. Samples were cut apart after completing the test procedure, clearly noting the lack of any sealant adhering to the substrate over the area of blister growth during the test.

After the sealant is applied and well cured to the sample, the constraining plate is placed over the sample. This constraining plate is separated from the top of the sealant by some predetermined distance. The spacer shim gives the appropriate spacing as seen in fig. 1.

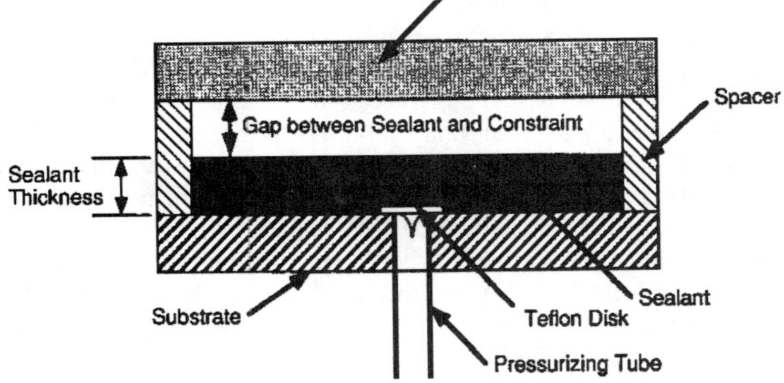

FIGURE 1 – Schematic of the Constrained Blister Adhesion Test

With the constraining plate clamped in place, the sealant is pressurized, a blister forms and pushes out until it flattens out against the constraining plate. A ruler is placed across the top of the clear restraining plate so that the growth of the blister can be more easily detected. In the tests carried out by these authors, the determination of the start of the pressure was visual and recorded from directly above with a video camera. A suggested modification of this test is to detect the blister growth electronically.

In the past, the blister was pressurized to a constant, sustained pressure, and the area of growth at a given pressure was noted. That technique is reminiscent of a creep rupture test for adhesives. We modified this test so that we increased the pressure at a steady rate and measured the force at which the blister began to grow. In essence the most significant modification of the test method is that we are using it much like a tensiometer, that is stressing the sealant at an ever increasing rate until a failure is noted.

In the test method, the following are the variables: the thickness of the sealant to be placed on the adherent, the distance of the constraining plate to the thickness of the sealant and the rate of pressure increase.

While an extensive study was not done, it was found that as the rate was increased from approximately 1 psi per minute to 5 psi per minute the values for adhesion got higher. At approximately 1 psi per minute the pressure at failure was about 40 psi, with the sealant tested. At a very steady 5 psi per minute rate of increase the same sealant produced values of 58 psi. The study did not pressurize faster than 5 psi per minute. Higher rates are attractive because they would speed the time it takes to run the test. The accuracy and reproducibility of the values generated at the faster rate would have to be studied.

We had an initial fear that if some very strong urethanes were studied they might produce values of 120 psi or greater and that was the limit of our equipment (pressure regulator). Thus we stayed with a very comfortable 5 psi and had the automatic pressure regulator set to increase at 5 psi per minute. We used that rate of pressure change for all of the data generated that will be presented as the paper progresses.

The above commentary does imply a very important restriction to the test data. The test is only useful in comparing values that were all obtained in identical procedure. Thus the modification of the constrained blister test as indicated here is only a comparative tool and not an absolute measure of bond strength.

An important part of the test development was determining the ideal sample thickness. In this work samples of 1/16 inch (1.6mm), 1/8 inch (3.2 mm), and 1/4 inch (6.4 mm) were studied. We would have liked the 1/16 (1.6 mm) sample to work since that gave the fastest cure rate and thus the overall test time could be significantly reduced. As it worked out, the configuration that gave the most reproducible data was with the 1/4 inch (6.4 mm) sample thickness.

In the study of the distance of the constraining plate to the top surface of the sealant we studied 1/32 inch (0.8 mm), 1/16 inch (1.6mm), 1/8 inch (3.2 mm) and 1/4 inch (6.4 mm). When the constraining plate was far above the sealant we saw 2 types of failure that were not beneficial. One undesirable mode of failure was when the sealant grew a very irregular blister and it essentially tunneled out one side and broke. This was due to the combination of thick samples and large distances to the constraint. This was not a constant but an occasional occurrence.

With thin sealants and large distances to the constraining plate (1/16 inch (1.6 mm) sealant and 1/8 inch (3.2 mm) to the constraining plate), we saw most typically a blister that would act like a balloon. The balloon would fold over and travel to the edge where it would break with the noise of a rifle shot. This was obviously not the way to run the test and did not give the adhesive failure mode desired.

The purpose of all of this work was to find conditions that would produce consistent adhesive failure (no cohesive failure), so we could actually study adhesion independent of the cohesive strength. The quest for consistent adhesive failure was finally achieved when the distance from the sealant to the constraining plate was 1/4 of the thickness of the sealant. This suggests the workable sample dimensions are 1/16 inch (1.6 mm) of sealant with a 1/64 inch (0.4 mm) gap between the top of the sealant and the under side of the restraining plate. 1/8 inch (3.2mm) sealant with a 1/32 inch (0.8 mm) gap; 1/4 inch (6.4 mm) sealant with a 1/16 inch (1.6mm) gap.

In the physical laying up of the sealant on the substrate surface it is quite difficult to achieve a uniform 1.5 mm thickness. In samples like this variations of 0.4 mm were common. While the actual variation is small the relative size of this variation, relative to the sample thickness, is large and not conducive to reproduceable data. These variations may contribute to the gap distance enough, so that ballooning is again a problem. Thus we settled on the 1/4 inch (6.4 mm) sample thickness as something that the lab technician, the scientist and the high school co–op would all do well. With this choice, sample preparation would not be a key variable in this test procedure, at least not in reference to thickness. Cure time will be longer with the 1/4 inch (6.4 mm) thick sample, but the curing characteristics of these samples will be most similar to that of the sealant in field applications. This is very important since early work, in an unreported study, has shown that changes in cure rate influence adhesion. Thus the 1/4 inch thick sample is a logical choice for several reasons.

Satisfied with the test configuration and the rate of pressure increase that we had chosen, the test method was checked for reproducibility. To do this we prepared three identical samples and ran each sample 3 times by pressurizing until blister growth is noted, relieving the pressure and repressuring. In this manner, 3 readings were completed on each sample. This study used an acid cure silicone on mill finish aluminum plates and the values are read directly from the pressure gauge in pounds per square inch. These are shown in Table 1:

TABLE 1
Reproducibility of a Sealant on an Aluminum Substrate

Sample No.	Pressure at Release Onset (psi)
1	58, 57, 58;
2	56, 60, 57;
3	58, 59, 58
Average	58.5 psi
Std. Deviation	1.4 psi

Obviously we were very pleased with the reproducibility of the data.

To determine what these results meant in terms of known adhesion tests, Peel adhesion ASTM C-794, the tensile-adhesion test ASTM C-1135, and ASTM D-3164 the lap shear test was run on the same sealant and same substrates. The substrates were cleaned in the same manner and sealant cured for the same time. The values are tabulated in Table 2:

TABLE 2
Adhesion Test Results for First Sealant Sample

Test Method	Failure	Tensile Strength	Elongation
Peel			
1/4" Thick	100% Cohesive Fail	12 psi	
1/8" thick	0% Coh (adhesive)	23 psi	
Tensile Adhesion			
Width x Depth			
1/2" Wide x 1/2" Deep	20% cohesive fail	95 psi	
1/4" Wide x 1/2" Deep	30% cohesive	94 psi	140%
1/2" Wide x 1/4" Deep	100% cohesive	77 psi	188%
Lap Shear			
1/16"	60% cohesive	156 psi	58%
1/8"	100% cohesive	102 psi	87%
1/4"	100% cohesive	83 psi	146%
Const. Blister			
Seal Thickness x Gap			
1/4" Seal. x 1/16" Gap	0% coh. (adhesive)	57.7 psi avg of 3	
	0% coh. (adhesive)	57.5 psi avg of 3	
	0% coh. (adhesive)	58.3 psi avg of 3	

One can attempt to analyze the data .

Peel Test

The peel test is well described by its title – in essence an edge of the sealant is held and pulled. The force that is takes to pull the sealant loose or break the sealant is recorded. In the peel test done here we see two grossly different pieces of data. That serves to remind one of how very imprecise the peel test is. The ASTM method suggests ± 20 % is the best one can expect in a given lab. This data is obviously outside that limit but since the thicknesses are different for the two samples, the apparent modulus of elasticity exhibited at the metal interface will be different and thus we expect a bit of a difference in values. More importantly, the peel test gives no indication of the sealant's ability to perform. That is the problem with peel test data. The number generated for adhesive strength does not predict the sealant ability to perform. By the most common rule of thumb used with peel data in the sealant industry, one would look at the peel data and almost ignore the values but look hard at the cohesive failure and suggest that the adhesion of the 1/4 inch thick sample is quite good. Similarily one could look at the 1/8 inch data and suggest that adhesion is bad because adhesive failure is seen. The most realistic conclusion is that the adhesion of the two samples (same sealant and same substrate and same cleaning process, cured for the same time in the same conditions) is the same and the most common way of interpreting the test data simply doesn't reflect the true adhesion characteristics

With sealants, the peel test is useful when a given sealant, is tested in a standard configuration, on a variety of surfaces. One can determine if the adhesion is comparable for each of the surfaces, but note well this comparison is only good to the point of cohesive failure. Or it can be used to determine if different batchs of the sealant adhere equally well to the same surface. Again, this can only be done up to the value of the cohesive failure, after which all are equal and no determination of one versus the other can be made. The cohesive failure limit is the highest adhesion value that can be obtained with this test. Thus we have a test with large accuracy problems and severe limitations in the range of adhesive strengths that can be determined. The beauty of this test is that it is simple to run. Manufacturers can and do use this test quite regularily to compare adhesion of different batches of the same sealant, as indicated above. The constrained blister looks to be a clear improvement on the peel test, with the exception that a larger sample is needed and different equipment is needed beyond that found in the typical sealant lab or testing lab.

Tensile – Adhesion

The tensile adhesion data shows how much movement the joint can take in a single catastrophic movement and how much force was exerted during that movement. This test has real value to sealant users and is probably the most important test for them that can be run relatively quickly. They can estimate the amount of movement the joint will see and determine from the test the force that will be generated at the bond line. The user can also get some idea of the sealant's adhesive ability to handle the stress introduced by that movement. With a minumum value in the structural glazing spec of T/A of 50 psi or greater (ASTM C–1184), it is obvious that the sealant used above, on the same substrate, would be satisfactory for structural glazing applications. This would be true in each of the sample geometrys used above.

The strength value has great significance only in structural glazing and is of little significance in general sealing applications. With sealants, unless a great deal is known about the rubber properties of the sealant being tested, the magnitude of the strength determined has no intrinsic value. The desired data for a sealant is relative to the adhesive strength being sufficient to handle the force generated with the movement of the sealant. The tensile–adhesion test does measure adhesion as a function of movement. Since the elongation value for the joints tested was greater than 100%, for all the joint configurations, there is some confidence, based on historical correlations, that this sealant will be able to handle ± 25% cyclic joint movement. This is a good test for sealants. It tests adhesion in the manner the sealant performs. For sealants used in sealing applications the notation of no failure as the sealant significantly exceeded the movement expected to occur in real life situations was the key value. The force exerted at that point is of no real value. In sealing applications the sealant that exerts a large force as it is strained is producing large forces at the bond line and large forces on the substrate to which it is attached. This is not desirable. The ideal situation is to have no failure adhesive or cohesive as the sealant exceeds the expected joint movement. It is more acceptable for the forces exerted at the bondline to be small. The real value of the tensile–adhesion test is to determine if the adhesion is adequate for the movement expected. The knowledge of the adhesion being adequate for twice the expected movement is also of interest. The actual force generated has little value in true sealing applications, except in structural glazing applications, which is more of an adhesive application. Thus, this is a good test for sealants in adhesive applications since it does give a joint strength value for the structural engineers and it is a good test for sealants in sealant applications because it does indicate the adequacy of the joint to handle movement. The idea is not to confuse the two applications of sealants and the two applications of this test.

The principal limitation to the tensile adhesion method is that it dosen't measure adhesion above cohesive failure and thus with sealants of equal strength adhesion to all surfaces that produce cohesive failures are reported as equal in adhesion. When using this test to study sealant applications (not structural glazing) it should be stated that the adhesion of sealants that allow the same movement are equal. Very importantly, this test method tests in a joint configuration and gives a direct measure of adhesion as a function of joint movement. Thus, this test method does give the ability to compare <u>necessary</u> adhesion (adhesion as a function of movement). When testing sealants for sealant applications, that is how data should be reported (adhesion to X% and not adhesion to Y psi). Regardless of how well the constrained blister test works in its final refinement, the tensile adhesion test will most probably be the most important test to the sealant user. It is expected that the constrained blister test and the tensile adhesion test will compliment each other with one measuring well the adhesive strength values and the other the stress – strain relationship and the movement ability of the joint, relative to the adhesion.

The tensile adhesion data generated above has a bit greater spread than anticipated but it does illustrate the dependence of strength on the joint geometry. As one can see from the above date there is so much variation on the actual adhesion values that the numbers have little if any meaning. What is clear from the tensile adhesion test that is not clear from the blister test is that in each of the joint configurations and the varying amount of stress put on each joint, the adhesion each time was more than adequate. It allowed the sealant to take more than six times the maximum allowed movement of this sealant (25% expansion) the required movement.

These authors find no relationship between the peel values and the tensile adhesion values.

Lap Shear

This is simply overlapping two pieces of substrate with sealant in the overlap area. This is cured and pulled in a tensiometer. The lap shear test, in thin sections, gives some strength values that are of interest if one is considering the adhesive strength of an adhesive. In thin sections it has little meaning to sealant applications. The real value of this test are the elongation numbers. These values give some indication of the sealants ability to handle shear movement. In thick sections the sample severely elongates and this test measures a combination of tensile and shear forces. That is quite good also, in that this combination of stresses is what actually happens to a sealant. Thus when the test specimen is of the same thickness as the proposed joint geometry, the elongation values of this test give relatively useful data to the architect or engineer. A lot to lot comparison of the sealants shear ability can be done and this is of value to the suppliers and the people who set QA specs. Often those who use the lap shear test do not measure the displacement but simply record the strength values and the mode of failure (adhesive versus cohesive) and miss the real value of this test to sealant users.

This test does not give values that are intrinsically useful to the sealant user. And this like all of the above does not test adhesion above the value of cohesive failure.

To the sealant supplier, the strength data generated by this can be important in that a lot–to–lot comparison of the sealant's adhesive strength can be done. The strength data from this test method is only of value up to the place of cohesive failure. At and above that cohesive strength value, all the adhesion is reported as comparable. In determining actual adhesive strength values, this test is better than the peel only in that it is a bit more reproduceable and a lot more discriminating. It is better than the peel for sealant users in that it does convey the useful shear movement value if the test is done with a sufficiently thick sample. The lap shear requires a more specific substrate shape for testing.

Constrained Blister

The constrained blister test values are not inherently useful to the sealant user, and from that aspect they are much like peel values. The constrained blister, like the peel test, measures adhesion rather independent of movement and does not indicate the sealants ability to adhere throughout the movement range in which the sealant will operate. However, it seems to measure adhesion values quite well. The first apparent advantage, as a gauge of adhesion, is that it seems to be very reproducible. The second and most unique advantage is that, when done well, it seems to produce only adhesive failures (There are some discrepencies that need resolution but at this point in test development, this does seem to be the case). It seems that lot–to–lot discrimination will be easily done and the comparison of adhesion to different surfaces is also easily done. When a sealant supplier is now asked by the curtain wall manufacturer to give a preferred surface, this should easily be done. The applications of this test can be many. This seems to be a true breakthrough in adhesion testing of sealants. However it is not the perfect test for the sealant user, the architect, and the specifying engineer. Putting a constrained blister value for adhesion into a sealant specification would be as totally useless as the present peel values called out in ASTM C–920 specification for construction sealants. While it is the opinion of these authors the peel test should be removed from C–920, it should not be replaced with the constrained blister test, but instead the tensile adhesion test or the test used by a similar test used by the International Standards Organization (ISO) that takes a sealant joint and holds it at a constant strain.

Sample preparation is a bit more difficult with the constrained blister test. Most of our work was done with a sufficiently thick sample to withstand the pressure of the test and the sample is flared so as to hold the flared rod that brings the air to pressurize the blister. It's still uncertain as to the technique to handle a tempered glass sample. Thin weight metal samples are adhered to more rigid metal plates and the hole is drilled in the composite. The thin metal samples were not difficult but did take a bit longer to prepare.

Our job here is not to find a replacement for the peel test or any other test but to show the advantages, limitations and possible uses for the constrained blister test. In doing this, there is one set of test results not yet reported that are very important. In further testing the sealant used in the above tests was modified with another adhesion promoter and to another the aluminum surface was primed. This all gave very good adhesion. There was a cohesive failure in one of the samples at 0.44 mPa (63 psi) and the other did not fail at the maximum of the pressure controller of 0.84 mPa (120 psi).

In the first sample the sealant was studied after failure and it is believed that an air bubble caused the premature cohesive failure giving a weak spot in the sealant.
However the cause is not certain and more investigation and further research is needed to say for certain this test will always give adhesive failures in well prepared samples. Is is rather certain that the degree of dependence of the test on the integrity of the sample must be known for this test to have general acceptance and applicability.

The second sample produced an inordinately high value for adhesion (0.84 mPa, (120 psi)). However this sample had a larger unbonded cross–section at the time the sample was read for the start of the blister growth. Here the unbonded cross–section had almost 32mm (1 1/4 in.) diameter while in all previous tests the unbonded diameter when the reading was made was very near 19mm (3/4 in.). The importance of the unbonded area to the reported adhesion values has yet to be determined in this version of the constrained blister test. There was a marked change in values in both the work of Moet and Dillard with blister size and perhaps this should be expected here as well. For this test to be of real value, with numbers relavent from sample to sample, the relationship between adhesion value and the unbonded diameter must be known.

Also noted earlier in the paper the rate at which the pressure was changed was a factor in the strength value adhesion noted in the test. This was expected and this is quite easily controlled. However this would caution one that values run at different pressurization rates cannot be compared unless some correlation equation is determined.

A very important conclusion to be drawn is that the constrained blister data does not seem to relate to any of the other of the above test methods. We could find no correlations and in fact we expected none. Each test method has some good uses and some not so good. We expect the constrained blister test to be a good study tool but not a good specification tool for sealants. For adhesives it may be an absolutely excellent test, even for a specification. However to be useful in the manner modified here, further research into the test method must be done.

To verify the above conclusions the above experiment was redone with a yet another sealant of different composition. All of this work was as identical to the previous work as the authors could make it. That data is listed in Table 3.

TABLE 3
Adhesion Test Results For Second Sealant Sample

Test Method	Failure	Yield Strength	Elongation
Peel Test			
1/4" Thick	0% Coh (100% Adh.)	28 pli	
1/8" Thick	0% Cohesive	5 pli	
Tensile/Adh. Test			
1/2" x 1/2"	0% Cohesion	48 psi	106%
1/4" x 1/2"	0% Cohesion	53 psi	128%
1/2" x 1/4"	0% Cohesion	93 psi	256%
Lab Shear Test			
1/16"	10% Cohesive	123 psi	60%
1/8"	30% Cohesive	90 psi	117%
1/4"	5% Cohesive	77 psi	220%
Const. Blister Test			
1/4" Seal. x 1/16" Gap	0% Cohesion	30 psi	
	0% Cohesion	33 psi	
	0% Cohesion	31 psi	

The data in Table 3 is very reminiscent of the previous set of experiments. The commentary about the advantages and disadvantages is again shown to be valid. The variability of the peel test is large, the reproduceability of the constrained blister is very good. This sealant was entirely different in its strength and adhesion, but the commentary on the varoius tests is the same.

The conclusion from the earlier data, that the data from the constrained blister test does not relate to the other test data, looks to be a good general conclusion.

In the lap shear test data on both of the above experiments, one can see the measured adhesive strength is greater as the sealant gets thinner. This thickness effect was tested with the constrained blister test as well. This study used the original sealant mentioned earlier and the general conclusion is the same as with the lap shear test. Table 4 contains the data for study:

TABLE 4
Sealant And Gap Thickness Data For Sealant One

Sealant Thickness	Distance To Constraint	Pressure at Release Onset
1/4 in. (6.4 mm)	1/16 in. (0.8 mm)	51 psi
1/8 in. (3.2 mm)	1/32 in. (0.8 mm)	59 psi

Here you see that the 1/8 inch sample was stronger than the corresponding 1/4 inch thick sample. The difference is not as great but it is real and expected.

Of special interest is that the distance between the sealant and the constraining plate, in the region studied, does not seem to be significant, as long as the constraint is sufficiently close as to prevent gross deformation of the blister and prevent cohesive failure. The two 1/4 inch thick samples noted below illustrate this:

TABLE 5
Constrained Blister Gap Thickness Test For Sealant One

Sealant Thickness	Distance To Constraint	Yield Strength
1/4 in. (6.4 mm)	1/32 in. (0.8 mm)	52 psi
1/4 in. (6.4 mm)	1/16 in. (1.6 mm)	51 psi

The sealants studied in all the work above were in the general classes of medium and high modulus silicones. These sealants were well cured. No investigation was yet done for sealants of very low modulus. However when a lower modulus version of the above sealants were tested in the precured state (very soft and stretchy), the sealant entrained air and took on a foam appearence. The test method will have some limitation on softness of sealants to be studied but the limit has not been established. However from this work we know that one will exist.

Another note should be made that the adhesion data is still sensitive to sample preparation procedures. Where there was air trapped between sealant and substrate, the experiment did not give a uniform bubble and premature failure was noted. This is a little different than the cohesive failure noted on the sample mentioned earlier which was also thought to be initiated by a flaw introduced by air.

A very important modification that we put into our test method right from the start that is different than that reported by either Dillard or Moet, is the use of a significantly large unbonded area at the start. We accomplished this with the use of the one inch diameter teflon disk set over the center of the inlet tube. There was a bit of difficuty in getting the disk to stay in place during the application of the sealant but this was solved by using a very low vacuum to hold the disk to the top of the inlet tube and we also used a silicone pressure sensitive adhesive to hold a small plastic spike to the center of the disk. In this case the spike fit easily down the tube and held the disk centered over the tube, in the middle of the test surface. Both techniques worked equally well and the spike caused no variation in data.

Conclusion

The constrained blister test as described in this paper seems to hold much potential but indeed still needs a good deal of work before being useful outside or inside the research lab. Hopefully the work indicated above will shorten the time and make the job of adopting this test to individual needs, easier.

Acknowledgements:

The authors would like to thank V. Johnson of the Electro–mechanical Development Group of Dow Corning for providing the assistance and manpower necessary to build the prototype test stand used for all the constrained blister tests outlined above. We would also like to thank B. Albe also of Dow Corning for setting up the computer control scheme, which allowed us to obtain uniform and reproducible pressure ramp rates. And we would like to thank Marcy Sewell, who was a summer college intern, and who was responsible for the majority of the data collected for this paper. Without her hard work none of this would have been possible.

References:

1. M.J. Napolitano, A. Chudnovsky, and A. Moet, J. Adhesion Sci. Technol., Vol. 2, No. 4, 311–323, (1988)

2. Y–H., Lai, D. A. Dillard, J. Adhesion, Vol. 31, 172–189, (1990)

3. Annual book of ASTM Standards, Section 4: Construction, vol. 04.07, (1989)

4. J.M. Klosowski, "Sealants in Construction", Marcel Decker, Inc, N.Y., (1989)

Voytek (W) S. Gutowski[1], Anthony Cerra[1] and Lee Russell[1]

THE INFLUENCE OF GLOSS LEVEL IN POLYESTER-POWDER COATINGS ON THE ADHESION OF SILICONE SEALANTS

REFERENCE: Gutowski, V. W. S., Cerra, A., and Russell, L., **"The Influence of Gloss Level in Polyester-Powder Coatings on the Adhesion of Silicone Sealants"**, Science and Technology of Building Seals, Sealants, Glazing, and Waterproofing: Second Volume, ASTM STP 1200, Jerome M. Klosowski, Ed., American Society for Testing and Materials, Philadelphia, 1992.

ABSTRACT: It has been frequently observed that the strength of adhesion between silicone sealants and polymer-coated substrates depends on the coating's gloss level.

The study described in this paper reveals that there seems to be a relationship between the gloss and the coating's surface properties as determined by the non-dispersive, $\gamma^{n\text{-}d}$, component of the total surface energy. It is shown that gloss increase is associated with an increase in the substrate $\gamma^{n\text{-}d}$. The exact pattern of this relationship cannot yet be established due to the narrow range of experimental data. Experiments carried out with several commercially available silicone sealants show that the higher the coating's gloss level, the stronger the adhesive bond between the sealant and the substrate. This finding is shown to be consistent with the earlier proposed mechanism of adhesion, according to which an increase in the value of the non-dispersive component of the substrate total surface energy leads to increased bond strength to the point where adhesive forces become stronger than the cohesive strength of the bulk sealant.

KEY WORDS: silicone sealant, polyester resins, powder coats, gloss level, pigments, surface energy

[1]Dr Gutowski is a Principal Research Scientist; Mr Cerra and Ms Russell are Experimental Scientists at the CSIRO Division of Building, Construction and Engineering, Melbourne, Australia. Postal address: PO Box 56, Highett, Victoria 3190, Australia

INTRODUCTION

Structural glazing which involves the attachment of claddings to building facades using silicone sealants, has become popular over the last decade. One of the framing materials used in this technique is aluminium coated with a range of organic coatings, e.g. polyester, PVF2, acrylic, etc.

A theory of adhesion involving the non-dispersive component of substrate surface energy is presented in references 1–3. This paper proposes to apply this theory to elucidate the effect of gloss level on the adhesion of silicone to polyester powder-coated aluminium.

EXPERIMENTAL

1. Substrate Materials

The range of polyester powder-coated substrate materials investigated in this project has been prepared using two commercially available formulations nominally described as standard and high durability grades. These were selected in a typical range of colours and gloss levels.

The products, as listed in Table 1, were formulated to obtain a range of colours and gloss levels. All white, black and grey colours were obtained by varying the concentration of the same white and black pigments. Gloss level was controlled by varying the type, fineness and quantity of fillers and othe compounds. The only coatings containing non-typical additives are P8, which contained an air-releasing agent, and P12 which additionally contains a different type of tinting agent.

Silicone Sealants

Twelve different commercial sealants were used for these tests consisting of neutral, alkaline and acid cure one-component (RTV1) systems, as well as a number of two-component systems. The exact nature of the sealants tested appears to be irrelevant since the results obtained are similar for all types of materials investigated.

Techniques

(a) *Surface preparation*. All substrates were cleaned by three wipes with ethyl alcohol prior to the preparation of lap-shear specimens, which were prepared within 2 hours of substrate cleaning.

(b) *Specimen cure*. All lap-shear specimens were allowed to cure for 4 weeks under controlled conditions i.e., 20°C and 50% relative humidity.

(c) *Bond Strength*. This was determined using special single lap-shear specimens prepared according to the procedure described in Reference [1]. After the sealant cure, as described in (b), the specimens were tested in an Instron testing machine at the rate of 10 mm/min.

(d) *Gloss determination.* The gloss level of the polyester powder-coated substrates was determined by the coatings' manufacturer using a 60° illumination angle. All measurements were carried out in accordance to ASTM D 523.

Table 1. Polyester powder-coated substrates investigated in this work

Symbol	Colour	Gloss level (60° angle)	$\gamma^{n\text{-}d}$ [mJ/m^2]#	Comments
P1	White	90	8.56 }	same formulation;
P2	White	40	6.40	different gloss
P3	Black	90	7.87	
P4	Black	45	5.90 }	same formulation; different gloss
P5	Black	35	5.43	
P6	Claret (special)	85	6.58 }	
P7	Claret	55	6.54	
P8	Grey	55	4.30	contains air-release agent
P10	White	90	8.55	as P1/different resin (high durability grade)
P11	Black	90	7.90	as P10/different pigment
P12	Greyish	35	3.87	as P10 + 0.5% tint A
P13	Dark grey	85	9.22	as P12 + extra 0.5% tint A

Note: Materials P1 to P8 are made from a 'standard' base polymer, and P10 to P13 are made from a 'high durability' grade polymer.
$\gamma^{n\text{-}d}$ is the non-dispersive component of the substrate total surface energy (see Ref. [2] for details).

(e) *Total surface energy (γ_l) of the substrate and its non-dispersivecomponent (γ^{n-d}).* This was experimentally determined using the procedure described in Reference [2].

RESULTS AND DISCUSSION

1. Relationship between the gloss level and γ^{n-d}, the non-dispersive component of substrate surface energy

The values of the experimentally determined gloss levels and non-dispersive component (γ^{n-d}) of surface energy of all substrates considered in this work are listed in Table 1 and illustrated in Figure 1 in the coordinates: 'gloss level *v.* γ^{n-d}. The data

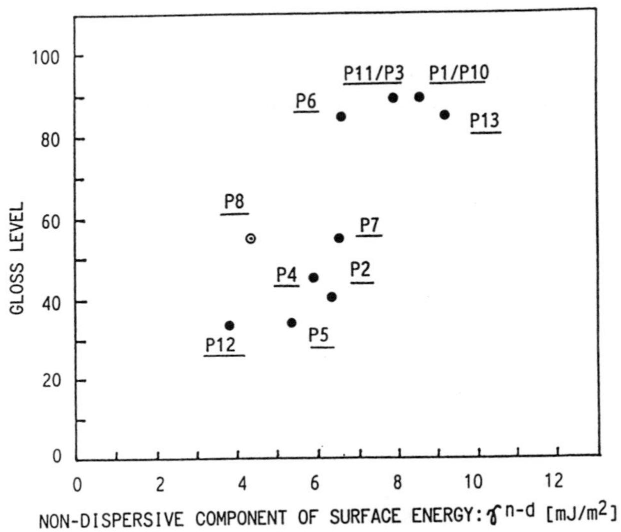

Figure 1. The relationship between the gloss level and the non-dispersive component, γ^{n-d}, of the total surface energy for polyester powder-coatings investigated in this work.

indicate that apart from coating P8 (which contains a waxy air-release agent resulting in a significant lowering of γ^{n-d} without decreasing the surface gloss level), all other points seem to fall into the following patterns:

(i) a high gloss level zone (about 80–90) for substrates exhibiting $\gamma^{n-d} \geq 6.6$ mJ/m^2, and

(ii) a low gloss level zone (about 35–55) for substrates exhibiting very low $\gamma^{n-d} \leq 6.4$ mJ/m^2.

A sharp gloss level decrease is observed at the boundary between these two zones, i.e. at about $\gamma^{n-d} \approx 6.4$ to 6.0 mJ/m^2.

Due to the shortage of experimental data it is difficult to determine the exact pattern of the gloss v. γ^{n-d} relationship. However, based on these observations, as well as on further data contained in Table 1, it can be concluded that both surface properties of polyester coatings investigated, i.e. gloss and the non-dispersive component of the substrate surface energy, γ^{n-d}, may be inter-related. The implications of such a relationship may be significant since it has been shown that the bond strength of the sealant/substrate combination is directly dependent on the numerical value of γ^{n-d}. [2]

Comparing the results for the black coatings P3, P4, P5 and P11, which are identical apart from gloss level, it appears that gloss level and γ^{n-d} may be related as shown in Figure 2. Results for the white colour coatings P1, P2 and P10 which differ from black polyester-coated substrates only by the type of the pigment used, show a similar trend, i.e. a decrease of γ^{n-d} with decrease in the gloss level. However, no definite conclusions on the profile of the 'gloss *v.* γ^{n-d} relationship should be drawn

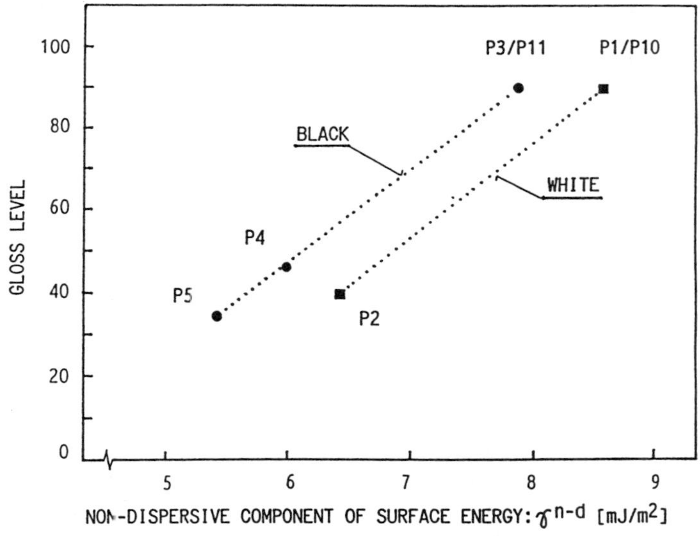

Figure 2. A trend observed in the relationship between the gloss level and non-dispersive component, γ^{n-d}, of the total surface energy in two groups of coatings with different colour: white and black.

due to the limited number of experimental points available. No comments can be made on materials P6 and P7 (claret) and P12/P13 (grey) since their colour is controlled by addition of non-identifiable tints. Some comments may also be offered (see Figure 2) regarding the influence of pigment type on the coating γ^{n-d}, although the data available is again limited to the white and black substrates only.

The white colour is obtained by using titanium dioxide, which has significantly higher $\gamma^{n\text{-}d}$ than the carbon black used in black colour substrates. The lower $\gamma^{n\text{-}d}$ of black-coloured substrates is in line with the lower value of $\gamma^{n\text{-}d}$ of carbon black pigment, compared with the relatively higher $\gamma^{n\text{-}d}$of white titanium dioxide.

2. Influence of gloss level on the adhesion of silicone sealants

In the earlier works of these authors on the mechanism of adhesion of silicone sealants [1,3] it has been shown that the strength of adhesion across the sealant/substrate interface can be related to the non-dispersive, $\gamma^{n\text{-}d}$, component of the substrate surface energy. It is shown in the present work that the value of $\gamma^{n\text{-}d}$ may be, in turn, related to the gloss level of the polyester powder coatings.

Figures 3 to 5 are typical examples of the relationship between the bond strength and $\gamma^{n\text{-}d}$ for the polyester powder-coated materials investigated in this work, bonded with the one-part silicone sealants A and C, and a two-part sealant B. All other sealants investigated in this work exhibit similar trends.

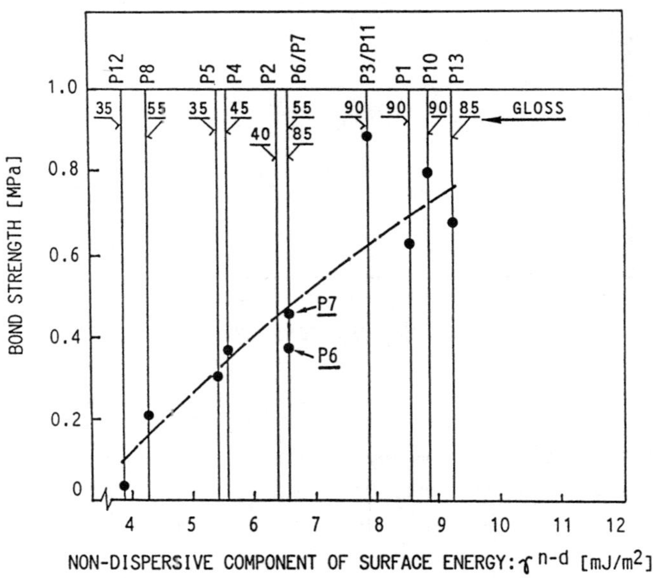

Figure 3. The relationship between the bond strength and surface $\gamma^{n\text{-}d}$ for all polyester powder-coated products investigated. Substrates bonded with sealant A. Gloss level of each coating is shown in the graph.

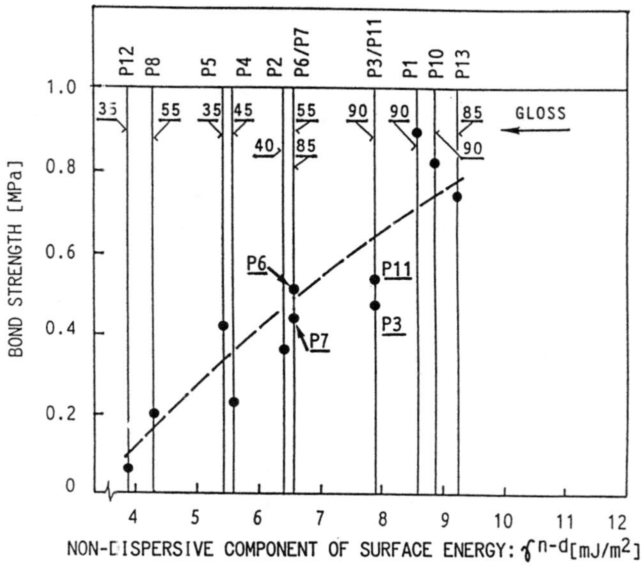

Figure 4. The relationship between the bond strength and surface γ^{n-d} for all polyester powder-coated products investigated. Substrates bonded with sealant B. Gloss level of each coating is shown in the graph.

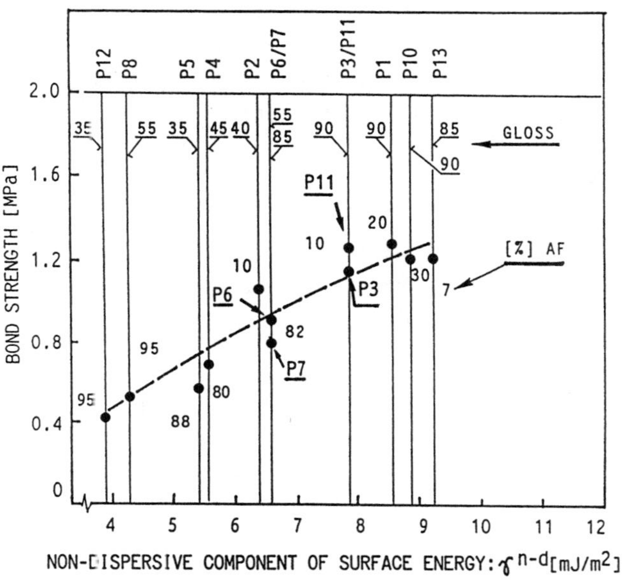

Figure 5. The relationship between the bond strength and surface γ^{n-d} for all polyester powder-coated products investigated. Substrates bonded with sealant C. Gloss level of each coating is shown in the graph.

A one-component sealant A (see Figure 3) exhibits almost entirely 100% adhesive failure with all polyester powder-coated materials analysed in this work. It can be observed that the increase in the substrate $\gamma^{n\text{-}d}$, is the factor responsible for an increase in the strength of adhesion. Although no typical pattern for the relationship between gloss level and bond strength can be established at this stage, it can be observed in Figure 3 that all substrates with low surface gloss (35 to 55) exhibit low bond strength, with the strength always increasing with increasing $\gamma^{n\text{-}d}$. The substrates with higher gloss level (85–90) are located in the region of relatively high bond strength.

A two-component sealant B (see Figure 4) again, exhibits almost 100% adhesive failure with all the polyester powder-coats studied in this work. The bond strength shows a power-dependence on the substrate $\gamma^{n\text{-}d}$, which in turn is related to the gloss level as illustrated in Figure 1. The general trend observed is that these substrate materials which exhibit low gloss level (35–45) produce low bond strength with sealant B, and those with higher surface gloss (85–90) exhibit relatively high bond strength.

Sealant C (one-component material), as Figure 5 illustrates, exhibits a generally better bonding capacity with polyester powder-coated materials than sealants A and B discussed above. The general trend observed in the strength v. $\gamma^{n\text{-}d}$ and gloss level relationship is as follows. Low gloss substrates (gloss level 35–45) exhibit low bond strength, increasing with $\gamma^{n\text{-}d}$ (within the range: $3.8 \leq \gamma^{n\text{-}d} \leq 6.5$ mJ/m^2). With further increase in the gloss level ($\gamma^{n\text{-}d} \geq 7$ mJ/m^2) there is a marked increase in the [%] cohesive failure within the sealant. It is noticeable again that materials with low gloss levels produce almost entirely 100% AF whilst those with high gloss level result in significantly greater bond strength and almost 100% cohesive failure within the sealant.

CONCLUSIONS

1. It has been shown within the scope of this experiment, that for a particular range of commercially available polyester powder-coated materials a relationship seems to exist between gloss level and the non-dispersive component, $\gamma^{n\text{-}d}$, of the substrate surface energy.

2. Based on this and other work [1–3] it may be concluded that: substrate non-dispersive component of surface energy, $\gamma^{n\text{-}d}$, is a factor controlling the adhesion quality of silicone sealants with polyester powder-coatings, and that the gloss level may be linked to the value of $\gamma^{n\text{-}d}$.

3. In the cases investigated in this project, (12 sealants with different cure chemistry; RTV1 and RTV2 types), substrates with low gloss level (typically 30–45) produced low strength bonds with 100% adhesive failure at the substrate/sealant interface. Materials with gloss level equal or greater than 90 always produced high strength bonds, frequently with 100% cohesive failure within the sealant.

ACKNOWLEDGMENTS

The authors wish to thank the Consortium of sponsors to the 'Structural Glazing Project' for the financial support of this work.

REFERENCES

1. Gutowski, W.S., Russell, L., Cerra, A., New Tests for Adhesion of Silicone Sealants, in Science and Technology of Building Seals, Sealants, Glazing and Waterproofing, Second Volume: ASTM STP 1142, (Jerome M. Klosowski, editor), Philadelphia, 1992.

2. Gutowski, W.S., Russell, L., Cerra, A., Adhesion of Silicone Sealants to Organic-Coated Aluminium, in Science and Technology of Building Seals, Sealants, Glazing and Waterproofing, Second Volume: ASTM STP 1142, (Jerome M. Klosowski, editor), Philadelphia, 1992.

3. Gutowski, W.S., in Building Seals – Materials, Properties and Performance, (ed. F. O'Connor), ASTM STP 1069, Philadelphia, 1991, p. 174.

William G. Webb[1]

MORTAR BLOCK SURFACE PREPARATION AND ASTM C 794

REFERENCE: Webb, W. G., **"Mortar Block Surface Preparation and ASTM C 794"**, Science and Technology of Building Seals, Sealants, Glazing, and Waterproofing: Second Volume, ASTM STP 1200, Jerome M. Klosowski, Ed., American Society for Testing and Materials, Philadelphia, 1992.

ABSTRACT: Historically, the results of ASTM C 794 Adhesion-In-Peel tests have been subject to high variation, especially in tests on mortar where results from different laboratories were compared. This experiment focuses on one potential source of error, the method used to prepare the surface of the mortar block. Five different techniques were evaluated with three generically distinct sealant types all intended for use on concrete and mortar. The procedures described in ASTM C 794 were followed as closely as possible. The results are evaluated for differences in precision which can be attributed to the substrate.

KEYWORDS: Adhesion-in-Peel test, elastomeric, sealant, mortar block, surface preparation

Despite carefully designed round robin tests, the results of ASTM C 794 Adhesion-In-Peel tests have historically been subject to high variation in the test results when tests are conducted in different laboratories. Adhesion tests on mortar blocks are especially susceptible to this variability, resulting in data ranges of +/- 100 percent of the mean result. This paper focuses on the method used to prepare the surface of the mortar test blocks and the storage conditions of the blocks before use as likely sources of the variation in between-laboratory test results.

---

1 William Webb is co-owner of Custom Concrete Company, 270 Stuart Road, Carlisle, PA 17013.

The purpose of this test program was to determine if any significant difference in the precision of the test results can be attributed to surface preparation and storage of the mortar test block, and if a less complicated preparation and storage procedure will produce equivalent or better precision. This experiment evaluated five surface preparation methods for mortar blocks prepared from the same batch of mortar. All the test blocks were sliced from a master concrete block measuring 152 by 76 by 600 mm (6 by 3 by 24 in.) The blocks were cut using a water cooled masonry saw with a diamond tipped blade to yield 10 by 76 by 152 mm (0.375 by 3 by 6 in.) test blocks. All the blocks were rinsed in clean tap water and scrubbed with a plastic bristle brush to remove the cutting residue. The blocks were then treated as listed below.

GROUP I. No further surfacing was performed. The blocks were stored dry until needed for use.

GROUP II. The cut blocks were surfaced on a cast iron lapping plate with number 60 aluminum oxide grit and water. Since the aggregate was already exposed, the grinding/surfacing was continued until the surface of the block appeared flat, with none of the peaks and valleys left by the saw blade. This set of blocks was scrubbed again in clean tap water to remove the grinding residue and stored dry

GROUP III. The saw cut blocks were stored in a bucket of clean tap water. No further surface preparation was performed.

GROUP IV. The cut blocks were surfaced on a cast iron lapping plate with number 60 aluminum oxide grit and water as in Group II. This set was stored in a bucket of clean tap water until ready for use.

GROUP V. The cut blocks were stored in clean tap water until needed. At that time they were removed from the water and surfaced on a cast iron lapping plate as described above. This group was prepared as instructed in the method and as such is the control group for the experiment.

As prescribed in the test method, just before test sealant was applied, the blocks were each scrubbed a second time under running water to remove any residual laitance or residue which may have formed during storage, and allowed to dry overnight in an oven at 93 C (200 F) They were allowed to cool to room temperature before the test sealant was applied.

Each preparation variable was evaluated using three different types of sealant. All the sealants are commercially available and all are recommended for use on concrete without the use of a separate primer. The sealant variables included a one part silicone, an acrylic latex, and a one part polyurethane sealant.

PROCEDURE

Sample Preparation

Peel specimens were prepared according to the method described in ASTM C 794, with some minor modifications. A layer of test sealant, approximately 1.5 mm (0.060 in.) thick, was applied to the surface of a prepared block. Next, two pieces of bright aluminum screen, typical window screen material, measuring 25 mm (1 in.) wide by 229 mm (9 in.) long were coated with the test sealant. These were laid on the first layer of sealant and covered with a second layer of sealant, also 1.5 mm (0.060 in.) thick. The 25 mm wide strips of screen are easier to handle than the 75 mm (3 in.) wide piece specified in the method and their use facilitates cutting and removing the excess sealant later in the test.

Sample Curing

As directed in the method, the test specimens were allowed to cure 7 days at 23 C (73 F) and 50 % relative humidity; 7 days at 37.8 C (100 F) and 95 % relative humidity; and finally 7 days at 23 C (73 F) and 50% relative humidity for a total cure period of 21 days. Immediately following the full cure period, the sealant on either side of the two pieces of screen was cut away to leave two 25 mm (1 in.) wide strips of sealant ready to be tested.

Test Procedure

Using a razor knife the sample was undercut to the concrete substrate then placed in the jaws of an Instron tester. Specimens were tested in 180 degree peel mode with jaw separation speed of 50 mm (2 in.) per minute. Ten specimens were tested per sealant, per mortar surfacing variable. Results were discarded if the screen broke or the mortar block broke before or during the test. The actual number of specimens tested, after attrition, was 145. After a specimen was tested, it was examined to determine the mode of failure of the sealant. Failure mode was determined to be adhesive, cohesive, or mixed. In the case of mixed failure, the percentage attributed to adhesive failure was estimated and recorded.

TEST RESULTS AND DISCUSSION

Table 1 shows the Mean, Range, Standard Deviation, and Data Spread for each set of specimens. The Mean is the arithmetic average of all the results for a particular set. The Range is the difference between the highest and lowest value for a particular set. The Standard Deviation is the measure of dispersion in the frequency distribution of a particular set, and is equal to the square root of the mean of the squares of the deviations from the arithmetic mean of the distribution. The Data Spread is the one-half the Range divided by the Mean and expressed as a percentage.

TABLE 1--Peel Test Statistics

Group	Sealant	Range	Standard Deviation	Mean	Data Spread Vs. Mean, %
I	ALL	27.09	5.88	11.80	114.79
	Latex Acrylic	10.29	2.78	5.80	88.23
	Silicone	7.00	2.09	12.10	29.05
	Urethane	18.50	4.82	17.50	52.80
II	ALL	10.52	2.83	8.16	64.49
	Latex Acrylic	4.55	1.48	5.49	41.49
	Silicone	3.26	0.89	7.99	20.39
	Urethane	4.59	1.69	11.70	19.62
III	ALL	14.97	4.23	8.68	86.20
	Latex Acrylic	4.44	1.36	5.60	37.49
	Silicone	7.78	2.88	7.78	56.81
	Urethane	10.50	3.43	13.20	39.81
IV	ALL	9.07	2.50	9.18	49.44
	Latex Acrylic	4.67	1.34	8.51	27.42
	Silicone	4.77	1.41	7.16	33.28
	Urethane	4.52	1.35	12.27	18.41
V	ALL	14.11	3.60	8.47	83.28
	Latex Acrylic	8.99	2.70	6.28	71.47
	Silicone	3.41	1.16	6.73	25.33
	Urethane	7.16	2.10	12.84	27.86
All	Latex Acrylic	10.29	2.31	5.34	79.21
All	Silicone	14.70	2.69	7.26	88.94
All	Urethane	21.30	3.75	14.04	77.94
ALL VALUES		27.09	4.25	9.30	145.69

na = not available

Table 2 illustrates the failure mode differences between the various types of block surfaces and the different sealants.

Table 2-- Failure Mode

Group	Sealant	100% CO	0-25% AD	25-40% AD	40-50% AD	50-99% AD	100% AD
I	Silicone	2	6	2			
	Latex Acrylic	1	2		1	2	4(a)
	Urethane	2	7	1			
	Total, %	16.6	50.0	10.0	3.3	6.7	13.3
II	Silicone	10					
	Latex Acrylic						10(a)
	Urethane	8					
	Total, %	64.3					35.7
III	Silicone	10					
	Latex Acrylic	4	1	3			2
	Urethane	4	4	1			1
	Total, %	62.1	17.2	13.8			10.3
IV	Silicone	9		1			
	Latex Acrylic	1		2		1	4
	Urethane	9	1				
	Total, %	67.9	3.6	10.7		3.6	14.3
V	Silicone	9		1			
	Latex Acrylic	1	2	4			3
	Urethane	9		1			
	Total, %	63.3	6.7	20			10
ALL	%	54.5	15.9	11.0	1.2	2.1	23.5

(a) A thin layer of concrete was removed from surface of block as sealant peeled. AD= Adhesive failure, CO= Cohesive failure
The numbers represent the number of specimens. Totals are shown as percentages since there were different numbers of test specimens among the various sets.

A cursory examination of the data developed in this experiment leads to the conclusion that the surface treatment can have a marked effect on the variability of test results obtained in the C794 Adhesion-In-Peel test. Grinding (or lapping) the surface of the mortar test block seemed to provide a significant improvement in the precision of the test. The manner in which the test blocks are stored prior to use in the test appeared to have a very minor effect on the variability of the results but no significant effect on the failure mode.

However, examination of Mode of Failure (see Table 2) shows that great variation occurs even when the samples do not peel off the substrate. In these cases, the substrate surface preparation has no effect on variability of the peel strength. In fact, of the 145 specimens tested, only 16.5 percent showed all adhesive failure, and only 2 percent more failed predominantly from the substrate with some cohesive failure. Of those that showed adhesive failure, the vast majority were latex acrylic sealant specimens.

Among the two groups of saw cut blocks, the greatest data spread, +/- 88.23 percent, was observed in the latex acrylic specimens of Group I, where the blocks were stored dry. Although 60 percent of this set showed predominantly adhesive failure, a thin layer of concrete peeled off the block surface with the sealant. The urethane, which failed cohesively, had approximately the same data spread. The silicone sealant which failed cohesively displayed a data spread of only +/- 29.05 percent. In Group III where the saw cut blocks were stored wet, only 20 percent of the latex acrylic specimens failed adhesively, together with 10 percent of the urethane samples. Here the data spread was between 37 and 40 percent versus nearly 57 percent for the silicone sealant which failed completely cohesively.

Turning to the post surfaced (lapped) blocks, in Group II all the latex acrylic specimens failed adhesively with a 41.49 percent data spread. In this group the silicone and urethane sealants, which failed cohesively, displayed a much narrower data spread of 20 percent. Group IV showed 50 percent of the latex acrylic group with adhesive failure but only +/- 27.42 percent data spread. In this set the silicone was more variable and the urethane less so. Finally, in the control Group V only 30 percent of the latex acrylic samples peeled adhesively, but the data spread was the second highest of all five experiments at +/- 71.47 percent of the mean. In this group the urethane and silicone sealants failed cohesively with much lower data spread of 25 to 28 percent.

If the results of the latex acrylic sealant are examined by themselves, some differences between the groups do emerge. Group I, where the blocks were not post surfaced after cutting and were stored dry, provided the most variable results. Group V, blocks that were stored wet and surfaced just before use, produced similar results. In the other three groups the variability was about the same, even in Group III, where saw cut blocks were also used. Interestingly, in this group the failure mode was predominantly cohesive. The differences in precision between the groups of latex acrylic specimens are illustrated in the normal distribution curves shown in Figures 1 through 5.

Examining the other statistics, the range and standard deviation for the latex acrylic are consistently lower than those of the urethane sealant although the urethane sealant failure mode was predominantly cohesive. The silicone was sometimes more variable and sometimes less variable than the latex acrylic sealant, but since its failure mode was always predominantly cohesive, the substrate had little if anything to do with the test results. The normal distribution curves in Figures 6 through 11 illustrate the variability of these test results.

It is interesting to note that while the results were more variable, the mean test result was generally higher on blocks that had been saw cut and stored dry. The exception to this statement is the result for latex acrylic samples. While this was the only group to show a large percentage of adhesive failure, and the data spread tended to be great, the average peel result varied very little from group to group. The variations in the peel results tended to cancel out, resulting in consistent averages, which are the most important statistic to the user of this test.

The mean result varied only +/- 7.40% between the groups of latex acrylic specimens. The mean peel value for the silicone sealant varied somewhat more between groups, +/- 36.98%, while the urethane peel strength varied +/- 20.66%. The variation in mean peel strength between the five sets of data was consistently lower than the data spread within any one group. Essentially this means that no matter what surface treatment or storage method was employed the average peel strength of the test specimens was about the same.

This experiment, while showing that the surface preparation of the block is <u>not</u> the cause of wide variation in the C794 test, does not explain why tests on mortar are more variable than tests on metal or glass. It is possible the natural porosity of concrete mortar causes it to retain moisture. This may affect the cure rate and state of cure of sealants which are affected by moisture. The natural alkalinity of the mortar may also play a role, interfering with the chemistry of curing sealants and catalyzing hydrolysis of bonds at the substrate/sealant interface during the immersion step. It is possible that during storage at least some of the soluble salts in the concrete are extracted and reach equilibrium, leaving a more uniform surface (chemically) than that of blocks which are stored dry, which would explain why the results are more consistent on mortar test substrates that were stored in lime water.

CONCLUSIONS

Clearly the substrate has little bearing on the variability in this test, and the way the surface is prepared is also not a significant factor. Use of saw cut versus carefully lapped (ground) blocks did not

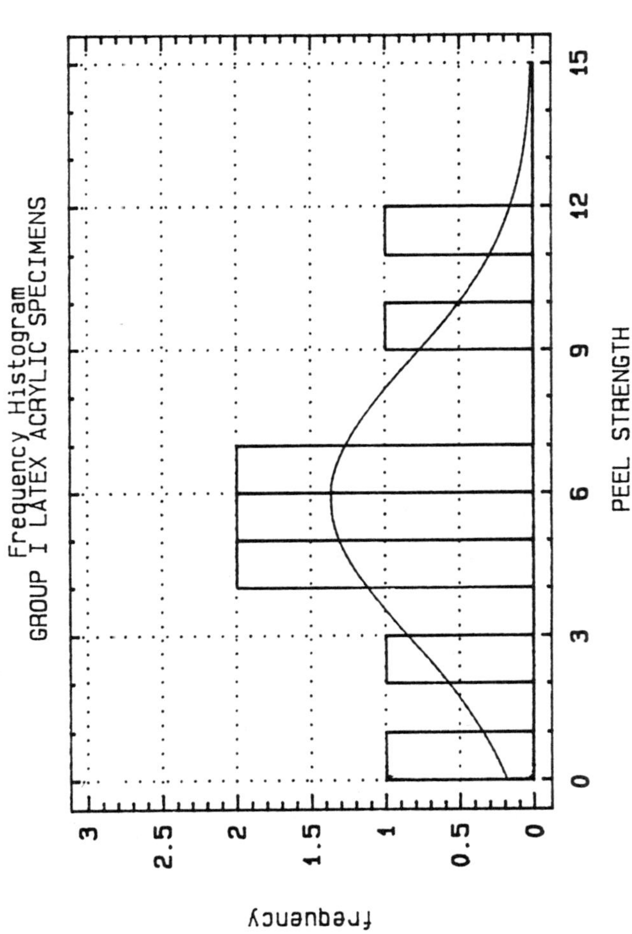

FIG. 1 -- Frequency distribution for Group I latex acrylic specimens

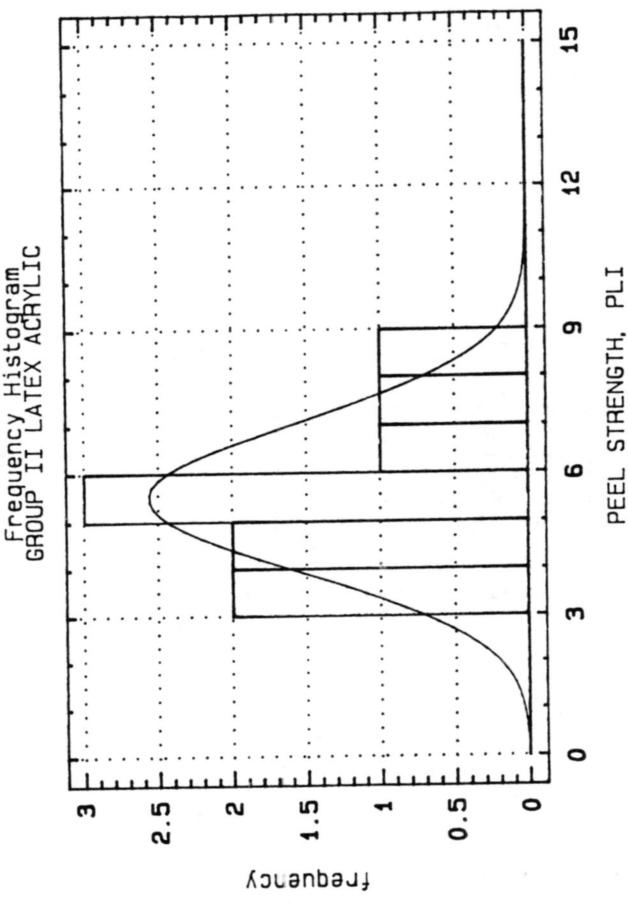

FIG. 2 -- Frequency distribution for Group II latex acrylic specimens

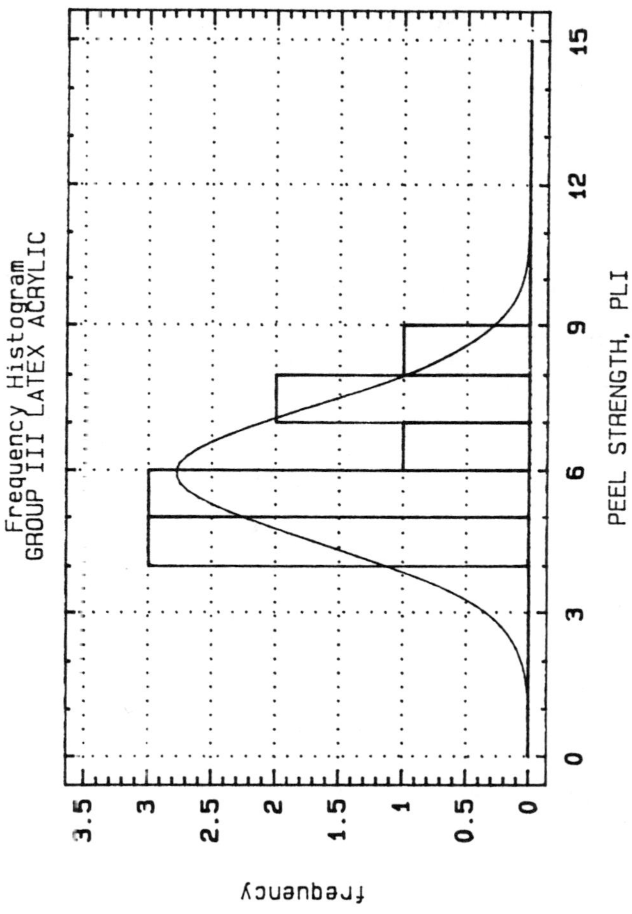

FIG. 3 -- Frequency distribution for Group III latex acrylic specimens

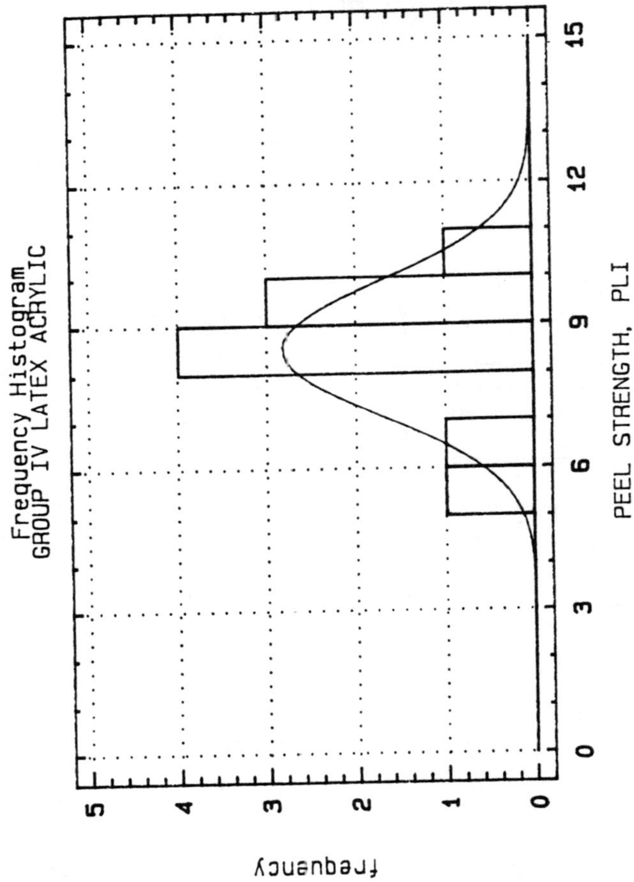

FIG. 4 -- Frequency distribution for Group IV latex acrylic specimens

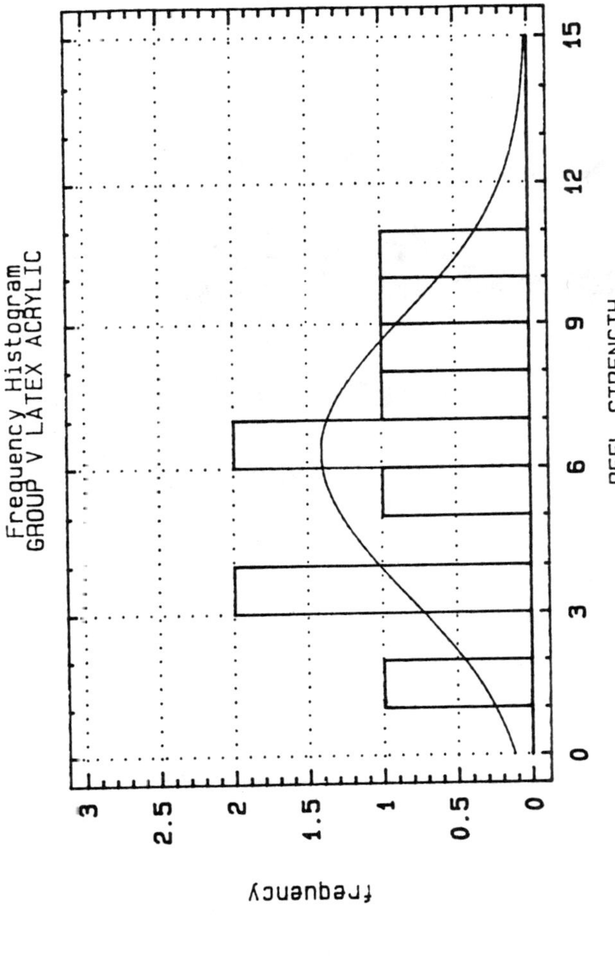

FIG. 5 -- Frequency distribution for Group V latex acrylic specimens

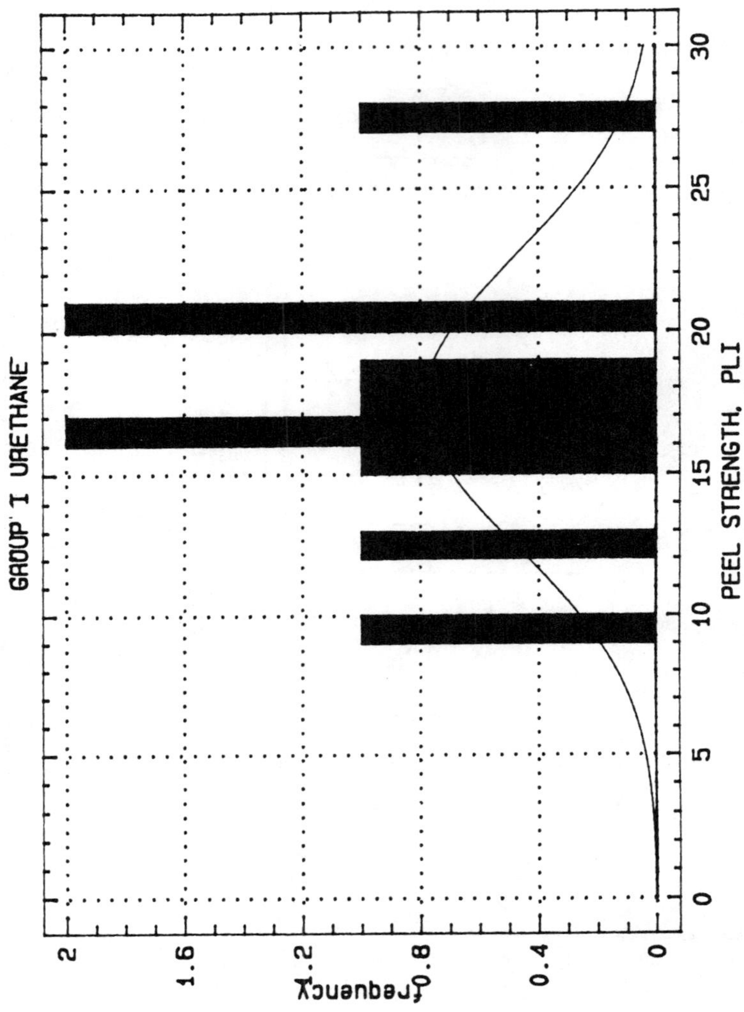

FIG. 6 -- Frequency distribution for Group I urethane specimens

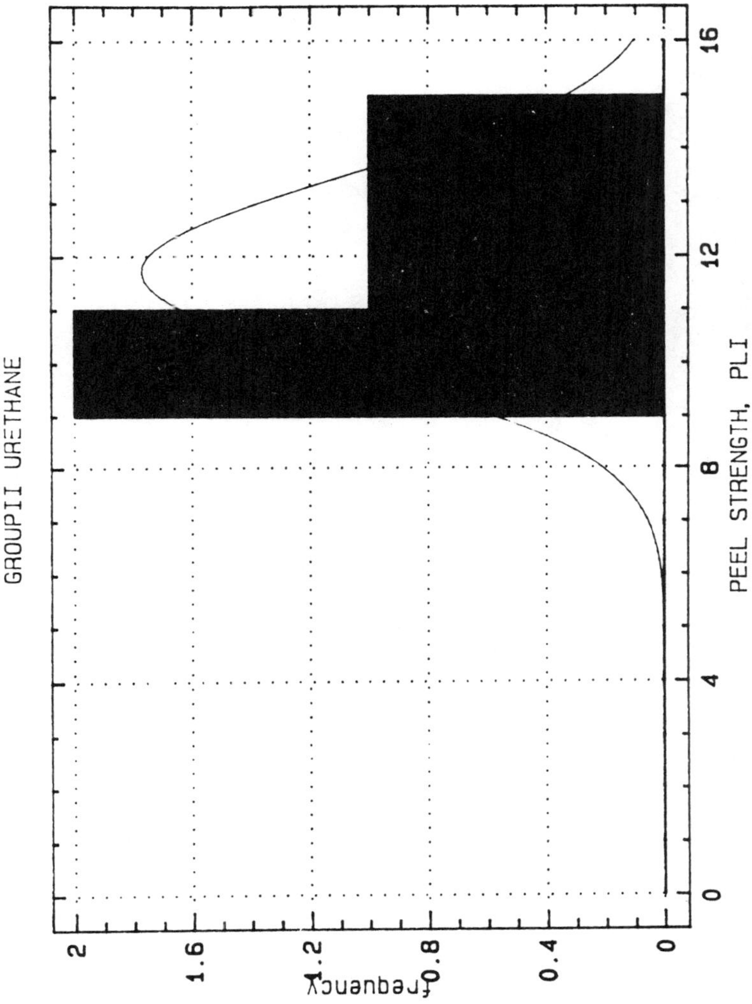

FIG. 7 -- Frequency distribution for Group II urethane specimens

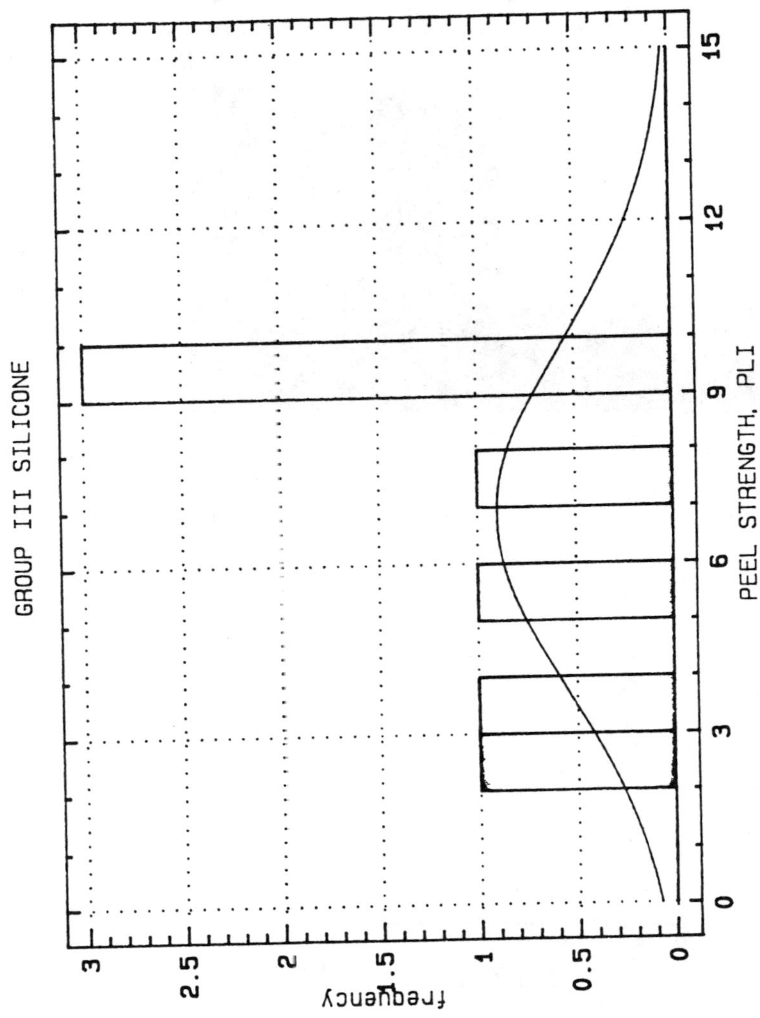

FIG. 8 -- Frequency distribution for Group III silicone specimens

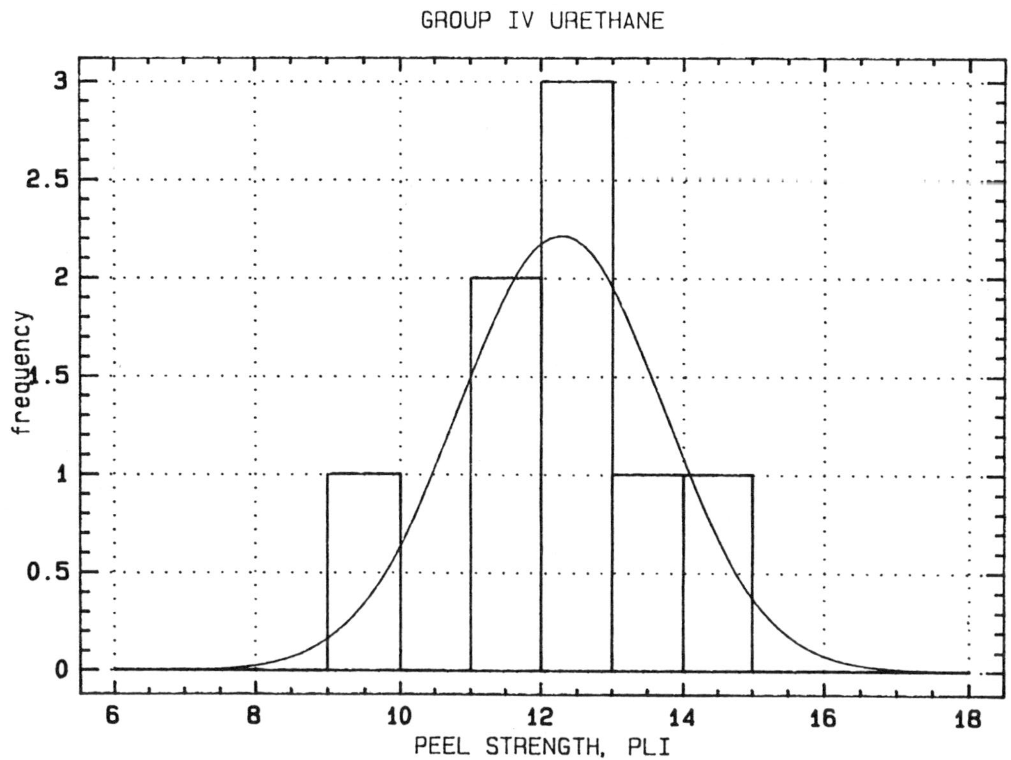

FIG. 9 -- Frequency distribution for Group IV urethane specimens

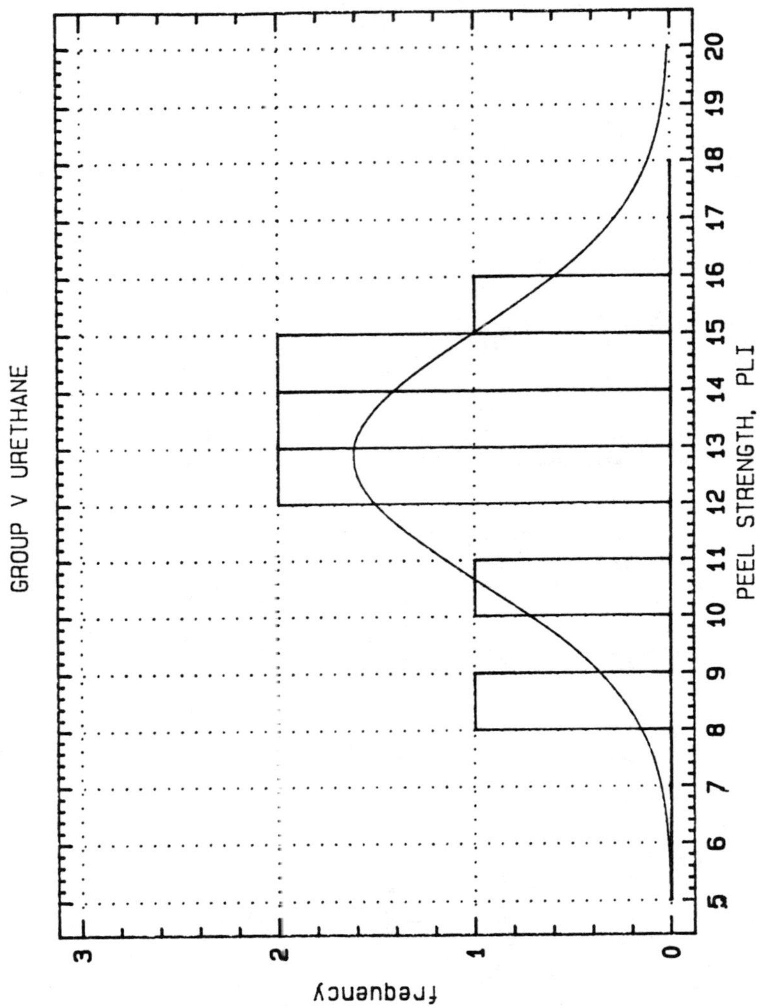

FIG. 10 -- Frequency distribution for Group V urethane specimens

significantly affect the precision of the method or the mode of failure. The practice of grinding the blocks just before use (Group V) may have contributed to wider variation in the data where adhesive failure was a factor.

In this set of tests no one sealant appeared to be significantly prone to variation. Overall they were all quite variable. The use of more specimens, 10 versus the recommended two, did not improve the overall precision, but that may be because no values were excluded from the statistical analysis.

RECOMMENDATIONS

Based on this experiment, the requirements in C794 for post-surfacing blocks on a cast iron lap with number 60 silicon oxide or silicon carbide grit seem too stringent. An equally serviceable surface would be achieved by saw cutting or sanding the blocks to uniformly expose the larger aggregate. Similarly, increasing the number of specimens to be tested is not recommended as it did not improve the precision of the method.

Wet storage of the concrete did have a positive effect on variability and should be retained as a requirement of the method.

Since concrete and mortar are prepared directly from a natural product, sand, an absolutely uniform test substrate may never be achieved. An interesting follow up to this experiment would be to prepare blocks from several sources of sand and evaluate the variability of those blocks versus the results of this experiment.

The concrete used in the construction industry will always be at least as variable as the test blocks constructed according to the ASTM C794 standard. A sealant manufacturer must take this into account and design a product for use on all kinds of concrete. Perhaps the +/- 100 percent variation in this method is necessary to factor in differing real world use conditions, and to build in a safety factor that will allow the sealant to continue to function on widely variable substrates.

ACKNOWLEDGEMENTS

The author gratefully acknowledges the assistance of Mobay Silicones, DAP, and Mameco for providing test materials, and to his family for patience and understanding during many evenings of preparation of this paper.

Voytek (W) S. Gutowski[1], Lee Russell[1] and Anthony Cerra[1]

ADHESION OF SILICONE SEALANTS TO ORGANIC-COATED ALUMINIUM

REFERENCE: Gutowski, V. W. S., Russell, L., and Cerra, A., **"Adhesion of Silicone Sealants to Organic-Coated Aluminum"**, Science and Technology of Building Seals, Sealants, Glazing, and Waterproofing: Second Volume, ASTM STP 1200, Jerome M. Klosowski, Ed., American Society for Testing and Materials, Philadelphia, 1992.

ABSTRACT: The integrity of a structurally glazed building envelope depends on the strength of adhesion at several interfaces, e.g. glass- glass coating, glass coating-silicone sealant, silicon sealant-substrate coating, substrate coating-primer, primer-substrate.

Frequently, the frame and cladding are aluminium alloys which are either anodised or finished with an organic coating to satisfy aesthetic and/or long-term corrosion resistance requirements.

In this paper, the mechanism of adhesion between organic-coated aluminium and silicone sealants is analysed. The sealants investigated are neutral-, alkaline-, and acid-cured, 1-part RTV silicones as well as 2-part RTV silicones, all commercially available. The organic coatings investigated are powder and wet-coated polyester, PVF2 (polyvinylidene flouride), acrylic and siliconised finishes, also all commercially available products.

It is shown that the strength of adhesion between any silicone sealant and the organic-coated substrate, as well as the extent of adhesive failure, depend on the substrate surface properties as quantified by the non-dispersive component of the substrate surface energy, γ^{n-d}. A typical relationship is shown to exist for strength *v.* γ^{n-d} and [%] adhesive failure *v.* γ^{n-d}.

[1]Dr Gutowski is a Principal Research Scientist; Mr Cerra and Ms Russell are Experimental Scientists at the CSIRO Division of Building, Construction and Engineering, Melbourne, Australia. Postal address: PO Box 56, Highett, Victoria 3190, Australia

Finally it is shown that the organic-coated substrates used in structural glazing can be assessed for their potential bondability with silicone sealants by determining the non-dispersive component of their surface energy. This technique is described in detail in this paper.

KEY WORDS:

Silicone sealant, adhesion, surface energy, acid-base interactions, polyester coatings, fluorinated coatings, acrylic coatings.

MECHANISM OF ADHESION OF SILICONE SEALANTS

From elementary physics, the interaction force, F, and hence the strength of an adhesive bond, can be defined as the first derivative of the interaction energy, U, with relation to the separation distance 'r' between the atoms or molecules, i.e.

$$F = -\frac{dU}{dr}\,. \tag{1}$$

In adhesion science, the (negative) energy of interaction between two materials in immediate contact is known as the thermodynamic work of adhesion, W_A, which can be estimated using the following fundamental Dupré equation:

$$W_A = -U = \gamma_1 + \gamma_2 - \gamma_{12}\,, \tag{2}$$

or the Young-Dupré equation:

$$W_A = \gamma_2(1 + \cos\Theta) \tag{3}$$

where γ_1 and γ_2 are the surface energies of materials 1 and 2 in contact, γ_{12} is the interfacial energy and Θ is the equilibrium contact angle exhibited by liquid (2) deposited on a smooth non-porous substrate 1.

An alternative expression for the estimation of work of adhesion was developed by Good and Girifalco [1], i.e.

$$W_A = 2\,\Phi(\gamma_1\,\gamma_2)^{1/2}\,, \tag{4}$$

where Φ is the interaction parameter given by

$$\Phi = (d_1\,d_2)^{1/2} + (p_1\,p_2)^{1/2} \tag{5}$$

and d_1, d_2 and p_1, p_2 are the dispersive and non-dispersive (frequently, but erroneously defined in the literature as 'polar') fractions of total surface energy (γ) of materials 1 and 2, i.e.,

$$d_1 = \gamma_1^d/\gamma_1 \qquad d_2 = \gamma_2^d/\gamma_2 \tag{6}$$

$$p_1 = \gamma_1^{n-d}/\gamma_1 \qquad p_2 = \gamma_1^{n-d}/\gamma_2 \tag{7}$$

and γ^d and γ_1^{n-d} are the dispersive and non-dispersive components of surface energy.

All materials, including polymers (with the exception of saturated hydrocarbons), can be categorised as basic or acidic [2,3] according to whether they are proton donors or acceptors. Thus, the adhesion of a polymeric sealant or adhesive to any organic or inorganic substrate can be considered in terms of acid-base interactions.

These interactions are associated with charge exchange between the materials in contact which may occur in the following alternative manner:

(a) An acid accepting an electron pair due to its incomplete electronic arrangements (Lewis theory), or proton donation according to the Brønsted theory, which leads to the formation of a strong covalent bond. Examples of materials acting in this manner include polymers such as PVC and PVDF, or inorganic materials like silica, Fe_2O_3 or Fe_3O_4.

(b) A base donating an electron pair (Lewis base) or accepting a proton (Brønsted theory). Examples of such materials include polymethylmethacrylate (PMMA), polycarbonate (PC), calcium carbonate (e.g. marble) and amorphous Al_2O_3.

(c) Amphoteric behaviour, i.e. acting as both electron acceptor and electron donor. Examples of materials in this category include amides (e.g. polyamides), amines and alcohols (e.g. polyvinyl alcohol).

According to Fowkes [2], the total work of adhesion, W_A, can be expressed in terms of the dispersive, i.e. W_A^d, and acid/base interaction, i.e. W_A^{ab}, components as follows,

$$W_A = W_A^d + W_A^{ab} . \tag{8}$$

The acid-base component of the thermodynamic work of adhesion can be calculated using the following expression:

$$W_A^{ab} = W_A - 2(\gamma_1^d \gamma_2^d)^{1/2} . \tag{9}$$

Interactions related to hydrogen bonds are a sub-set of acid-base interactions.

It has been shown earlier [4] that the strength of adhesion between a silicone sealant and a range of materials such as engineering plastics, milled aluminium and glass depends on the energy of acid-base interactions, W_A^{ab}, as expressed by Eqn (9).

The relationship, illustrated in Figure 1, shows that depending on the acid-base character of interacting materials, the following three zones can be distinguished in the relationship between bond strength and W_A^{ab} as follows:

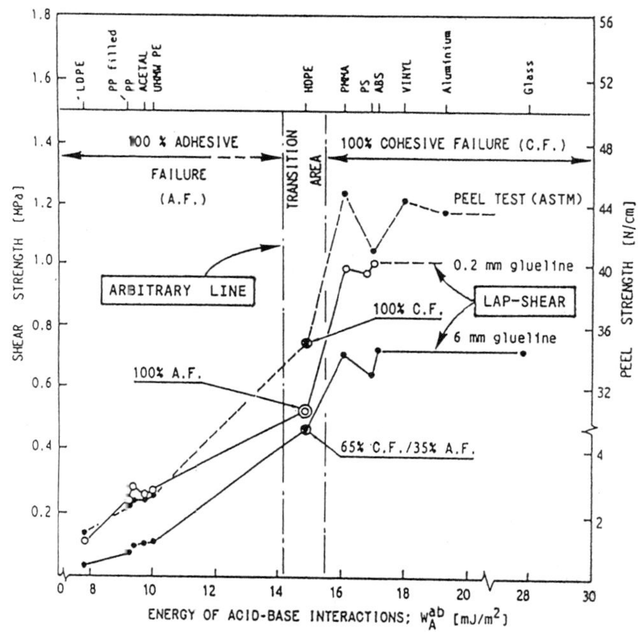

Figure 1. The relationship between the strength of an adhesive bond and the energy of acid-base interactions. All bonds made with the Dow Corning Silicone Sealant 795 and a range of engineering plastics, milled aluminium and glass.

- a 100% adhesive failure zone where bond strength increases linearly with W_A^{ab},
- a mixed failure mode zone.
- a 100% cohesive failure zone in which the strength of the adhesive forces exerted across the sealant/substrate interface exceed the cohesive strength of the sealant thus resulting in cohesive failure within the sealant.

The concept of acid-base interactions requires recognition of the fact that the acidic sites of one of the interacting phases, e.g. sealant, must interact with the basic sites of the other phase, e.g. substrate. That is to say that if both phases (e.g. sealant and substrate) are of the same nature, i.e. if they both exhibit acidic only (or basic only) character, there will not be any acid-base interaction across the sealant/substrate interface.

Experimental determination of the acid-base character of materials [5–7] requires relatively sophisticated and expensive techniques such as inverse gas chromatography, flow microcalorimetry or other procedures.

Therefore, in this paper we investigate a somewhat simpler concept according to which the strength of the adhesive bond across the sealant/substrate interface may be found to be related to a value of the energy of non-dispersive interactions,

$$W_A^{n-d} = W_A - W_A^d. \tag{10}$$

All parameters of equation (10) can be determined by relatively simple wettability studies, as described in detail in a further part of this paper. The magnitude of the energy of non-dispersive interaction, W_A^{n-d}, depends on the values of the non-dispersive components of the surface energy of the substrate, γ_1^{n-d}, and that of the silicone sealant, γ_2^{n-d}, i.e.:

$$W_A^{n-d} = 2(\gamma_1^{n-d}\gamma_2^{n-d})^{1/2}. \tag{11}$$

It follows from expression (11) that for the interaction of a silicone sealant with a range of substrates of varying polarities, $\gamma_{1\,(i)}^{n-d}$, the energy of the non-dispersive interaction is

$$W_A^{n-d} = 2(\gamma_2^{n-d})^{1/2}\,[\gamma_{1\,(i)}^{n-d}]^{1/2}\text{, or} \tag{12a}$$

$$W_A^{n-d} = C\,[\gamma_{1\,(i)}^{n-d}]^{1/2}, \tag{12b}$$

where

$$C = 2(\gamma_2^{n-d})^{1/2}. \tag{12c}$$

Thus, it can be concluded from expression (12b) that according to the concept analysed in this paper, the energy of non-dispersive interactions of a silicone sealant with a range of substrates is directly proportional to the substrates' non-dispersive component of its surface energy.

Based on the above, the following two hypotheses are to be tested in this work:

(a) the strength of adhesion between a silicone sealant and a range of organic-coated substrates is proportional to the energy of non-dispersive interaction exerted across the substrate/sealant interface, and thus

(b) the strength of adhesion between a silicone sealant and a range of organic-coated substrates is proportional to the substrate non-dispersive component, γ_1^{n-d}, of its total surface energy.

We would like to emphasise here that W_A^{n-d} by definition is not equal to the energy of the acid-base interactions since values of γ_1^{n-d} and γ_2^{n-d}, in the form presented in this paper do not take into account the inherent acidity or basicity of interacting materials. It may therefore occur in some cases that the experimental points relating bond strength to W_A^{n-d} will not satisfy the relationship described by Eqn (12b). This may be due to the already mentioned possibility that the materials of a given value of γ_1^{n-d} may differ in terms of their acid-base character.

The following describes the results of the work undertaken so far to verify the above hypotheses.

EXPERIMENTAL

1. Substrates

The following types of organic coatings were investigated in this work:

(a) Polyester – powder coats
(b) Polyester – wet coats
(c) Fluorocarbon (PVF2)
(d) Acrylic
(e) Hybrid materials with PVF2 base polymer.

All substrate materials were commercially prepared using coatings manufactured by Australian manufacturers. The base alloy material was aluminium **6063**.

2. Silicone Sealants

Silicone sealants investigated in this work were of the following type:

(a) One-component (RTV1) sealants: neutral-, alkaline-, and acidic cured,

(b) Two-component (RTV2) sealants.

In total, there are 10 RTV1 and 2 RTV2 sealants investigated and analysed in this work. All sealants are commercially available and are commonly used either in structural or weatherseal applications.

3. Techniques

Surface preparation. All substrates were thoroughly cleaned using high purity ethyl alcohol. After triple wiping the specimens were allowed to dry for about 1 hour. Lap-shear specimens were prepared typically within 2 hours after cleaning.No primers were used.

Adhesive bond strength. This was determined using the special lap-shear specimens (in duplicate) as described in detail in Reference [8]. The glueline thickness was 0.2 mm. All bonded specimens were allowed to cure for 4 weeks under controlled conditions (temperature 20°C and relative humidity 50%). The cured specimens were tested using an Instron machine at the rate of 10 mm/min.

Adhesive bond quality assessment. All lap-shear specimens were assessed for two parameters (i) the actual bond strength calculated by dividing the force required to break the specimen by the bond area, and (ii) the percentage of adhesive failure [%AF] which was determined using a high powered optical microscope.

Determination of the surface energy of the substrate and its dispersive/non-dispersive character. The surface energy (γ_1) of the substrate and its non-dispersive (γ_1^{n-d}) and dispersive (γ_1^d) components are determined experimentally by wettability studies. In this technique, sessile or suspended drops of a series of test liquids are deposited on the substrate's surface [see Figure 2a].

Once the liquid drop has established equilibrium (after some initial geometry

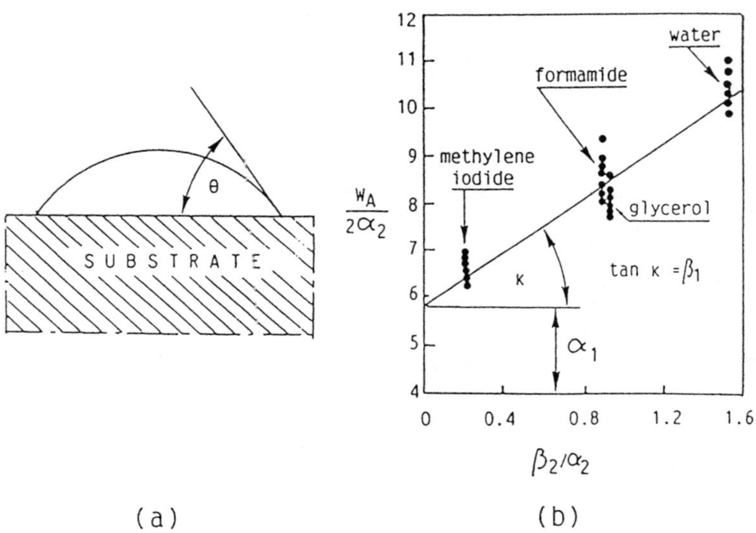

Figure 2. Determination of the surface energy of a solid; (a) sessile drop of a test liquid exhibiting a contact angle θ_{21} at the substrate/liquid interface; (b) typical graph $W_A/2\ \alpha_2$ v. β_2/α_2 for determining surface energy of a solid. The slope is the square root of the non-dispersive component, γ^{n-d}, and the intercept is the square root of the dispersive component, γ^d.

changes), the equilibrium contact angle θ is determined for each test liquid. The test liquids, as listed in Table 1, must have known total surface energy (γ_2) as well as the dispersive (γ_2^d) and non-dispersive (γ_2^{n-d}) components.

The surface energy of the substrate is then determined by applying the following equations [6]

$$0.5\,\gamma_2^{(1)}\,(1 + \cos\theta^{(1)}) = (\gamma_1^d\,\gamma_2^{d(1)})^{1/2} + (\gamma_1^{n-d}\,\gamma_2^{n-d(1)})^{1/2} \qquad (13)$$

$$0.5\,\gamma_2^{(2)}\,(1 + \cos\theta^{(2)}) = (\gamma_1^d\,\gamma_2^{d(2)})^{1/2} + (\gamma_1^{n-d}\,\gamma_2^{n-d(2)})^{1/2} \qquad (14)$$

which are solved for the unknown parameters (γ_1^d) and (γ_1^{n-d}). The total surface energy of the substrate may be calculated as follows

$$\gamma_1 = \gamma_1^d + \gamma_1^{n-d} \qquad (15)$$

The parameters $\theta^{(1)}$ and $\theta^{(2)}$ in Eqns (13) and (14) are the equilibrium contact angles exhibited by the test liquids (1) and (2) deposited on the substrate's surface, whilst the known properties of the test liquids are γ_2, (γ_2^{n-d}) and (γ_2^d). The values of $\gamma_1 = \gamma_1^d + \gamma_1^{n-d}$ are estimated as the average from the results of tests involving several of the test liquids listed in Table 1.

Table 1. Total surface energy (γ_2) and the dispersive (γ_2^d) and non-dispersive (γ_2^{n-d}) components for selected test liquids (t = 20°C).

No.	Test liquid	γ_2 (mJ/m^2)	γ_2^d (mJ/m^2)	γ_2^{n-d} (mJ/m^2)
1	Water	72.8	21.8	51.0
2	Glycerol	63.4	37.0	26.4
3	Formamide	59.2	39.5	18.7
4	Methylene iodide	50.8	48.5	2.3
5	Trichlorobiphenyl	45.3	44.0	1.3
6	α-Bromonaphthalene	44.6	44.6	0
7	Tricresyl phosphate	40.9	39.2	1.7
8	n-Hexadecane	27.6	27.6	0

An alternative method for determining the substrate surface energy currently used in this work is explained in Figure 2b. In this technique, a series of test liquids, selected from Table 1 are used for measuring the equilibrium contact angles Θ for substrates of unknown surface energy. It has been shown by Kaelble [9] that the following expression can be derived from Eqns (2) to (7):

$$\frac{W_A}{2\alpha_2} = \alpha_1 + \beta_1 \left(\frac{\beta_2}{\alpha_2}\right) \tag{16}$$

where:

$$\alpha_1 = (\gamma_1^d)^{1/2} \qquad \beta_1 = (\gamma_1^{n-d})^{1/2} \tag{17}$$

$$\alpha_2 = (\gamma_2^d)^{1/2} \qquad \beta_2 = (\gamma_2^{n-d})^{1/2} \tag{18}$$

It can be seen that Eqn (16) represents a straight line with the coordinates $[W_A/2\alpha_2]$ and $[\beta_2/\alpha_2]$, as illustrated in Figure 2b. The intercept of this line, i.e. α_1 represents the square root of the dispersive part of the substrate's surface energy, γ_1^d, and the slope, tan κ, represents the square root of the substrate's non-dispersive component, as given by Eqn (17).

RESULTS AND DISCUSSION

1. Non-dispersive component, γ_1^{n-d}, of substrate's surface energy

Table 2 contains the results on the numerical value of the non-dispersive component of the total surface energy, i.e. γ^{n-d} for all organic-coated substrates investigated in this work.

The scope of this paper does not allow for a detailed analysis of the effects of all the 'processing variables' on the value of γ_1^{n-d} of the organic-coated substrates. For the needs of this paper, the following observations regarding the γ_1^{n-d} of the substrates will suffice.

(a) The numerical value of γ^{n-d}, the non-dispersive component of the substrate's total surface energy, depends on several processing variables, e.g.: the type of base polymer, air-reducing compound, gloss-control additive, cure conditions, etc.

(b) The value of γ_1^{n-d} of the polyester powder-coated products investigated varies within the range: 3.87 to 9.22 mJ/m^2.

(c) The base polymers used in the 'standard' and 'high durability' versions of the polyester powder-coated products exhibit similar surface 'non-dispersivity' (materials P1 and P10 in Table 2).

(d) All fluorinated coatings investigated in this work exhibit non-dispersive component of total surface energy within the narrow range γ^{n-d} = 7.1 to 8.8 mJ/m^2.

Table 2. Non-dispersive component of total surface energy (γ_1^{n-d}) for all organic-coated substrates investigated in this work

Substrate	γ^{n-d} [mJ/m^2]	Substrate	γ^{n-d} [mJ/m^2]
Polyester: powder-coated		Polyester: wet-coated	
P1	8.56	P14	6.76
P2	6.40	P15	8.95
P3	7.87	P16	9.54
P4	5.90		
P5	5.43	PVF2	
P6	6.58	D1	8.80
P7	6.54	D2	8.30
P8	4.28	PVF2 hybrid	
P9	7.76	M2	8.60
P10	8.85	M3	7.10
P11	7.90	M4	7.70
P12	3.87	Acrylic	
P13	9.22	M1	11.80

(e) The acrylic coating (M1) investigated exhibits the highest value of $\gamma^{n-d} = 11.8$ mJ/m^2.

2. Influence of the non-dispersive component of total surface energy of the organic-coated substrate on the strength of adhesion

As already mentioned, there were a total of 12 one- and two-component silicone sealants of various cure chemistry investigated in this work. All of these sealants whatever their chemistry and curing mechanism exhibit similar performance in terms of the relationship between the bond strength and a value of non-dispersive component, γ_1^{n-d}, of the substrate's surface energy. In order to avoid unnecessary repetitions, the authors have therefore selected five typical examples to demonstrate the *'strength v.* γ^{n-d}' relationship. For commercial reasons the type of sealants discussed and their cure chemistry cannot be disclosed in this paper.

In order to validate the proposed hypothesis i.e.

$$(\text{bond strength}) \propto W_A^{n-d} \propto C[\gamma_{1\ (i)}^{n-d}]^{1/2} \quad (19)$$

the analysis must first consider those systems which produce predominantly adhesive failure (~100% AF) at the sealant/substrate interface. From all the sealants tested, those denoted A and B satisfy this requirement. The relationship between the bond strength and γ_1^{n-d} for the range of substrates bonded with sealants A and B has been illustrated in Figures 3 and 4.

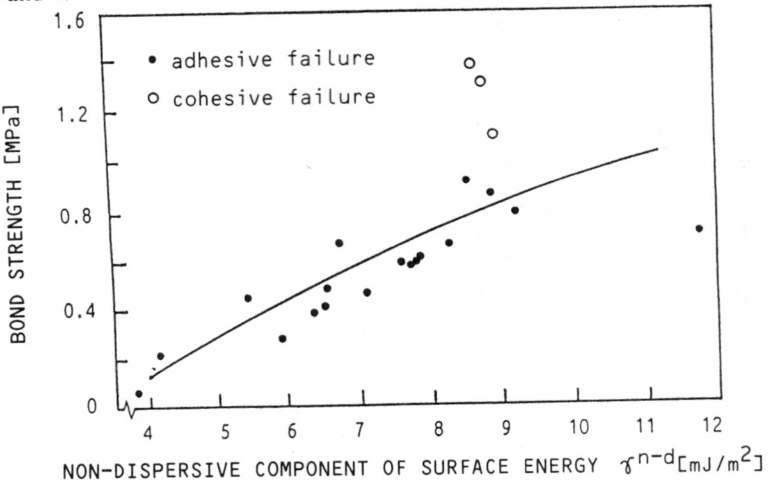

Figure 3. Influence of the non-dispersive component of substrate's surface energy on bond strength. Substrates bonded with sealant A.

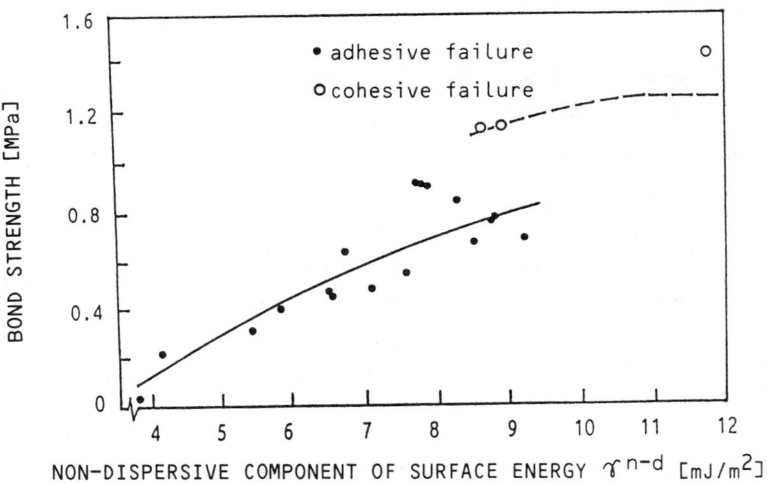

Figure 4. Influence of the non-dispersive component of substrate's surface energy on bond strength. Substrates bonded with sealant B.

The results indicate that an increase in substrate's γ_1^{n-d} results in an approximately power-related increase in bond strength, in a fashion similar to that suggested by Eqn (19). It also appears that the two substrates with values of γ_1^{n-d} about 9 mJ/mm^2, i.e. M2 and P15 deviate from the typical trend by exhibiting 100% cohesive failure with sealants A and B, in contrast to other substrates, probably due to the formation of stronger, chemical bonds resulting from strong acid-base interactions across the substrate/sealant interface. Another substrate, i.e. M1 exhibits significantly lower bond strength (0.67 MPa) with sealant A than it would be predicted from Eqn (19) for a material with a relatively high γ^{n-d} = 11.8 mJ/mm^2. In this case, this deviation can be attributed to relatively weak acid-base interaction between sealant A and this material.

Analysis of the results involving other sealants reveals that the pattern of '*strength v.* γ_1^{n-d}' relationship is somewhat more complex than that illustrated in Figures 3 and 4. For example see sealants C, D and E as illustrated in Figures 5 to 7.

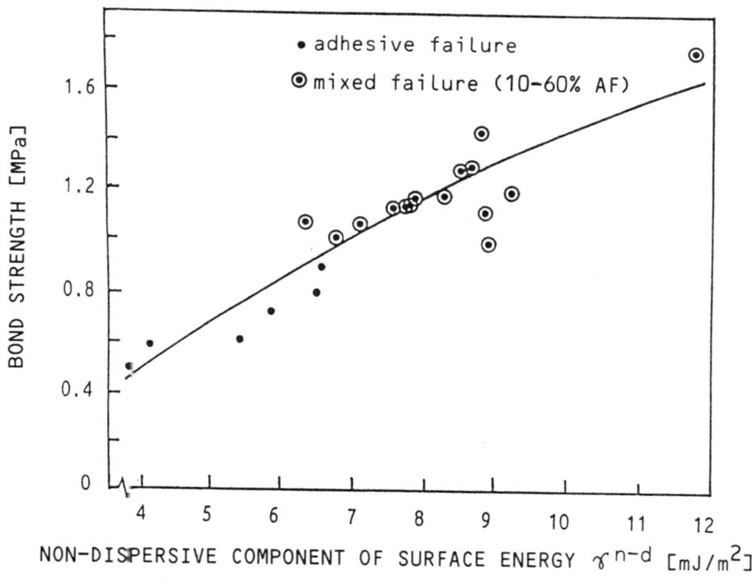

Figure 5. Influence of the non-dispersive component of substrate's surface energy on bond strength. Substrates bonded with sealant C.

The following general observations can be made from the data presented in these Figures:

(a) An increase in the substrate's γ_1^{n-d} results in an approximately power-related increase in bond strength [in a fashion similar to that suggested by Eqn (19)] up to a threshold value which is related to the cohesive strength of the sealant. Any further increase in γ_1^{n-d} does not influence the bond strength.

(b) there are three zones in the '*strength v.* γ_1^{n-d}' relationship, i.e.

- *very low non-dispersive component of substrate surface energy:* $\gamma_1^{n-d} \leq 6$ mJ/m². In this region most sealant/substrate combinations exhibit entirely 100% adhesive failure. An increase of γ_1^{n-d} in this region results in an

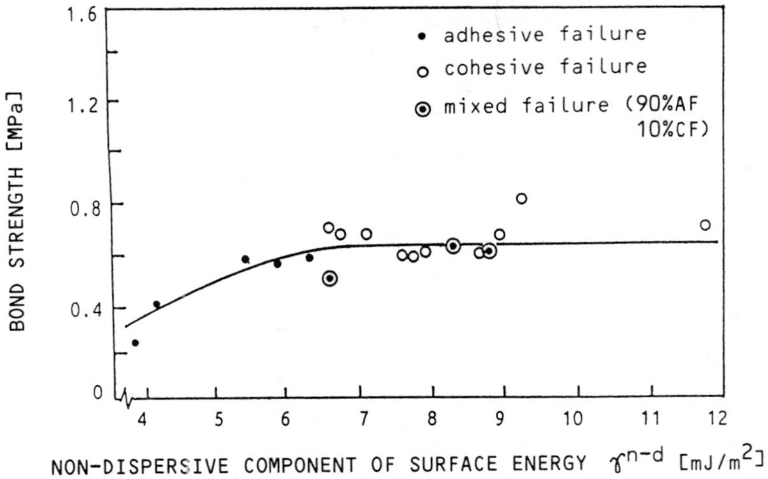

Figure 6. Influence of non-dispersive component of substrate's surface energy on bond strength. Substrates bonded with sealant D.

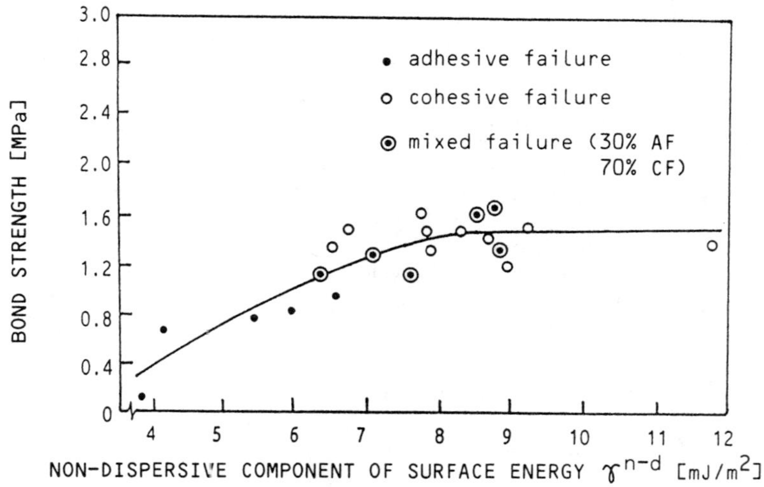

Figure 7. Influence of non-dispersive component of substrate's surface energy on bond strength. Substrates bonded with sealant E.

approximately power related increase in the bond strength, as expressed by Eqn (19).

- *low non-dispersive component of substrate surface energy:* 6 mJ/m$^2 \le \gamma_1^{n-d} \le 9$ mJ/m^2. In this zone there is a transition from 100% AF at the lower limit, through the combination of adhesive and cohesive failure (AF/CF), up to predominantly 100% CF in some sealants. Despite the occurrence of both failure modes i.e. mixed 'AF/CF' or '100% CF' in this zone, the strength difference between the bonds that failed 100% cohesively or in the mixed AF/CF mode is relatively small. The possible reasons for the occurrence of both types of failure mode, i.e. 'mixed AF/CF' or 100% cohesive failure for different substrates exhibiting similar values of γ_1^{n-d} in this zone may be due to the different levels of acid-base interaction of these materials exhibiting different donor/acceptor character despite similar values of the non-dispersive components of substrate surface energy. This concept will be investigated further by the authors of this paper. It may also be attributed, but to a lesser degree, to the local differences in the sealant thickness within the specimen's overlap area (see Reference 8 for more comments relating to the latter).

- *medium non-dispersive component of substrate's surface energy:* 9 mJ/m$^2 \le \gamma_1^{n-d} \le 12$ mJ/m^2. In this zone most sealants investigated exhibit 100% cohesive failure, since the strength of the adhesive forces exerted across the sealant/substrate interface exceeds the strength of the elastomeric sealant.

CONCLUSIONS

1. It has been shown in this work that the strength of adhesion between silicone sealants and typical organic-coated substrates as used in facade design depends on the energy of non-dispersive interactions, W_A^{n-d}, across the sealant/substrate interface, which in turn is directly related to the substrate's non-dispersive component, γ_1^{n-d}, of its total surface energy.

2. An increase in the substrate γ_1^{n-d} results in an approximately power-related increase in the bond strength up to a threshold value corresponding to the cohesive strength of the sealant. A detailed study regarding the analytical form of the relationship between the bond strength and γ_1^{n-d} [based on Eqn (19)] will be presented in a future paper.

3. Some deviations from the typical behaviour (in terms of the relationship between the bond strength and γ_1^{n-d}), as observed in this work are attributed to different levels of acid/base interactions for materials exhibiting similar values of γ_1^{n-d} but exhibiting different donor/acceptor character.

4. It is proposed that the suitability of a given substrate for building facade applications should be assessed by first determining the substrate's non-dispersive component of surface energy followed by the preparation and testing of specimens for determining the strength of adhesion and failure mode. It is suggested at this stage that all substrates exhibiting $\gamma_1^{n-d} \leq 6$ mJ/m^2 should be treated cautiously.

ACKNOWLEDGMENTS

The authors wish to thank the Consortium of sponsors involved in the 'Structural Glazing Project' for the financial support of this work.

REFERENCES

1. Good, R.J., Girifalco, L.A., A Theory for Estimation of Surface and Interfacial Energies. III. Estimation of Surface Energies of Solids from Contact Angle Data. Journal of Physical Chemistry, **64**, 1960, p. 561–565.

2. Fowkes, F.M., Acid-Base Interactions in Polymer Adhesion in Physico-Chemical Aspects of Polymer Surfaces, 2, (ed. K.L. Mittal) Plenum Press, New York and London, 1978, p. 583–663.

3. Gutman, V., Donor-Acceptor Approach to Molecular Interaction, Plenum Press, New York and London, 1983.

4. Gutowski, W.S., "Adhesive Properties of Silicone Sealants" in Building Seals – Materials, Properties and Performance, (ed. F. O'Connor) ASTM STP 1069, Philadelphia, 1991, p. 174–192.

5. F.M. Fowkes, Role of Acid-Base Interfacial Bonding in Adhesion, J. Adhesion Sci. Tech. **1**(1), 1987, p. 7–27.

6. J. Schultz, L. Lavielle, C. Martin, The Role of Interface in Carbon Fibre-Epoxy Composites. J. Adhesion, **23**, 1987, p. 45.

7. Lloyd, D.R., Ward, T.C., Schreiber, H.P. (Editors), Inverse Gas Chromatography: Characterisation of Polymers and Other Materials. American Chemical Society, ACS Symposium Series No. 391, Washington DC, 1989.

8. Gutowski, W.S., Russell, L., Cerra, A., "New Tests for Adhesion of Silicone Sealants", Science and Technology of Building Seals, Sealants, Glazing and Waterproofing, Second Volume, ASTM STP 1142, Jerome M. Klosowski, editor, American Society for Testing Materials, Philadelphia, 1992.

9. Kaelble, D.H., "Dispersion-Polar Surface Tension Properties of Dynamic Solids", Journal of Adhesion, **2**, 1970, p. 66–81.

Andrew Charles Yanoviak, AIA, CSI

ACCOMMODATION of MOVEMENT within BUILDING SEALS and SEALANTS for HIGHRISE TOWER ARCHITECTURE and CONSTRUCTION

REFERENCE: Yanoviak , A. C., "**Accommodation of Movement within Building Seals and Sealants for Highrise Tower Architecture and Construction**", Science and Technology of Building Seals, Sealants, Glazing, and Waterproofing: Second Volume, ASTM STP 1200, Jerome M. Klosowski, Ed., American Society for Testing and Materials, Philadelphia, 1992.

ABSTRACT: This manuscript includes an overview of architectural design research on the subject of cyclical and non-cyclical 'movement systems' encountered in the design and construction of highrise building towers. Furthermore, it takes into consideration the need for structural 'accommodation' and 'resistance' of these multiple movements and 'movement systems' in the design, manufacture, application and maintenance of 'building seals and sealants'.

This presentation includes a basic review of architectural and structural design factors, building codes and standards for the effects of both 'live load' and 'dead load' on the structural building frame and 'exterior building envelope'. It also addresses the application of appropriate construction materials and building systems including 'cladding and glazing systems'. Due consideration is given to the development of design standards and the standard design details required, to accommodate these movements. The accumulative single and multiple effects of wind and seismic forces, as well as gravity and thermal loads are also addressed.

Practical examples of water infiltration problems and inadequate design details, specifications, material applications, waterproofing methods and challenges are noted. References to 'building seals and sealants', including the phenomena of 'curing', 'creep', 'elongation', 'elasticity', 'plasticity', 'yield limits', 'electrolysis' 'stress', 'strain', 'fatigue', 'adhesion', 'cohesion', and 'bonding', are also presented. Maintenance factors to be considered in remedial repair specifications, design details and installation methods, are also referenced in this manuscript.

Design challenges by architects, developers, and building owners for contractors and manufacturers of 'building seals and sealants' are also noted.

KEYWORDS: building movement, building seals, sealants, highrise, structure, exterior wall panels, applications, specifications, design details, building standards, deficiencies, quality control, testing, technical literature

Andrew Charles Yanoviak, AIA, CSI is an architect-planner and construction litigation/expert witness consultant with Andrew Charles Yanoviak,AIA; 1188 Bishop Street / Suite 3011; Honolulu, Hawaii 96813

The analysis of 'building movement' for highrise towers in the range of 10 to 100 stories, i.e. 100 to 1000 feet (30 to 300 meters) tall and above is quite complex. In fact, 'Computer Aided Drafting and Design' (CADD) programs had their origin in tall building design problem analysis about 15 years ago. Now, CADD computer terminal networking is fundamental among interdisciplinary design professionals in large multidisciplinary architect/engineer ("A/E") firms. CADD generated three-dimensional modeling studies are now paramount to the effective creation of innovative custom designed highrise building towers (Figure 1).

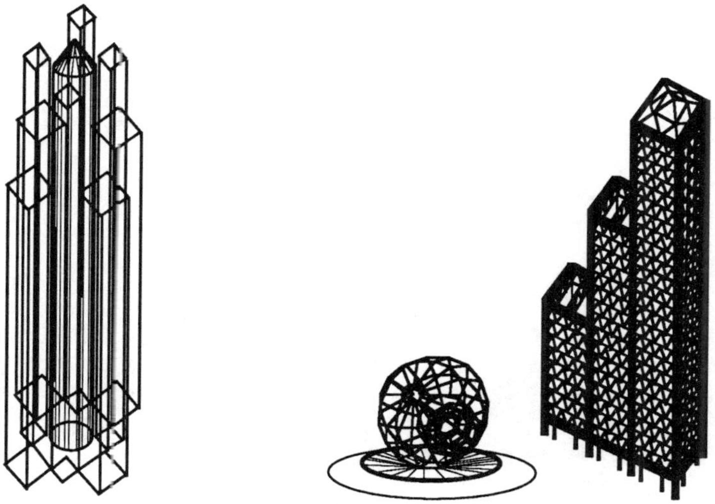

Figure 1 : 'CADD' GEOMETRICAL ANALYSIS OF 'TUBULAR' AND 'TRUSSED BRACED FRAME' HIGHRISE TOWER STRUCTURAL SYSTEMS

Pioneering structural engineers ("SE"), mechanical ("ME"), electrical ("EE"), civil ("CE"), and soils engineers ("S"), along with architects who are challenged with finding economical solutions to highrise building design, require a concerted team effort (Figure 2). This is necessary to address the integration of various 'building movement' systems which incorporate advanced construction and design technology. In addition to the owner/ developer ("DEV") and architect, both the structural engineer ("SE"), and other engineers are intensely involved at the conceptual design, design development, and design detailing stages.

Highrise towers with complex and transformational geometric shapes and differential heights require specialized design and construction consultants who address 'building seals and sealants' as well as windows and cladding systems.

In the area of 'building seals and sealants', during construction there are glazing ("GL"), caulking, cladding, coatings, and other specialty subcontractors ("SUB"). And, during both the design and construction phases of highrise tower development, the inorganic chemists, physical chemists, and biochemists engaged by the manufacturers and suppliers ("MFR/ SUPPL") of building materials are very much needed in research and development of new products and applications (Figure 2). Not only the contractors and engineers, but also the scientists and consultants, are constantly being challenged by innovative highrise designs, sculptural architectonic shapes and forms, and technological and scientific inventions [1, 2, 3].

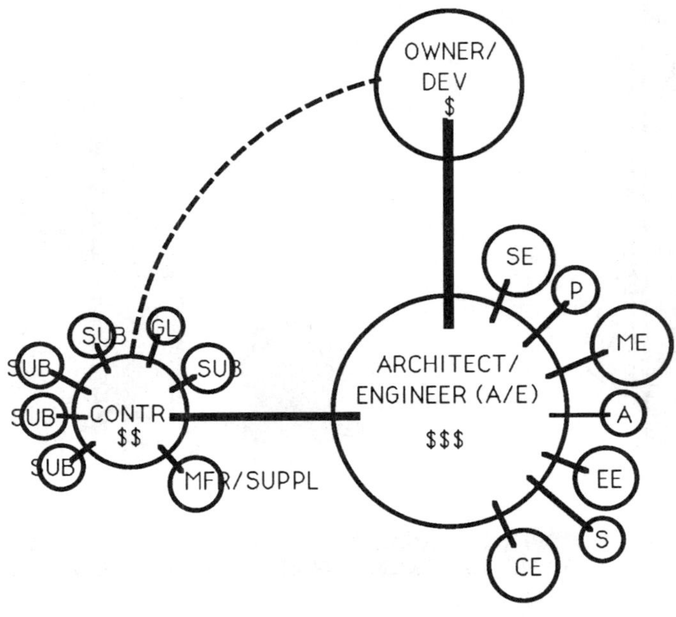

Figure 2 : TYPICAL 'DESIGN-BUILD' CONTRACTURAL RELATIONSHIPS

Simplistic highrise towers, formed of basic rectangular prisms in the 'upended cigarette carton' vernacular, which were predominant in much earlier periods and styles of Modern Architecture, are no longer in vogue. In our Post Modern Architecture era, they are being superceded by complexly intertwined geometrical 'signature' configurations that challenge the most skilled technologists in the Construction Industry [2, 3, 4, 8] . (See also 'Photo 1' of the "Century Square" high-rise office tower in Downtown Honolulu designed by Architect Jo Paul Rognstad, AIA; and, 'Photo 2' of the "One Waterfront Plaza" residential highrise tower in Honolulu designed by noted Architect Warner Gayle Boone, AIA).

Photos 1 and 2 : TYPICAL 'MODERN ARCHITECTURE' HIGHRISE TOWER FORMS in HONOLULU (on Right) CONTRASTED with 'POSTMODERN ARCHITECTURE' TOWERS

There are many different types of 'building movement' that must be considered in highrise tower design and construction. Foremost among these, are those affected by simultaneously resisting and accommodating the overturning and sliding forces generated by the wind and earthquakes. Generally, either wind or seismic design factors govern. These determinations are made on the basis of criteria such as type of construction, structural system selected, building height, geographical location, and seismic intensity and frequency zones as designated in the building codes [4, 8, 9, 11, 21, 22, 23, and 24].

The 'seismic zones' illustrated (Figure 3) are based on the Honolulu Building Code. This particular code established higher 'design standard' criteria for 'seismic zones' in the Hawaiian Islands prior to their eventual incorporation in the Uniform Building Code (UBC) which is published by the International Conference of Building Officials (ICBO). The higher numbered 'seismic zones' ('3', '2B', '2A') require greater overturning and shear design resistance in the core of the structure and compatible behavior in the exterior building envelope.

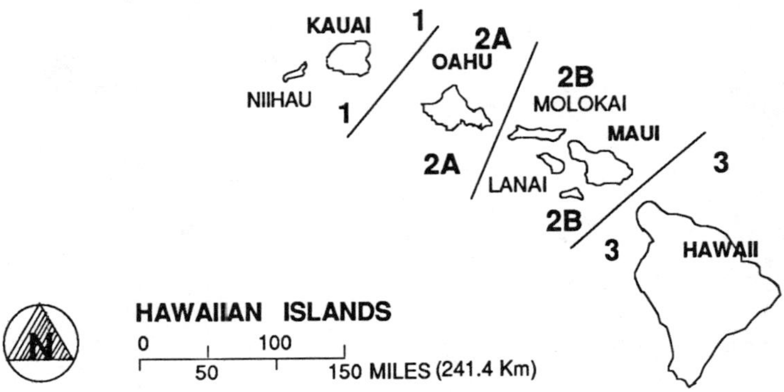

Figure 3 : 'SEISMIC ZONES' of the HAWAIIAN ISLANDS in the HONOLULU BUILDING CODE and the UNIFORM BUILDING CODE

It is interesting to note in the the Uniform Building Code (UBC) 'Wind Rose Diagram' for the United States that Alaska and Fort Lauderdale share comparable hurricane and gale force wind velocities (Figure 4). In the design of highrise towers in these geographical locations and others including Honolulu, both the American National Standards Institute (ANSI) and the Architectural Aluminum Manufacturers Association (AAMA) design standards must be consulted in conjunction with 'minimum requirements' of the building codes [10, 12, 14, 15].

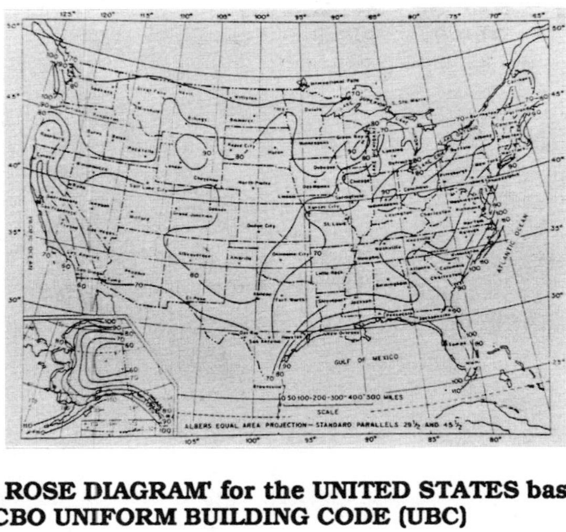

Figure 4 : 'WIND ROSE DIAGRAM' for the UNITED STATES based on the ICBO UNIFORM BUILDING CODE (UBC)

Most highrise building towers are designed to accommodate downward gravity forces (Figure 5), attributed to code specified 'dead loads' (weights of building materials incorporated within the structure) and some 'live loads' (such as building occupants, furniture, equipment, stored products, etc.). However, other building code specified 'live loads' (such as lateral wind loads and seismic forces) are accommodated by designing the highrise tower as a giant cantilevered beam, literally bracketed off a vertical wall rather than a horizontal ground plane (see Figure 6).

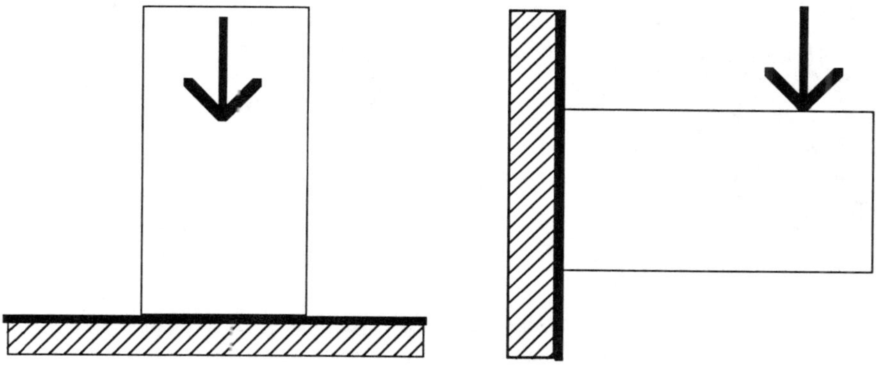

Figures 5 and 6 : 'FORCE ANALYSIS' of HIGHRISE TOWERS for GRAVITY and LATERAL LOADS ('DEAD' and 'LIVE')

There are many configurations comprising highrise tower designs which are a blend of architecture and engineering by necessity of their height, weight and economics. As the exterior architectural forms and interior spaces have become more complex in recent times, likewise the structural engineering concepts have evolved into a much higher level of sophistication. In addition to the 'cantilevered beam' principle illustrated above, several highrise towers have been designed and constructed in accordance with the 'tubular shell' concept; e.g., the 1350 feet (or 400 meters) tall twin towers of the 'World Trade Center' in New York City designed by the late American Architect Minouru Yamasaki, FAIA. Others incorporate the 'bundle of tubes' concept exemplified by the 1500 feet (or approximately 450 meters) tall 'Sears Tower' by Skidmore, Owings and Merrill (SOM) in Chicago (Figure 1). There are also 'trussed frames' such as the 'Bank of China' in Hong Kong ('Photo 3') designed by AIA Gold Medalist Ieoh Ming Pei, FAIA of New York City and world famous Structural Engineer Leslie Robertson, P.E. [2]. And also, 'braced frames' such as the 'John Hancock Center' in Downtown Chicago ('Photo 4') designed by Bruce Graham, FAIA of SOM, and 'ductile moment frames' which are more common, but still exciting, in their expressions of ingenious resolutions of structural forms and forces (Figures 1 and 7) and [5, 6, 7, 8, 10, 21, and 22].

Photos 3 and 4 : 'SPIRALING TRUSSED FRAME' HIGHRISE TOWER FORM in HONG KONG (on Left), COMPARED with 'BRACED FRAME' HIGHRISE ARCHITECTURE in CHICAGO

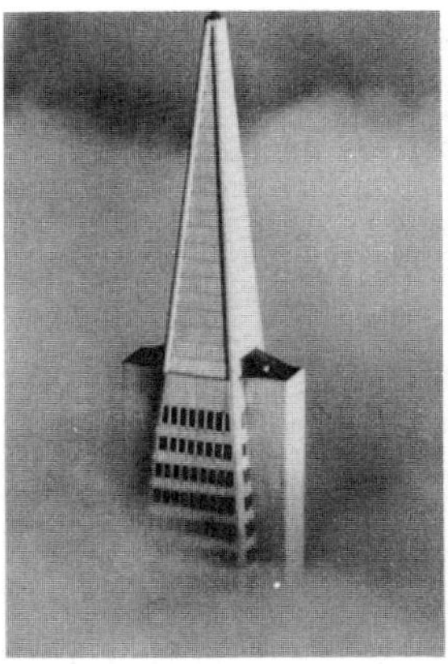

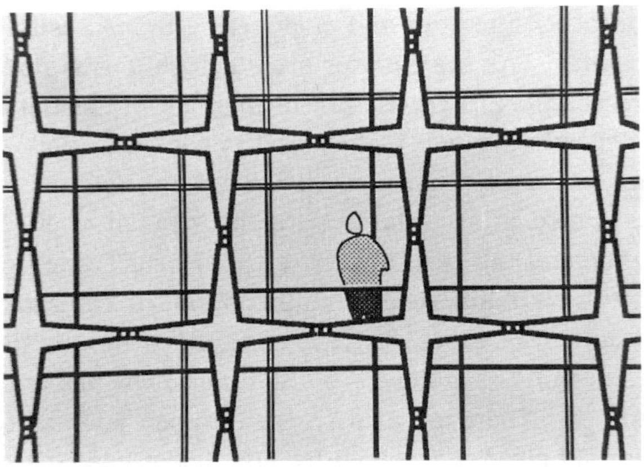

Photo 5 and Figure 7 : 'SPIRE' of TRANSAMERICA TOWER above CLOUDS in SAN FRANCISCO' & 'DUCTILE MOMENT FRAME' of EXTERIOR BUILDING ENVELOPE in HIGHRISE ARCHITECTURE

Highrise building tower occupants on the uppermost floors with their heads above the clouds, may need to call downstairs to receive information on local weather conditions regarding precipitation levels ('Photo 5'). However, those with aural equilibrium sensitivity to 'building movement', are well aware when strong winds are blowing outside. Side-sway and flexure (or 'horizontal displacement' in structural engineering terms of 3 to 6 feet or 1 to 2 meters) may be common in the tallest of highrise architecture (Figure 8). In seismic conditions, the highrise tower may also be subjected to the convoluting forces of horizontal 'p' waves, vertical 's' waves, and diagonal ground motions resulting in sinusoidal contraflexure (Figure 9), 'story-racking' (Figure 10), or even a rotational 'torque' twisting force (Figure 11). 'Base isolation' theories including 'shock absorbers' are being applied to highrise tower foundation design and construction to mitigate 'building movement' due to earthquake forces.

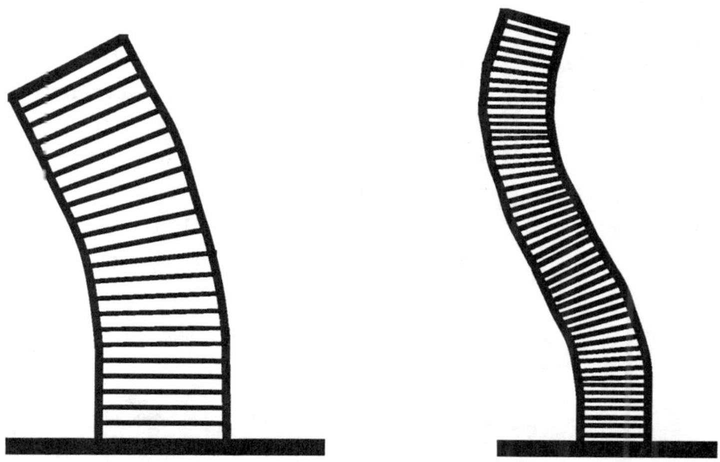

Figures 8 and 9: DIAGRAMS of 'FLEXURE' (on Left) and 'SINUSOIDAL CONTRAFLEXURE' in HIGHRISE TOWER FORMS BEING SUBJECTED to HIGH WIND and EARTHQUAKE FORCES

Depending on the building 'footprint' layout on the site, number and proximity of highrise towers, volumetric configuration of their masses, etc. 'building movement' may result in not only 'story racking' but even flexural 'collision' between towers or portions of towers, depending upon their natural 'frequency' and innate 'period' in resisting earthquake vibrations. These phenomena have been very clearly demonstrated on seismic 'shake tables' in private and university laboratories. Rarely, if ever, are 'building seals and sealants', cladding, and glazing designed for such calamitous and catastrophic events. In some U. S. city centers, it has been predicted that in a severe earthquake of long duration at over 6.0 intensity on the Richter scale, the fallen glazing and cladding materials from highrise towers will be piled several stories in height on the surrounding streets and rooftops. 'Minimum requirements' in building codes with seismic provisions, are satisfied by calculations based on empirical formulas and equations derived from an assumed maximum earthquake intensity on the Richter scale of only 8.0; although, we have experienced several more severe seismic events up to 9.5 on the Richter scale during this twentieth century A.D. Consequently, both building owners and developers, as well as design professionals have the judgmental responsibility to exceed the 'minimum requirements' provided in building codes whenever and wherever necessary [6, 7, and 13].

In geographical areas where the severity of earthquake hazards are not clearly established by building performance provisions specified in even the the 'minimum requirements' of the building codes, progressive architects and engineers have demonstrated the need to provide for the public health, safety and welfare in their innovative highrise architecture, engineering and construction. One outstanding example of such accommodation of 'building movement' and prevention of destruction to 'building seals and sealants', glazing, and cladding systems, is the 'Citibank' building tower in Manhattan, New York City designed by architect Hugh Stubbins, FAIA, where rooftop mechanical 'dampening devices' were installed to counteract the earthquake forces [2, 3, 8, 10, 16, and 17].

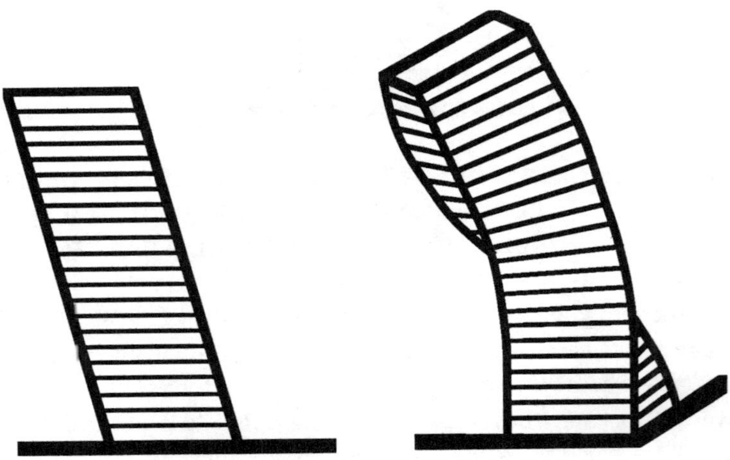

Figures 10 and11: DIAGRAMS of 'STORY-RACKING' (on Left) and 'TORQUE' FORCES on HIGHRISE TOWER FORM BEING SUBJECTED to HIGH WIND & EARTHQUAKE FORCES

The extent of resistance to 'building movement' manifested in lateral drift, flexure, contraflexure, and torque is also dependant on many variables. The type of structural systems selected for different building parts and configurations including variable tower heights, can vastly affect the resulting behavior of the highrise in resisting static and dynamic wind and earthquake forces. Architects and engineers are now mixing precast and poured-in-place concrete structural elements, composite steel and concrete construction, and even designing very tall highrise towers with several bottom stories of reinforced concrete and top stories of skeletal steel beam and column construction. Each of these portions of composite building elements adopts a different 'period' and their overall 'building movement' is quite complex; and therefore, invariably requires alternate design details at various building height levels for successful glazing, cladding, and 'building seals and sealant' systems [2, 3, 4, 5, 6, 8, 9, 10, 21, 22, 23, and 24].

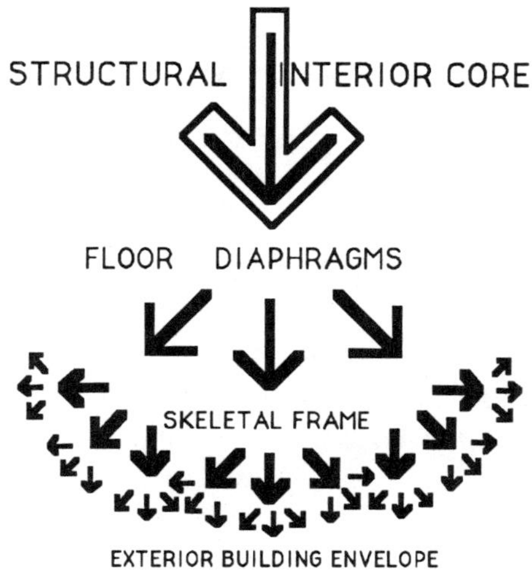

Figure 12: SCHEMATIC DIAGRAM of 'TRANSLATIONAL FLOW of FORCES' between 'INTERIOR STRUCTURAL CORE' and 'EXTERIOR BUILDING ENVELOPE' in HIGHRISE TOWERS

Engineering management of the interrelationships and translational flow of resistance forces between the internal structural core of highrise towers and the exterior building envelope ('Photo 6' and Figure 12), as well as vice versa (see Figure 13), are enhanced by the computerized technique known as 'finite element analysis' based on Newtonian physics. These 'Base isolation' and 'damper' theories and practices in response to dynamic seismic forces, have now complicated engineering design resistance to highrise tower overturning and sliding.

Principles applied to bridge design are being used on buildings. Furthermore, 'intermediate trussed structural frames' to receive opaque claddings, beyond direct attachment to columns and beams are now quite common ('Photos 6 and 7').

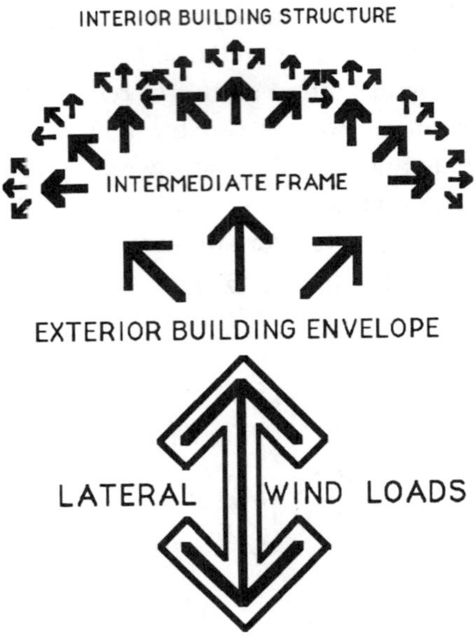

Figure 13: SCHEMATIC DIAGRAM of 'TRANSLATIONAL FLOW of FORCES' between the 'EXTERIOR BUILDING ENVELOPE' and the 'INTERIOR STRUCTURAL CORE' in HIGHRISE TOWERS

A growing number of building owners and their architects and structural engineers are specifying the installation of 'seismic recording instruments' in accordance with recommended provisions contained within the UBC Appendix. This is being done, in order to learn more about the actual behavior of the 'interior structural core' and 'exterior building envelope' in high rise towers during earthquakes of different magnitudes, frequencies, and durations. Although such provisions are not yet contained within the 'minimum requirements' of the UBC, this ICBO building code recommends practices similar to those already being implemented in Japan. That is, installation of these 'seismic recording instruments', which are about the size of a loaf of bread', at the lowest, highest and intermediate floor levels in a highrise tower at minimal cost. In this manner, structural engineers will be better able to obtain reliable data on the periodic behavior of the structural frame in earthquakes, and thereby balance the forces requiring both resiliency and ductility as well as stiffness and rigidity.

Photos 6 and 7 : 'INTERIOR CORE', 'EXTERIOR STRUCTURAL FRAME', 'INTERMEDIATE TRUSSED BUILDING ENVELOPE FRAME', and CLADDING and GLAZING on HIGHRISE TOWER FORMS in DOWNTOWN HONOLULU, HAWAII

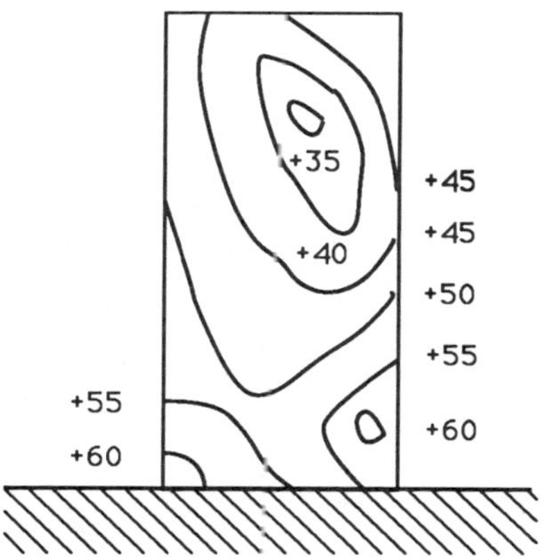

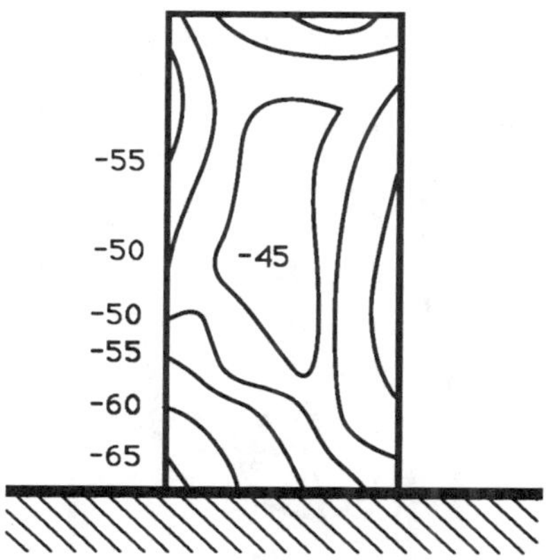

Figures 14 and 15: DIAGRAMS of 'LABORATORY WIND TUNNEL SIMULATED FORCES' on HIGHRISE TOWER FACADE (POSITIVE PRESSURES on Left and NEGATIVE PRESSURES on Right)

Building codes contain provisions for 'minimum requirements' for 'building movement' due to averaged wind and seismic 'live loads' that are primarily applicable to the building structural frame, and secondarily to the exterior building envelope. Therefore most architects and engineers involved in highrise tower design (shape and form as well as selection of structural system) and construction, subject their preliminary design concepts to environmentally simulated wind tunnel modeling studies. Wind tunnel laboratory analyses of the highrise tower facades can determine the maximum positive and negative pressures (Figures 14 and 15) affecting the design details and specifications for glazing, cladding, and building seals and sealant materials [13 , 21, and 22].

The amount of distortion experienced in 'building seals and sealants' and within cladding and glazing anchorage systems subjected to 'building movement', must be contained within established design standards and tolerance limits to prevent failure. There are building systems and construction materials laboratories involved in not only static but also dynamic load testing of the exterior building envelope. To a certain extent, they can simulate the positive and negative pressure differentials being experienced within the exterior building envelope of cladding and glazing systems.

Anchorage systems in seismic zones for the attachment of curtain walls, cladding and glazing systems, are generally designed for at least four-way 'building movement'; i.e., to accommodate in and out, up and down, sideways, and even diagonal displacements. There is a certain amount of inherent design redundancy in the structural frame, and it is generally more flexible and ductile than the rigid infill panels enclosing the building envelope (Figures 16 and 17).

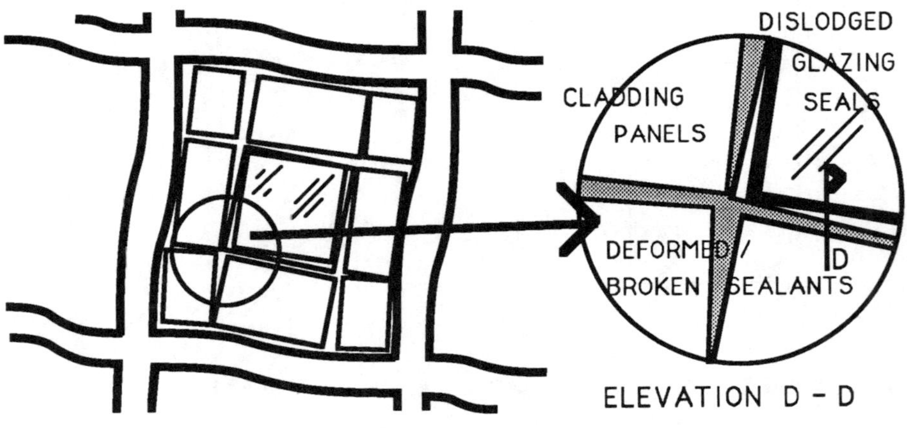

Figures 16 and 17: 'CONTORTED STRUCTURAL FRAME ' DUE to SEISMIC GROUND MOTION FORCES and EFFECTS on BUILDING SEALS and SEALANTS

In dealing with wind, seismic and other 'live loads' on highrise tower architecture, architects and engineers are constantly involved in balancing forces of 'resistance' with those of 'accommodation' ('Figures 9 and 10'). The translational flow of forces between the exterior building envelope and the interior structural core, contain many of the same phenomena in microcosm that are experienced within the larger structural frame. Most highrise tower forms are clad with curtain wall systems which are also subjected to the phenomena of creep, deflection, elongation, elasticity, plasticity, yield limits, stress, strain, fatigue, etc. The specific type of curtain wall system selected and specified, such as 'stick', 'split-mullion', 'panel', etc. has a definite effect on 'building movement' and the 'building seals and sealants' selected and specified, as well as the design of thermal movement and expansion joints ('Figures 17 and 18' and 'Figures 19 and 20').

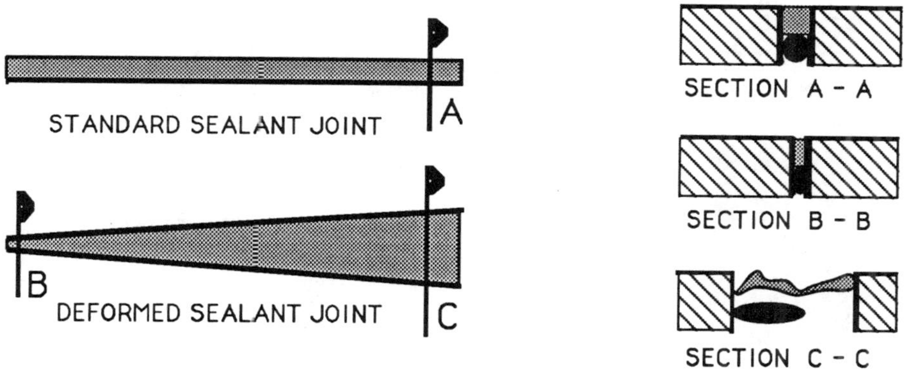

Figure 18 and 19: DETAILED SECTIONS of STANDARD and FAILED and DEFORMED SEALANT JOINTS'

Substantive technical literature, specifications, and design standards address the idealized and standard design details for 'building seals and sealant' applications. Volumes have also been written on methods and means to avoid failure in 'building seals and sealants'; especially, in relation to water infiltration problems [16, 17, and 18]. However, there is very little technical literature or design standards re-search, which addresses the composite interface between various construction materials incorporated within building systems, and their different manufacturing and installation or application tolerance levels. Furthermore, there is almost no design research or standards development work being done on the overall com-prehensive interrelationships between tolerance levels and ranges for building performance of construction materials, with different rates of moisture absorption swelling and shrinkage, or thermal expansion and contraction.

To a limited extent, there have been some laboratory and field experiments employing photoelastic analysis and synthesis on exterior building envelope systems; including, glazing, cladding, and building seals and sealants [19, 20, and 25]. This and other materials research has led to the use of structural load resisting and accommodating 'building seals and sealants' (Figures 19 and 20). Most cladding materials employ butt joints rather than shiplap or overlapping shingle joints; therefore, 'building seals and sealants' are often over-challenged to deal with both 'building movement' resulting in adhesion cohesion, bonding, elongation, and fatigue failures with ensuing water infiltration problems ('Photo 8' and Figures 16, 17, 18 and 19). Electrolysis and premature corrosion may result when dissimilar metals are used in curtain-wall systems, which can also impact building seals and sealants in a negative manner and cause water leaks.

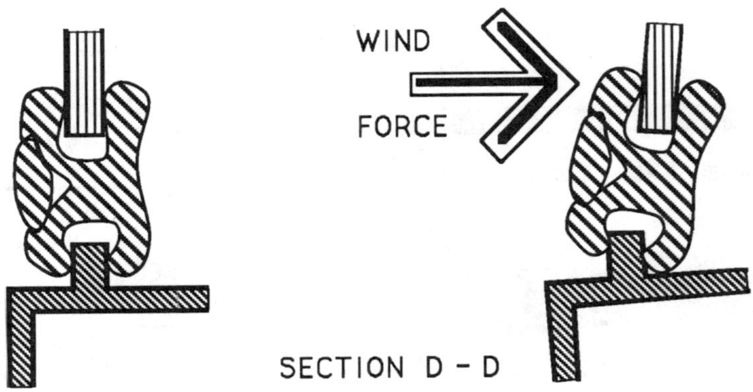

Figures 20 and 21: STANDARD and DEFORMED/OVERSTRESSED 'STRUCTURAL LOCKSTRIP GASKET SEALS'/GLAZING SYSTEM with EXCESSIVE HIGH WIND FORCES

Most glazing and cladding systems consist of panelized building products that are rigid in their composition as a structural diaphragm in their length and width dimensions. Therefore, a rectangular or square infill panel will not become a parallelogram with building movement as will the structural frame [Figures 16 and 17]. However, in the cross-sectional thickness dimension, most panelized building products (including glazing) are subjected to flexure, and can impose undue stresses on 'structural lockstrip gasket seals' [Figures 20 and 21] and other 'building seals and sealants' [Figures 16, 17, 18, and 19]. Given the geometry of roofless highrise towers with spires, skylights, and spaceframes, and the windy working conditions associated with the installation of panelized glazing and cladding systems at higher elevations, the challenges to produce watertight seals between moving parts are formidable.

Photo 8: HIGHRISE ARCHITECTURE in HONOLULU with BUTT JOINTS between MASONRY INFILL PANELS REQUIRING SEALANTS

As we move from 'water resistive' and 'fire resistive' exterior building envelopes toward higher expectations of building owners for 'waterproof' and 'fireproof' highrise towers, we are also moving from 'wind resistive' and 'earthquake resistive' structures towards 'windproof' and 'seismicproof 'highrise architecture. However the applications and specifications of 'building seals and sealants' and their design standards continue to be challenged by roofless highrise 'signature' towers exposed to urban biochemical environments and other foreign elements, for which there are no experimental laboratory precedents.

Obviously, there is much work to be done in addressing design standards development not only for 'building seals and sealants', but also for 'building movement' at macrocosmic and microcosmic levels of building design. The integration of analog and digital computers employing CADD and CAD/CAM with not only detailed building materials and systems component parts, but also incorporating sophisticated structural engineering design data and mathematical simulation equations, has been quite helpful. However, there is no substitute for upgrading quality performance and prescriptive specifications delineating testing and evaluation procedures as well as design standards. In this regard, through its voluntary consensus adoption process, ASTM is to be lauded for its major contribution to the construction industry and to the design of distinguished and innovative highrise architecture in our major metropolitan centers.

ACKNOWLEDGMENTS:

Inspiration for the subject matter of this timely manuscript on 'building movement' and the concomitant effects on 'building seals and sealants' is wholly attributed to Thomas F. O'Connor, AIA, FASTM. As 1991 ASTM/C-24 Committee Symposium Chairman, Mr. O'Connor offered this particular challenge following my presentation on "Building Seals and Sealants for Parking Garage Decks in Hawaii" [17]. The author also acknowledges with deep appreciation, the able technical computer graphics assistance of Andre C. Yanoviak in the preparation of this illustrative manuscript.

REFERENCES:

[1] Yanoviak, A. C., "The Development Process and Building Failures", HAWAII ARCHITECT, Journal of the Hawaii Society / The American Institute of Architects, Vol.16, No.1, Honolulu, January 1987, pps. 18 - 25.

[2] Yanoviak, A. C.,"Architectural Design Challenges for Sealant Technology and Design Standards"; Building Sealants: Materials; Properties; and Performance, ASTM STP 1069, Thomas F. O'Connor, editor, American Society for Testing and Materials; Philadelphia, Pennsylvania 1990, pps. 334 - 346.

[3] Yanoviak, A. C., "Universe: City 2000; Geometrical Explorations of Urban Form and Content", Tall Buildings and the Growth of Cities; Pan Pacific Tall Buildings Conference Proceedings, Ernest N. L. Chiu and Ernest T. Yuasa, editors, Joint Committee on Tall Buildings:International Association for Bridge and Structural Engineering; American Society of Civil Engineers; The American Institute of Architects; American Institute of Planners; International Federation for Housing and Planning; International Union of Architects; Honolulu, Hawaii, January 1975, pps. 76 - 89.

[4] Guise, D., Design and Technology in Architecture, John Wiley & Sons; New York, N.Y. 1985, pps. 1 - 275.

[5] Yanoviak, A. C., "Quality Control in Building and Architecture"; Part I, HAWAII ARCHITECT, Journal of the Hawaii Society / The American Institute of Architects, Vol.17, No. 5, Honolulu, May 1988, pps. 36 - 39.

[6] Yanoviak, A. C., "Quality Control in Building and Architecture"; Part II, HAWAII ARCHITECT, Journal of the Hawaii Society / The American Institute of Architects, Vol.17, No. 6, Honolulu, June 1988, pps. 34 - 35.

[7] Yanoviak, A. C., "Hawaii/Pacific CIRIES"; HAWAII ARCHITECT, Journal of the Hawaii Society / The American Institute of Architects, Vol.16, No. 3, Honolulu, March 1987, pps. 38 - 40.

[8] Mainstone, R. J., Developments in Structural Form, The M. I. T. Press; Cambridge, Massachusetts 1983, pps. 5 - 350.

[9] Forell, N. F., "Structural Engineering -- Codes and Design", Earthquake Fundamentals: The Earthquake Engineering Research Institute Regional Conference Proceedings; Honolulu, Hawaii, October 1986, pps. 1 - 33.

[10] Yanoviak, A. C., "Building Codes and Design Standards: Who's Responsible?"; HAWAII ARCHITECT, Journal of the Hawaii Council / The American Institute of Architects, Vol.19, No. 10, Honolulu, October1990, pps. 25 - 29.

[11] Yanoviak, A. C., 'Architects, Architecture and Building Codes"; HAWAII ARCHITECT, Journal of the Hawaii Society / The American Institute of Architects, Vol. 17, No. 2, Honolulu, Hawaii, February 1988, pps. 22 - 24.

[12] Yanoviak, A. C.,"A Once in a Triennial Opportunity"; HAWAII ARCHITECT, Journal of the Hawaii Chapter / The American Institute of Architects, Vol. 18, No. 2, Honolulu, February1989, pps. 26 - 28.

[13] Yanoviak, A.C.,"Giving Credit Where Credit is Due"; HAWAII ARCHITECT, Journal of the Hawaii Council / The American Institute of Architects, Vol. 19, No. 7, Honolulu, July1990, pps. 24 - 26.

[14] Yanoviak, A. C., "Water Infiltration Codes and Standards"; HAWAII ARCHITECT, Journal of the Hawaii Council / The American Institute of Architects, Vol.19, No. 5, Honolulu, May 1990, pps. 26 - 27.

[15] Yanoviak, A. C., "Windows and Doors: Professional Practice Alerts"; HAWAII ARCHITECT, Journal of the Hawaii Council / The American Institute of Architects, Vol.19, No. 12, Honolulu, December1990, pps. 34 - 36.

[16] Yanoviak, A.C., "Water Infiltration in Hawaii and Ensuing Construction Litigation", Water in Exterior Building Walls: Problems and Solutions, ASTM STP 1107, Thomas F. O'Connor, editor, American Society for Testing and Materials; Philadelphia, Pennsylvania 1992, pps. 182 - 200.

[17] Yanoviak, A.C., "Building Seals and Sealants for Parking Garage Decks in Hawaii", Building Sealants: Materials; Properties; and Performance, ASTM STP 1168, Charles J. Parise, editor, American Society for Testing and Materials; Philadelphia, Pennsylvania 1991, pps. 151- 169.

[18] Parise, C. J., editor Science and Technology of Glazing Systems, ASTM STP 1054, American Society for Testing and Materials; Philadelphia, Pennsylvania 1990, pps. 1- 150.

[19] Klosowski, J. M., Sealants in Construction, Marcel Dekker, Inc., New York, N.Y. 1989.

[20] Cook, J. P. and Panek, J. R., Construction Sealants and Adhesives: Second Edition; John Wiley and Sons, New York 1984.

[21] Scanlan, R. H. and Simiu, E., Wind Effects on Structures: Second Edition; John Wiley and Sons, New York 1986.

[22] Ambrose, J. and Vergun, D., Design for Lateral Forces; John Wiley and Sons, New York 1987.

[23] Seed, H. B. and Idriss, I. M., Ground Motions and Soil Liquefaction during Earthquakes; Earthquake Engineering Research Institute, Berkeley, California 1982.

[24] Yanev, P., Peace of Mind in Earthquake Country; Chronicle Books, San Francisco, California 1980, pps. 123 - 219.

[25] Donaldson, B., editor Exterior Wall Systems: Glass and Concrete; ASTM STP 1034, American Society for Testing and Materials; Philadelphia, Pennsylvania 1991, pps. 1- 224.

Kurt R. Hoigard and Robert J. Kudder[1]

PERFORMANCE OF A MODIFIED SEALANT FILLET DETAIL

REFERENCE: Hoigard, K. R., and Kudder, R. J., **"Performance of a Modified Sealant Fillet Detail"**, Science and Technology of Building Seals, Sealants, Glazing, and Waterproofing: Second Volume, ASTM STP 1200, Jerome M. Klosowski, Ed., American Society for Testing and Materials, Philadelphia, 1992.

ABSTRACT: A high-rise building experiencing curtainwall leakage was targeted for remedial repairs to correct the problem. The cost of a conventional repair using a new mullion cover and sealant detail was judged to be prohibitively high, and an alternate design was sought. The alternate design selected required no new extrusions and incorporated a modified fillet bead with a foam bond-breaking tape. As this was an unproven detail, a test program was initiated to compare its performance to that of the mullion cover repair. Both repair designs were subjected to the equivalent of five years expansion and contraction at twice the anticipated extension range. Both designs performed without adhesive or cohesive failure of the sealants observed during the test period. The alternate design was judged to be comparable in extension performance and a viable substitute for the more expensive mullion cover repair.

KEYWORDS: sealant, joint design, fillet, extension, testing

INTRODUCTION

A high-rise building experiencing curtainwall leakage was targeted for remedial repairs to correct the problem. Investigation of the glass and aluminum curtainwall revealed a source of the infiltration to be a gasketed joint between the exterior structural mullion and the jambs of the stack windows (Figure 1). The gasket incorporated in the mullion had failed, allowing rain water to penetrate the building facade, causing damage to the interior finishes.

[1]Mr. Hoigard is an Associate, and Dr. Kudder is a Principal, with Raths, Raths & Johnson, Inc., 835 Midway Drive, Willowbrook, Illinois 60521.

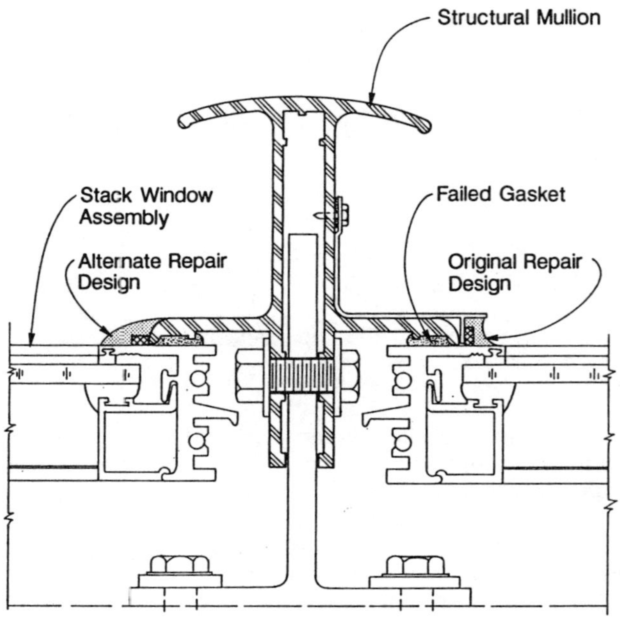

FIG. 1 -- Failed gasket at structural mullion/stack window interface.

The first repair design considered by the authors followed conventional sealant theory, and required modification of the interface through the addition of a new aluminum mullion cover extrusion and a sealant bead. The new extrusion would have to be coated with a thermal setting, silicone acrylic coating whose color matched that on the original aluminum components. Attachment of the new aluminum to the old would be time consuming, waterproofing of the required screws difficult, and the aesthetic effects questionable. After reviewing bids from three contractors, this design was considered prohibitively costly, and alternative repair concepts were sought.

The second repair design considered was the application of a conventional fillet-shaped bead of sealant. The temptation to use this type of detail can be very strong in a curtainwall repair project. The economy and expediency of a conventional fillet-shaped cap bead is an attractive way to fix an existing metal-to-metal interface which is known to leak, and it is normal practice to use this method for joints which will not undergo relative movement of the constituent parts over the life of the joint. Unfortunately, if the parts move, it is a folly, and will only provide a short-term solution to the leakage problem. The theoretical strain at the heel of a moving fillet joint is infinite, causing sealant cohesion failures which originate at the heel and propagate through the body of the fillet. As it could not provide long-term performance, this design was dismissed.

The third repair design considered was to use an unconventional, modified, fillet-shaped bead, designed in accord with good sealant practice. Sealant manufacturers' literature and industrial standards do not discuss or encourage the use of modified fillet-shaped sealant beads for moving joints. The purpose of the study reported in this paper was to determine if a modified fillet-shaped sealant bead (hereafter identified as the Alternate Repair Design) could provide in-place performance equivalent to a conventional sealant bead (hereafter identified as the Original Repair Design) when conditions include a moving interface. An apparatus and testing program was developed to subject both joints to the same joint movement and to the same number of movement cycles. The ultraviolet light exposure conditions are not the same for the two joint configurations, but this parameter was not part of the study.

JOINT DESIGNS

The center-to-center spacing of the structural mullions on this project was typically 1.595 m (5 feet 2-3/16 inches). Assuming a temperature range of 78° C (140° F), and using a coefficient of linear expansion for aluminum of 0.0027 per 100° C (0.0015 per 100° F), the calculated maximum change in length of a horizontal stack window component is 3.1 mm (0.123 inch), assuming a 44 mm (1-3/4 inch) lap at each end of the window. This is approximately 1.6 mm (1/16 inch) movement on each end of the horizontal member. Allowing for an unknown temperature at the time of application of the sealant, the joints were designed and tested for a plus or minus 1.6 mm (1/16 inch) range of movement.

The geometry for the Original Repair Design is shown in Figure 2. The joint accommodates left/right movements by placing the sealant in shear. The 7.9 mm (5/16 inch) depth of the new aluminum extrusion was dictated by the geometry of the structural mullion. The overall joint thickness of 7.9 mm (5/16 inch) results in a plus or minus 20 percent shear strain in the sealant at the limits of the designed movement. The 3.2 mm (1/8 inch) thick foam strip at the heel of the joint acts as both a bond breaker and a cushion to prevent three-sided loading of the sealant.

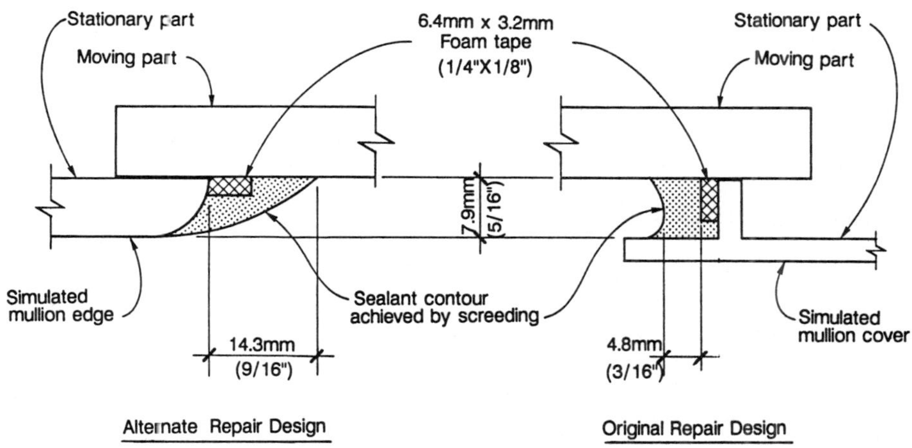

FIG. 2 -- Cross-sections of Original and Alternate Repair Designs.

Figure 2 also shows the geometry selected for the Alternate Repair Design. The joint accommodates left/right movements by placing the sealant in compression and tension, respectively. A maximum allowable sealant extension or compression of 25 percent was assumed, resulting in an unbonded sealant length of 6.4 mm (1/4 inch) and a thickness in the unbonded zone of 3.2 mm (1/8 inch). A foam strip 6.4 mm (1/4 inch) wide and 3.2 mm (1/8 inch) thick was selected to act both as a bond breaker and as a strain relief area at the heel of the joint.

MATERIALS

The sealant used for the testing program was a one-part, gun grade, medium modulus silicone recommended for use on coated aluminum, with an advertised axial movement range of ± 50 percent. No primers were used for the test.

The foam strips used as bond breakers were 6.4 mm (1/4 inch) by 3.2 mm (1/8 inch) adhesive backed, open cell weatherstripping. This material was packaged for residential use and sold on the retail market. Commercial equivalents are available.

The aluminum pieces used to simulate the curtainwall components were custom machined by a local machine shop, as original components were not available. These pieces were then coated with the same thermal setting silicone acrylic coating that had been specified on the curtainwall shop drawings and in the project manual.

APPARATUS

Structural mullion/stack window interfaces like that shown in Figure 1 are normally subjected to in-plane linear displacements to the left and right as temperature changes cause thermal expansion and contraction of the stack window units. In order to simulate these movements, the apparatus shown in Figure 3 was built for this test program. A moveable arm was pivoted on roller bearings and a solid shaft by a link connected to an adjustable eccentric on a gear motor shaft. Adjustment of the eccentric allowed variation in the stroke of the pivoting arm assembly, thus varying the range of motion at the sealant interfaces. While the motion produced by a pivoting arm assembly is actually rotary in nature, the extremely small arc traversed by the arm approximates linear travel.

The 457 mm (18 inch) sealant beads were cut into three 152 mm (6 inch) segments after curing, and sealant extension measured at the center of the top 152 mm (6 inch) segment as indicated by the reference height shown in Figure 3. Sealant segments closer to the pivot point were subjected to proportionately lower strains. The sealant movements were limited to five extension/compression cycles per minute by adjusting the gear motor speed to five RPM through the use of a voltage controller.

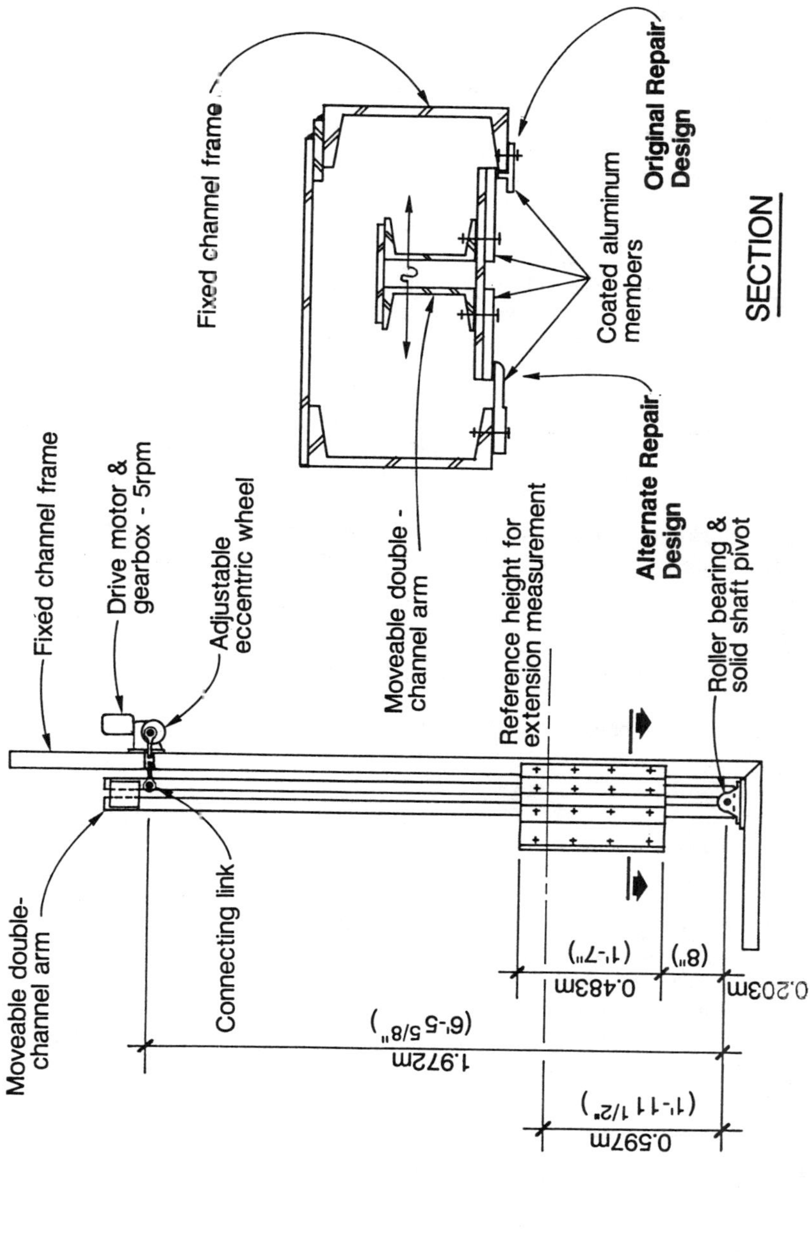

FIG. 3 — Schematic of test apparatus.

SEALANT INSTALLATION

The procedure for installing the sealant in the test areas for the two repair designs was as follows:

1. Clean adhesion areas with acetone.
2. Apply foam tape tight to the interface between the moving parts (Figure 4).
3. Apply sealant using a standard gun and cartridge. The bead for the Original Repair Design was installed with a beveled tip on the gun. The bead for the Alternate Repair Design was installed with a fan-tail tip on the gun.
4. Use carved wooden screeds to tool the sealant into the desired contours (Figure 5). Screeding was also considered necessary to remove any air bubbles and to ensure good contact at the adhesion surfaces, serving the same purpose as tooling in conventional joints.
5. After the sealant forms a surface skin, remove excess sealant from the screeding process with a wooden scraper, taking care not to damage the coating on the aluminum.
6. Trim the tops and bottoms of the beads back 12.7 mm (1/2 inch) to allow observation of the sealant contours (Figures 6 and 7).
7. Cut each sealant bead into three 152 mm (6 inch) lengths.

TEST PROGRAM

The test program was designed to nominally approximate five years of expansion and contraction at twice the anticipated extension range. Table 1 summarizes the ranges of movement tested and the number of cycles performed for each range. In the table, extension is given for the center of the topmost sealant segment, as indicated by the reference height shown in Figure 3. The "Percent of Maximum" column was calculated relative to the anticipated 0.8 mm (1/32 inch) maximum extension used in the design of the repair joints.

TABLE 1 -- Test extension summary

Number of Cycles	Extension	Percent of Maximum
250	±1.6 mm (1/16 inch)	200
500	±0.8 mm (1/32 inch)	100
750	±0.4 mm (1/64 inch)	50

FIG. 4 -- Foam tape installed at the interface between the moving parts of the Alternate Repair Design.

FIG. 5 -- Wood screed used to tool the sealant and produce the surface contour for the Alternate Repair Design.

FIG. 6 -- Trimmed cross-section of the Original Repair Design.

FIG. 7 -- Trimmed cross-section of the Alternate Repair Design.

Three series of tests were run. The parameters for each test are shown in Table 2. In each test, the sealant was applied by a different staff member in an attempt to average the effects of workmanship. For the first two tests, the sealant was applied indoors and allowed to cure seven days at an average temperature of 22° C (72° F) before testing. For the third test, the sealant was applied outdoors and allowed to cure for only one day at an average temperature of -7° C (20° F). 50 cycles of movement at ± 0.4 mm (1/64 inch) were then applied to test the effect of movement on the uncured joint. It was then allowed to cure an additional six days at an average of 3° C (37° F) before completing the test.

CONCLUSION

Both the Original and the Alternate Repair Designs performed without adhesive or cohesive failure of the sealants during any of the test series. The modified fillet sealant repair detail was judged to be comparable in extension performance and a viable cost-saving substitute for the more expensive mullion cover repair.

TABLE 2 -- Test series summary

		Test 1	Test 2	Test 3
Applicator		Kudder	Hoigard	Habel
Test Location		Laboratory	Laboratory	Outside
Application Temperature		22° C (72° F)	22° C (72° F)	-6° C (22° F)
Run 1A	Cure Time	7 Days	7 Days	1 Day
	Temperature	22° C (72° F)	22° C (72° F)	-7° C (20° F)
	Extension Range	±1.6 mm (1/16 inch)	±0.4 mm (1/64 inch)	±0.4 mm (1/64 inch)
	Cycles	250	750	50
	Result	No Failure	No Failure	No Failure
Run 1B	Cure Time			7 Days
	Temperature			3° C (37° F)
	Extension Range			±0.4 mm (1/64 inch)
	Cycles			700
	Result			No Failure
Run 2	Cure Time	7 Days	7 Days	7 Days
	Temperature	22° C (72° F)	22° C (72° F)	3° C (37° F)
	Extension Range	±0.8 mm (1/32 inch)	±0.8 mm (1/32 inch)	±0.8 mm (1/32 inch)
	Cycles	500	500	500
	Result	No Failure	No Failure	No Failure
Run 3	Cure Time	7 Days	7 Days	7 Days
	Temperature	22° C (72° F)	22° C (72 ° F)	3° C (37° F)
	Extension Range	±0.4 mm (1/64 inch)	±1.6 mm (1/16 inch)	±1.6 mm (1/16 inch)
	Cycles	750	250	250
	Result	No Failure	No Failure	No Failure

David H. Nicastro

DIFFICULT SEALANT JOINTS

REFERENCE: Nicastro, D. H., **"Difficult Sealant Joints"**, Science and Technology of Building Seals, Sealants, Glazing, and Waterproofing: Second Volume, ASTM STP 1200, Jerome M. Klosowski, Ed., American Society for Testing and Materials, Philadelphia, 1992.

ABSTRACT: Many premature sealant failures occur in joints which differ from the "archetype" -- the common parallel-sided, uniform, panel-to-panel expansion joint, with an hour-glass sealant profile. Parameters are discussed for obtaining good long-term performance from non-classical types of weatherseal joints. The conditions which are addressed include problems with substrate surface profile, intrinsic properties of substrates, joint shape, and dimensions. Also presented is a summary of the considerations in preventing premature sealant failure which are common to all types of exterior weatherseal joints.

KEYWORDS: Sealant joint failure, substrate, fillet, dimension, surface profile.

INTRODUCTION

Most of the technical literature concerning the use of elastomeric sealants for building weatherproofing is directed toward the common parallel-sided, uniform, panel-to-panel expansion joint, with an hour-glass sealant profile (Figure 1). Probably this "archetype" joint represents the majority of high-performance sealant application in commercial construction. However, many premature sealant failures occur in "specialty" joints which are significantly different from the archetype.

David H. Nicastro, P.E., is a Senior Materials Engineer for Law Engineering, Inc., 13831 Northwest Freeway, Suite 500, Houston, Texas 77040.

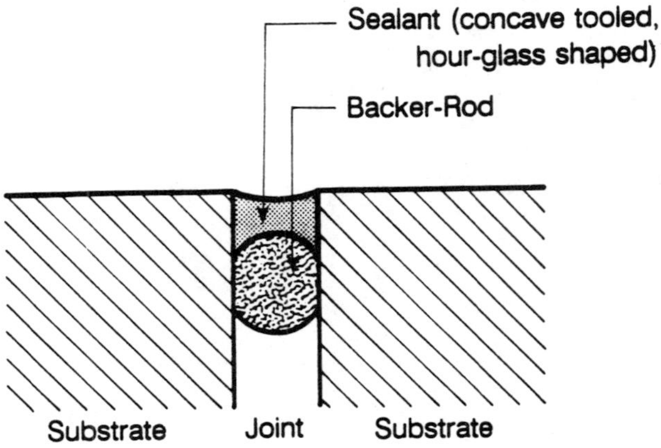

Figure 1 - "Archetype" Sealant Joint

The behavior of the archetype joint is well understood. Even new technical discussions of the behavior of non-classical joints are typically related back to the archetype for comparison, to put the discussion in a comprehensible context.

This paper describes criteria for obtaining good performance in non-standard joints. Because the scope of the paper is limited to exterior weatherseal joints, only high-performance sealants are considered. These typically include silicone-, polyurethane-, and polysulfide-based sealants.

The joint types addressed occur frequently in exterior building envelope weather-sealing, for both new and remedial construction. These include:

1. Expansion joints.
2. Perimeter joints around windows.
3. Closure details, such as caulk troughs and narrow fillets.

Although several specific recommendations are given in this paper concerning product types or installation procedures, it is generally assumed that the reader is familiar with proper sealant industry practice, especially those items which are common to all types of weatherseal joints. Such items are discussed in detail in previous publications [1,2,3]. The following list briefly summarizes some of the common considerations in preventing premature sealant failure in weatherseal joints.

Proper Practice

1. Product Selection: Every sealant product has a proper and an improper range of applications, despite sealant manufacturers' product literature, which often makes a product appear to be suitable to any application [4]. The performance parameters for selecting a product include:

SUBSTRATES

Certain types of substrates pose a challenge for joint sealing because of their surface profile, and others because of their intrinsic properties. Examples of each type follow.

Surface Profile

The most common example of poor surface profile is exposed-aggregate precast concrete (and similar exposed-aggregate cladding panels). When a joint is formed between these rough surfaces, it is impossible to tool the sealant bead completely (Figure 2). At the critical interface between the sealant and substrate, the tool bounces off each stone as it is dragged along the joint, creating small voids next to each stone. A joint formed in this way may appear visually to be weathertight, but still admits water readily during testing.

During original construction, this problem can be prevented by casting the joint surfaces smooth. Note that the face of the panels adjacent to the joints must also be smooth (not just the sealant bite area), so that the tool can be dragged smoothly along the joint.

As a remedial measure for panels with rough substrates, it is often possible to "float" the joint surfaces before applying sealant, using a cementitious grout or plaster. This reconstitutes the surface to a profile similar to an originally smooth-cast face. Of course, the float material must adequately bond to the substrate.

Another profile problem is posed by brick masonry construction with raked joints. This mortar joint profile causes a change in joint width at each course (Figure 3). Even with meticulous workmanship, it is not possible to tool the sealant into each raked mortar joint, resulting in voids between the sealant and mortar.

Preferably, the sealant joints in unit masonry should be formed between flush faces. Again, one remedy is to float the raked joints with mortar before applying sealant, to form a flush masonry face.

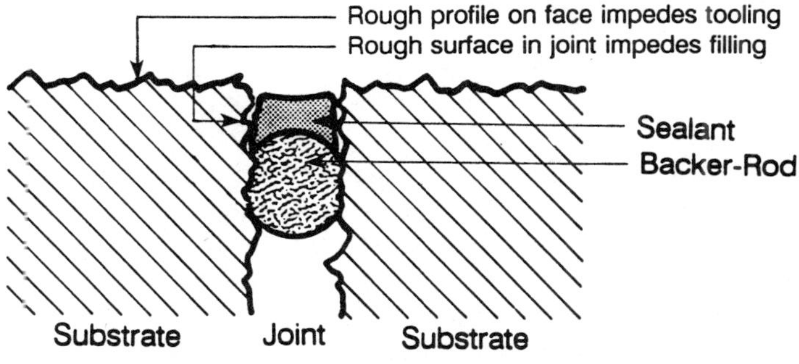

Figure 2 - Rough Surface Profile

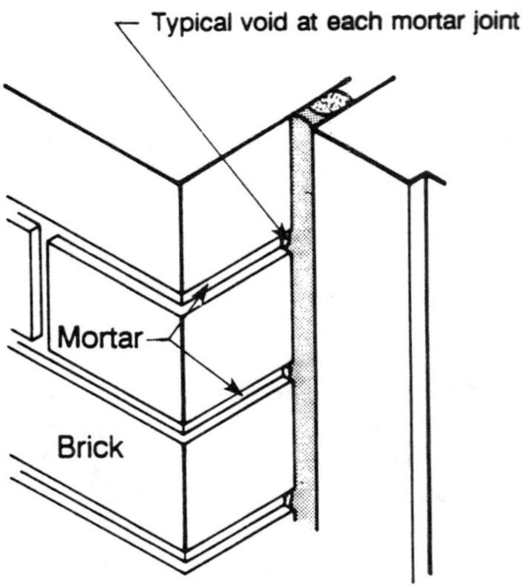

Figure 3 - Raked Mortar Joints

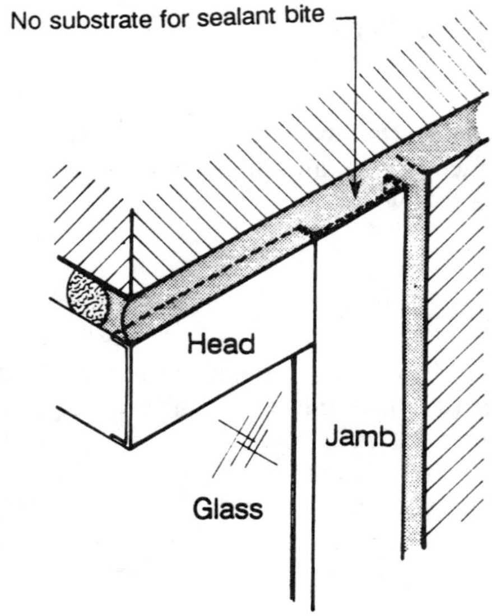

Figure 4 - Lack of Substrate

A disturbingly common example of inadequate substrate surface profile is found at aluminum window frame perimeter joints. Often the horizontal or vertical extrusions terminate at a sealant joint, with no return leg, effectively resulting in no substrate for backer installation or for the sealant to adhere to (Figure 4). This condition is analogous to applying sealant in a joint formed between a cladding panel and a hollow pipe; the bite width is only the thickness of the extrusion (or pipe). This problem can be seen at the corners of many windows, where the sealant has failed adhesively from the ends of the mullions, jambs, sills, or heads.

Proper design of the glazing would provide a mullion cap or other integral return as a substrate for the sealant. As a remedial measure, it is usually possible to use a small piece of break-metal to form a cap. This results in an offset in the joint width equal to the thickness of the flashing metal, and therefore a localized stress concentration occurs; however, this detail will provide much longer service than sealing just to the end of the extrusion.

Substrate Properties

Most cases of premature sealant failure related to substrate properties are simply due to incompatibility. Proper sealant product selection (or priming, in some cases) can prevent compatibility failures between the sealant and substrate. While the products are usually selected to be compatible with the principle cladding components, it is often overlooked that other materials contact the sealant, and therefore serve as substrates as well.

A. Modulus of elasticity, hardness, and movement ability (class), to ensure that the sealant is not overstressed.

B. Compatibility with substrates and incidental materials which will contact the sealant, and ability to adhere to the substrates.

C. Tolerance for environmental conditions, such as temperature range, chemical exposure, or water immersion.

2. Joint Design: The spacing and size of joints must be calculated based on more than a simplistic thermal expansion calculation [5]. Most joint sealants are stressed in more than one direction, including a combination of shear and tensile stresses [6].

3. Workmanship: A properly selected sealant in a properly designed joint still may fail prematurely for a variety of reasons related to improper installation practices. The critical considerations during execution include:

A. The substrate and ambient temperature during installation and curing must be within the allowable limits of the product, to prevent overstressing prior to obtaining full cure. The weather can also be a factor if frost, condensation, or rain wets the substrates.

B. Cleaning and priming to obtain good bond to substrates.

C. Mixing multi-part sealants, to obtain a homogeneous mixture without excessive air entrainment.

D. Depth control using backer-rod, and joint filling, to obtain a uniform cross-sectional area.

E. Tooling to force substrate wetting, compress the surface of the sealant for durability, and provide the desired surface profile (typically concave, in conjunction with a round backer-rod, to complete the hourglass cross-section).

In short, for proper joint performance, a sealant must be able to be installed, then must be able to develop its desired properties, and finally must remain intact in the joint for a reasonable service life.

The remainder of this paper addresses conditions which are inherently difficult for obtaining good sealant joint performance. Assuming that the product selection and workmanship are proper, difficulties can arise during execution due to a variety of existing conditions, most of which can be prevented during the design phase. The discussion is organized into three common categories of difficult conditions: substrates, joint shape, and joint dimensions.

For example, the perimeter joints around windows are typically visualized as being formed between the window frame and the adjacent cladding panel. However, the perimeter joint may intersect other sealant joints in the aluminum frame or in the cladding panel, which are sealed with incompatible products. This problem is especially common at the corners of aluminum-framed windows, where the internal frame joint sealant has exuded and contacted the perimeter joint sealant, resulting in a point defect (typically characterized by one sealant curling away from the other where they meet). Point defects at each corner of each window can result in substantial water infiltration.

Of course, choosing a compatible sealant is the simplest solution to this problem. In cases where other factors dictate using a non-compatible sealant, experimenting with primers applied over the existing sealant (between the non-compatible sealants) often will yield a satisfactory solution.

The properties of Exterior Insulation and Finish Systems (EIFS) as substrates have led to numerous premature sealant failures. In addition to the typical problems with joint design and sealant installation, polymer-based (PB) EIFS have been associated with a unique failure mechanism:

> Finish coat softening leads to delamination from the base coat at sealant joints. One hypothesis, supported by materials laboratory testing, is that, once softening has occurred, the bond between the finish coat and sealant is stronger than the bond between the finish and base coats. The sealant retains its bond to the finish coat; however, the finish debonds from the base coat. This is known as cohesive EIFS failure [7].

This phenomenon is actually a substrate failure, but it may appear at first observation to be adhesive sealant failure. Williams and Williams [7, 8] have presented recommendations for sealing to PB EIFS; these recommendations, which are alternatives to sealing to the finish coat, are only briefly summarized here:

1. Seal to base coat.
2. Seal to accessories.
3. Install double seals.

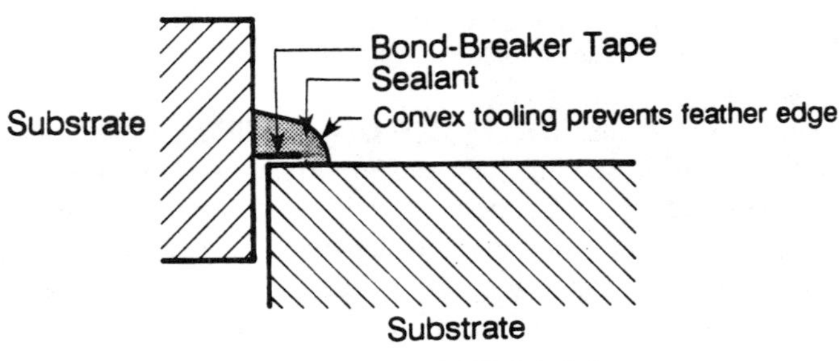

Figure 5 - Typical Fillet Joint

The problems with EIFS sealant joints are exacerbated by the typically high coefficient of thermal expansion of these synthetic systems, which can lead to large joint movements. This can be especially problematic if high cyclic movement occurs during the early life of the sealant, before it is fully cured. Proper design and application can accommodate this type of movement.

SHAPE

Joints shaped differently from the archetype suffer a disproportionately high number of premature sealant failures. This indicates that the behavior of these non-classical joints is not well understood; yet they comprise a large fraction of weatherseal joints. Discussions of two common types follow: fillets and caulk troughs.

Fillets

Fillet joints (triangular sealant bead between perpendicular substrates) have been studied extensively by Myers [9], and therefore will not be discussed in detail here (Figure 5). Myers' recommendations (with which the author concurs) can be summarized as follows:

1. Use a bond-breaker in all fillet joints.

2. Preferably, use diagonal bond-breakers that omit the sealant at the root of the fillet. If a flat bond-breaker is used (which is more common than diagonal), position it on the surface parallel to the direction of movement.

3. At the minimum cross-section (throat), the sealant depth should be 6 mm (1/4 in.).

Myers recommends tooling fillet joints concave, but also presents the conclusion that the shape of the concealed (inner) face of fillets plays a larger role in determining peak stresses than the shape of the exposed face [9]. The Sealant, Waterproofing and Restoration Institute recommends tooling fillets convex [1, p. 45]. Tooling a joint convex requires more skill, and possibly a special tool, to compress the sealant adequately.

In the author's experience with UV-sensitive sealants, convex fillets perform better because they eliminate a feather-edge of sealant adhered directly to a substrate. Spreading a thin layer of sealant over a substrate (as results from concave tooling) can cause "baking," which is similar to performing an accelerated weathering test.

Myers' work focused on large fillet joints, which would typically be used much like the archetype joint configuration, but in applications where the two substrates are perpendicular. Another type of fillet is a closure detail between two butted components, such as between a window frame and cladding panel, which are anchored together with no open joint between them. These small fillets are often treated as non-working joints, and therefore may not use high-performance sealants, backing, or adequate dimensions.

Even if such fillets do not have to accommodate large movements, sealant material properties require that the joints be detailed like working joints. As mentioned above, a thin layer of sealant spread over a substrate does not behave as an elastomeric material bonded between two separated surfaces. Some sealants in this type of closure fillet will "revert," losing cohesive strength and cured properties.

Typically, a larger fillet can be used, with proper backing, to avoid this detail. If space does not allow for a larger fillet, a low-modulus silicone can often be used successfully in an unbacked, narrow fillet.

Another type of fillet joint is a bead of sealant applied around the exterior perimeter of a lite of glass, typically as a remedial measure to alleviate water infiltration through gaskets or glazing tape. There are many mis-applications of this technology, which have resulted in its being considered only a temporary solution. However, when properly done, this detail provides long service life. Following are some of the important considerations in designing this type of remedy, which is often termed "wet-sealing."

1. For most applications, silicone must be used for long-term adhesion to glass exposed to UV radiation.

2. Bond-breaker tape should be used to separate the new sealant from the existing gasket or glazing tape, to prevent compatibility failures. Also, the bond-breaker improves the performance of these joints, as with any fillet.

3. Adequate bite must be provided on each substrate (glass and metal), typically specified as 6 mm (1/4 in.). Also, a minimum specified throat thickness (the thinnest part of the unbonded width) of 6 mm (1/4 in.) is necessary, so that the installed depth will be greater than the minimum allowable (typically 3 mm (1/8 in.)). These dimensions result in a relatively large fillet, which can be unsightly if not applied neatly; this requires skill beyond many applicators.

4. In addition to alleviating water infiltration, wet-sealing can improve the structural performance of glazing (whereas gasket replacement may negatively impact the structural integrity of the glazing). However, it also can have several drawbacks:

A. Some insulated glass unit (IGU) manufacturers void their warranties if water is trapped within the glazing system, because "fogging" (loss of hermetic seal) can occur. Wet-sealing should reduce the amount of water entering a glazing system, and should therefore reduce the potential for fogging. However, plugging weep holes often accompanies wet-sealing, and may in fact trap water in the glazing.

B. After wet-sealing, it will typically be more cumbersome (and therefore more expensive) to replace a lite of glass, because the sealant bead must be cut out and replaced as well.

C. Wet-sealing increases the edge shadowing of a lite of glass. Although this pattern and small amount of shading is not typically associated with glass breakage, it should be considered when other shadowing conditions might also contribute to thermal shock.

Caulk Troughs

A caulk trough is another common type of joint with a non-classical shape (Figure 6). Many buildings use this termination detail as a closure between a surface-mounted flashing and a wall. If the trough is wide enough and the angle between the wall and the flashing is not too acute, a backer-rod can be used in the bottom of the trough to approach the archetype joint configuration.

If the existing caulk-trough dimensions will not permit a backer-rod, a bond-breaker tape can be used to give an unbonded width between the substrates. Without an unbonded width, the stress at the base of the joint becomes excessive, typically resulting in premature failure.

DIMENSIONS

Due to material variations, poor erection tolerances, or misalignment of components, the actual field dimensions of some joints are not adequate to perform as weatherseals. In the author's experience, the most common cause of premature sealant failure is dimensional control (width or depth).

A common example of this in modern construction is a perimeter joint between an aluminum window frame and the adjacent cladding panels. Many of these joints vary in dimension more than the specified tolerance. As the joint dimension narrows, there is a tendency to leave out the joint backing and apply a solid, unbacked fillet over the aluminum. As discussed previously, this joint profile has shorter service life than the specified archetype joint. Unlike a panel-to-panel joint, the joint width cannot simply be widened by cutting or grinding.

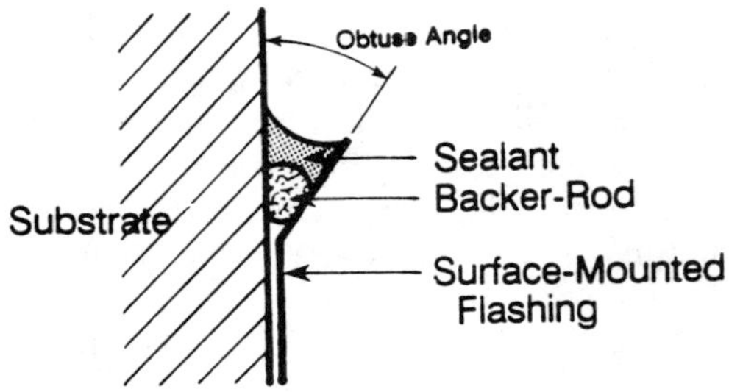

Figure 6 - Typical Caulk Trough

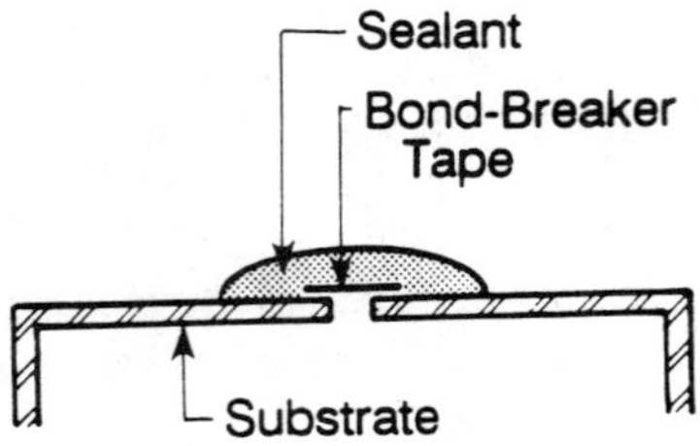

Figure 7 - Typical Butt-Joint

With slight modification, the fillet can work well to solve this problem. A bond-breaker tape can be used to maintain a uniform unbonded width, matching the specified joint width. As with other fillets, tooling the sealant convex instead of concave eliminates a feather-edge on either substrate.

Another application of the convex sealant bead over bond-breaker tape is the butt-joint (or "band-aid," which it visually resembles; see Figure 7). Whereas a convex fillet is used over a narrow gap between two perpendicular substrates, a butt-joint is used over a narrow gap between two abutting components in the same plane.

A technique to achieve long service life from a butt-joint is to center bond-breaker tape over the narrow joint, and "mound" a bead of sealant over the tape, tooling to a convex profile. A minimum bite of 6 mm (1/4 in.) to each substrate and a minimum throat thickness of 3 mm (1/8 in.) are recommended. The size and shape of these joints can be consistently and neatly achieved by striking the joints with a tool fabricated from a putty knife with the end ground to match the finished joint dimensions. Another method is to apply two pieces of thick tape as screeds, strike off the bead flush with the tape surfaces, and then remove the tape.

Many waterproofing designs call for the perimeter sealant joint beneath a window sill to be recessed to allow weep holes to drain from the underside of the sill extrusion (Figure 8). This joint configuration requires greater-than-average skill to seal during original construction, when the substrates are relatively clean. However, it is even harder to reseal; depending on how far recessed the sealant is, it may not be possible to properly clean the joint to achieve adequate adhesion.

In this type of design, a silicone sealant should be used (assuming it meets the other design requirements) to maximize the service life of the joint. Also, silicone is typically easier to clean out of a joint than other aged high-performance sealants, when resealing is finally necessary, and it bonds well to itself if the old sealant is not able to be completely removed.

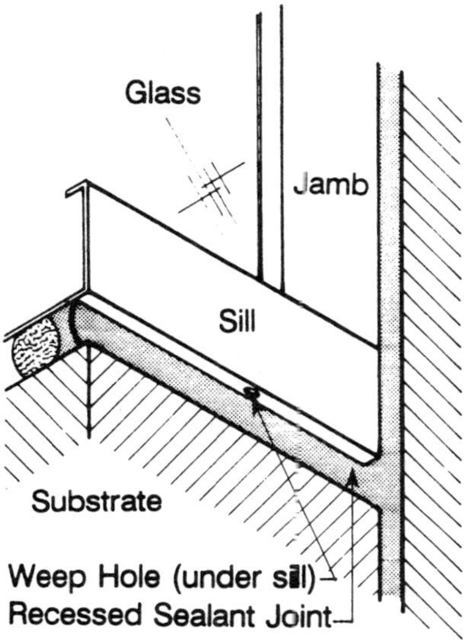

Figure 8 - Recessed Sealant Joint

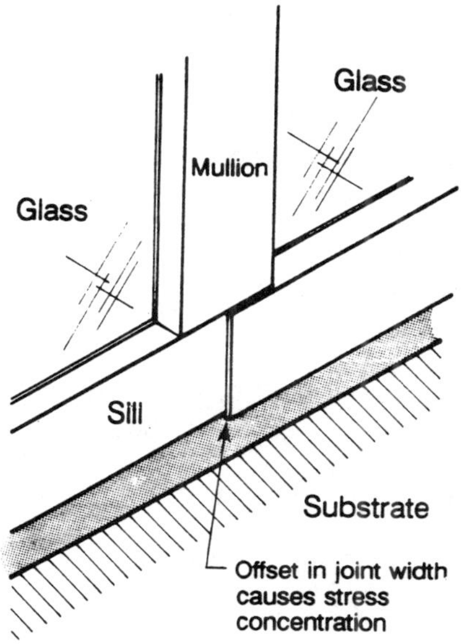

Figure 9 - Joint Offset

A final example of a dimensional problem which leads to early sealant failure is a sudden offset in a joint, caused by misalignment of two adjacent components (Figure 9). This is especially common in window perimeter joints, at the connection between aluminum extrusions. Typically, these joints are sealed without attention to the offsets, which can cause stress concentrations at the complex geometry.

Ideally, the distance over which the offset occurs should be extended to minimize the stress concentration. This might be accomplished by widening the joint or making the transition more gradual by grinding the edges of the substrates. If it is still anticipated that a sealant failure will occur at such a location, a shield can be formed in front of the offset to protect the immediate area from impinging rainfall. The shield can be fabricated from sheet metal, and installed as a small surface-mounted flashing.

CONCLUSION

The archetype joint configuration -- the common parallel-sided, uniform, panel-to-panel expansion joint, with an hour-glass sealant profile -- is not only used most frequently in construction because its behavior is well understood, but because it works extremely well. Deviations from this classical design generally do not perform as well. Therefore, the solutions to the various difficult conditions presented above can be summarized by the principal of approximating, as closely as possible, the archetype joint configuration.

ACKNOWLEDGEMENTS

The author wishes to thank John Fairchild for his excellent illustrations which accompany this paper.

REFERENCES

[1] Sealants: The Professional's Guide, Sealant, Waterproofing, & Restoration Institute, Kansas City, Missouri, 1990.

[2] Warseck, K. L., "Why Construction Sealants Fail--An Overview," Building Sealants: Materials, Properties, and Performance, ASTM STP 1069, T. F. O'Connor, Ed., American Society for Testing and Materials, Philadelphia, 1990.

[3] Hall, A. R., Ed., Sealant Manual, Flat Glass Marketing Association, Topeka, Kansas, 1990.

[4] Nicastro, D. H., "Parameters for Comparing High-Performance Sealants," The Construction Specifier, Vol. 42, No. 3, April 1989, pp. 122-127.

[5] O'Connor, T. F., "Design of Sealant Joints," Building Sealants: Materials, Properties, and Performance, ASTM STP 1069, T. F. O'Connor, Ed., American Society for Testing and Materials, Philadelphia, 1990.

[6] Shisler, F. W. III and Klosowski, J. M., "Sealant Stresses in Tension and Shear," Building Sealants: Materials, Properties, and Performance, ASTM STP 1069, T. F. O'Connor, Ed., American Society for Testing and Materials, Philadelphia, 1990.

[7] Williams, M. F. and Williams, B. L., "PB EIFS Claddings and Sealant Joints," The Applicator, Sealant, Waterproofing, & Restoration Institute, Kansas City, Missouri, Vol. 13, No. 1, Spring 1991, p. 13.

[8] Williams, M. F. and Williams, B. L., "Sealant Usage for Exterior Insulation & Finish Systems," Building Sealants: Materials, Properties, and Performance, ASTM STP 1069, T. F. O'Connor, Ed., American Society for Testing and Materials, Philadelphia, 1990.

[9] Myers, J. C., "Behavior of Fillet Sealant Joints," Building Sealants: Materials, Properties, and Performance, ASTM STP 1069, T. F. O'Connor, Ed., American Society for Testing and Materials, Philadelphia, 1990.

Issam Minkarah, John P. Cook, and A.S. Rajagopal

APPLICABILITY OF ARTIFICIAL WEATHERING TESTS FOR PERFORMANCE EVALUATION OF ELASTOMERIC SEALANTS

REFERENCE: Minkarah, I., Cook, J. P., and Rajagopal, A. S., "Applicability of Artificial Weathering Tests for Performance Evaluation of Elastomeric Sealants," Science and Technology of Building Seals, Sealants, Glazing, and Waterproofing: Second Volume, ASTM STP 1200, Jerome M. Klosowski, Ed., American Society for Testing and Materials, Philadelphia, 1992.

ABSTRACT: A number of sealant samples, Silicone, D1190, D3405 and other designations, were subjected to the standard battery of ASTM tests used in Ohio for acceptance of sealant materials. In addition, non-standard tests such as artificial weathering, cyclic shear/tension, stress relaxation, creep and modulus of elasticity were carried out on samples of the same materials. The results of the investigation revealed that the standard tests are not sufficient by themselves to characterize the long time behavior of the sealants. This paper presents the results of the artificial weathering tests with a brief description of the others. The artificial weathering tests were carried out for 608 hours and the samples were examined by way of color, surface characteristics, movement of sample, flexibility and hardness. Most of the samples passed the standard tests. However, despite the short duration of exposure in the weathering tester, there were significant differences in the ability of different materials to withstand the artificial weathering. In general, the D3405 materials survived the weathering test well, but they did show some scaling after 608 hours. The D1190 materials were uniformly poor, with some flowing off the exposure panel. Silicone gave the best performance, showing very little effect after 608 hours. The authors recommend continued studies to determine an appropriate exposure time so that the weathering test may be added to the standard bank of tests for elastomeric sealants.

KEYWORDS: Elastomeric Sealants, Weathering, hardness, adhesion, elasticity

INTRODUCTION

Untimely failure of pavements, particularly at joints and cracks, has placed an increasing pressure on highway engineers to devise appropriate technology for controlling the deterioration. Although failure is attributable to several factors viz., pavement composition, thickness, traffic, environment, construction quality etc., it is observed that, in concrete pavements the distress often starts at the joints and/or cracks. A proper maintenance measure taken at an appropriate time would

---

Professor, Professor Emeritus and Post Doctoral Fellow, Dept. of Civil and Environmental Engineering, University of Cincinnati, Cincinnati, OH 45221.

result in a better performance and in turn considerable savings. Preventive maintenance techniques for concrete pavements generally involve one or more of the following actions -- joint sealing, subsealing, subdrainage and restoration of load transfer. This paper focuses on the use of joint sealants, types, desirable properties, factors affecting their performance and various methods of testing them in the laboratory with emphasis on artificial weathering in a QUV weathering tester.

JOINT SEALANTS AND THEIR DESIRABLE PROPERTIES

Current practice allows the use of a range of materials for filling joints and cracks in pavements. Selection of a filler material depends primarily on regional, environmental and traffic conditions. Several agencies are continuously engaged in evaluating the requirements of these materials for use in various situations. The selection procedure calls for identifying available materials and testing their adaptability under various anticipated field conditions.

In general, the filler compounds are expected to have good adhesion to concrete, a required elasticity and be stiff enough at all times to resist the ingress of foreign matter. The important functions of the fillers include (i)prevention of accumulation of incompressible that cause spalling and blow-ups in hot weather due to expansion, (ii)minimizing the infiltration of surface water into the subbase and foundation which would lead to severe pumping, faulting, loss of support, breakup and frost heave. Sealing materials are unable to transmit traffic loads. Instead they merely prevent or minimize the ingress of water and dirt into the joint. In addition, these materials must be able to absorb edge movements and joint displacements, be water repellent, and remain always in full contact with the concrete surface. It is desirable that the joint filling compounds possess and retain these properties for a long time. It is often noticed that, in a relatively short time the performance of filler materials starts to decline. They become brittle with age and under such conditions hinder slab movements. A range of factors affect the performance characteristics of the sealants. Figure 1 depicts the primary factors that govern the performance of sealants. Standard ASTM testing procedures are directed toward determining the effects of one or a combination of these factors. However, it is recognized that these standards are not sufficient by themselves to predict fully the long term effectiveness of the sealants. Hence, additional tests like, artificial weathering, cyclic shear/tension, modulus of elasticity, stress relaxation and creep should be carried out. While all these tests were investigated, the emphasis in this paper will be on the artificial weathering tests with a brief report on the others.

SCOPE OF THE STUDY

The present study is a part of a research project that was carried out to determine the validity of the standard ASTM tests as they relate to sealant materials used in Ohio, and to investigate and develop other test methods. The testing program involved fourteen different tests on twenty different materials. In all, the laboratory investigation consisted of testing over 1600 specimens. This paper will briefly present the results of some of these tests and highlight the importance of the Accelerated Weathering test.

LABORATORY TESTING OF SEALANTS

The laboratory tests carried out include Bond (D3407), Resilience (D3407), Penetration (D3407), Flow (D3407), Weathering (G53-83 and D3583-85), Modulus of Elasticity, Creep, Stress Relaxation and Cyclic Shear and Tension at low temperature.

In the standard ASTM test for bond, concrete blocks are used as the substrate material. Since most of the sealants considered here are typically applied to asphalt pavement, the ASTM bond test was carried out using asphalt blocks of the same dimensions as the concrete blocks.

Standard specifications exist for artificial weathering, but such tests are seldom used for testing pavement crack and joint sealants. In the present study, five samples of each material were tested in

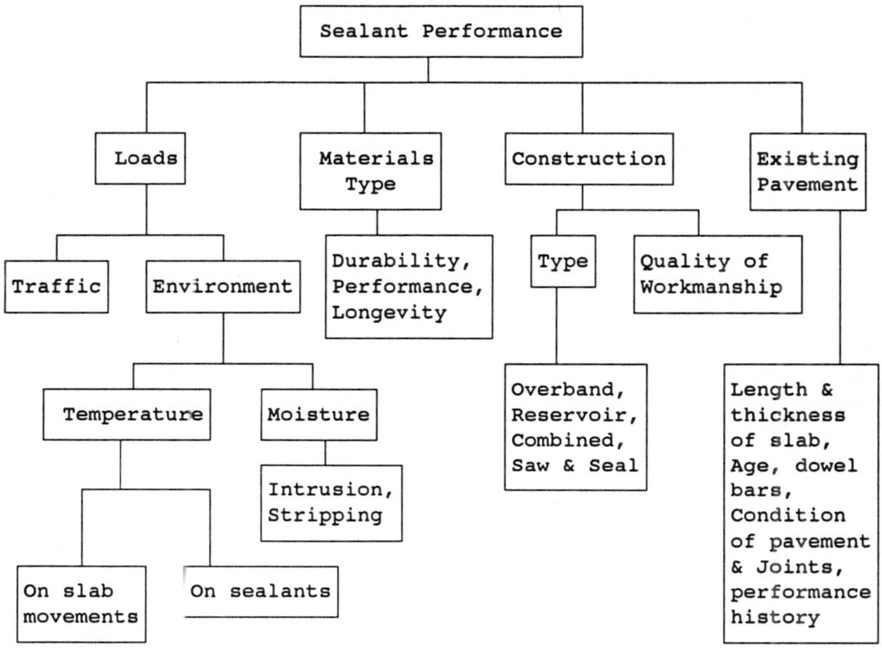

Figure 1. Factors Affecting Performance of Sealants

a QUV Accelerated Weathering Tester, using cycles consisting of four hours of UV exposure at 50^{0}C followed by four hours of condensation exposure at 40^{0}C. UVA-340 lamps were used in the weathering unit. The total exposure time was 608 hours which is the time when 20% of the samples showed significant degradation. At mid-point of the test (304 hrs.), the specimens were inspected visually for surface degradation and checked for flow and hardness. Photographs were taken at the beginning of the test and subsequently at 304 hours of exposure. Typical photographs are presented in Figure 2. The samples were visually inspected for surface degradation at 0, 304 and 608 hours. A list of the materials used in the test is shown in Table 1. The placement of the samples in the weathering tester is shown in Table 2.

Two forms of tests to determine the modulus of elasticity of the sealants were used. The first form of the test employed specimens which are identical to those used in bond tests. The blocks are drawn apart at a rate of 3.2 mm per minute and the force required is recorded continuously. From the load-displacement record, a stress-strain curve may be plotted. The second form of modulus test employed specimens which are three inches long and have a cross section 12.7 by 12.7 mm. Such specimens have a stress state which is much less complex. However, these specimens are difficult to prepare, handle and test accurately due to their shape. So an alternate form is being developed.

In the cyclic shear/tension test, specimens which are identical to those used in the bond test are put into tension and then subjected to a cyclic shear load. The test apparatus is placed in a chest freezer which maintains a temperature of -24^{0}C. This attempts to model the action of truck traffic passing over joints at cold temperatures. The number of cycles for failure and the appearance of the specimen at the end of the test are recorded as the results.

In the stress relaxation test, the specimen is subjected to a suddenly applied extension and then the decay in stress magnitude with time is recorded. This test closely simulates the field conditions.

The Creep test is, in a sense, the inverse of the stress relaxation test because here the load is held constant and the variation in deformation with time is recorded. The usefulness of this test is in identifying any permanent set experienced by the specimen under load.

TEST RESULTS AND DISCUSSION

Table 3 presents the ASTM designation tests carried out and a brief summary of the results. Results of the weathering tests are shown in Table 4.

The bond failure of all the specimens formed on concrete blocks appeared to initiate at the edges of the contact area whereas, some of the specimens on asphalt blocks displayed a cohesive failure in the asphalt substrate near the center. It is possible that the latter phenomenon is due to softening of the substrate at the center of the specimen owing to greater heat retention of D3405 materials in this area. Results of flow and penetration tests were more encouraging with 88% and 83% of the specimens passing the tests. This, however, was not the case with resilience tests in which only 58% of the specimens passed.

A majority of the samples (80%) appeared to withstand the weathering test. Only a few samples actually oozed or ran down the test panel. This behavior appeared to be caused by the sample melting from the operating temperature of the weathering tester. In some samples, bubbles or pinholes were present indicating that the surface of the samples couldn't withstand the operating temperature. This may have caused the surface to bubble up. All of the samples lost their shiny appearance over the test cycle. Also at 0 hrs. all the samples were tacky to the touch. After 608 hrs. all the samples lost their tackiness. On many samples a scale-like surface developed which could also be viewed as many small cracks. This condition appeared in over 50% of the samples and may have been caused by the UV light directly or the operating temperature of the UV lamps. The exposure time of 608 hours is not long enough to produce more pronounced deterioration of the samples, but

Table 1. Material Type and Specimens for Weathering Test

Material	ASTM Desig.	Batch	Specimens
Crafco	D3405	002	5
Koch	D3405	003	5
Neyra (PLS)	no spec	002	5
Neyra Hi Temp	D3405	001	5
Neyra Spec Plus	D3405	002	5
Prismo	D1190	001	5
Sealtight Hispec Polym Sealing Compound	D3405	001	5
Sealtight #164 Polymeric Compound	D1190	001	5
Sealtight Sofseal LM	D1190	001	5
Sealtight Sofseal XLM	no spec	001	5
Trumbull	D3405	002	5

Table 2. Sample Placement in Weathering Tester

Back Panel		Front Panel	
Number 13	Blank	Number 26	Blank
Number 12	Koch 003	Number 25	Blank
Number 11	Koch 002	Number 24	Neyra Hi Temp 001
Number 10	Neyra PLS 001	Number 23	Crafco 003
Number 9	Superseal 001	Number 22	Crafco 002
Number 8	Trumbull 002	Number 21	Sealtight #164 Polymeric Seal Compound
Number 7	Thermocouple	Number 20	Sealtight Hi Spec Polymer
Number 6	Trumbull 001	Number 19	Sealtight S.S. LM
Number 5	Prismo 001	Number 18	Sealtight S.S. XLM
Number 4	Neyra Spec PLS 001	Number 17	Dow Corning Silicone
Number 3	Dow Corning Silicone	Number 16	Neyra PLS 002
Number 2	Neyra Spec PLS 002	Number 15	Blank
Number 1	Blank	Number 14	Blank

Table 3. Brief Summary of Test Results on Various Sealant Materials

Test	Brief Summary of Results
Bond (D3407)	50% percentage of specimens failed. Experienced some difficulty in reproducing results. D3405 materials performed better.
Flow, Penetration and Resilience (D3407)	Flow - 88% and Penetration - 83% specimens passed. Resilience - only 58%passed
Weathering (G53-83 & D5383-85)	In general, D3405 materials survived but showed some scaling after 608 hours. Silicone gave best results
Modulus of Elasticity	Test illustrates high sensitivity of materials to temperature. Glass transition temperature exhibited by polymers which is undesirable. Several materials were subjected to impact loading at low temperature which exhibited a brittle fracture and some of the materials shattered
Cyclic Shear and Tension	None of the D1190 or D3405 materials survived more than a few thousand cycles. All specimens failed in adhesion. Silicone specimens survived more than 25,000 cycles with no apparent damage
Stress Relaxation	All materials tested exhibited high rates of stress relaxation. Their behavior is dependent largely on temperature
Creep	Very few tests were run. No practical conclusions could be drawn

Table 4. QUV WEATHERING TESTER DATA

SAMPLE NAME	VISUAL OBSERVATIONS	
1) Dow Corning Silicone	0 Hrs	Light gray color, flexible surface.
	304 Hrs	No observable differences.
	608 Hrs	Same as above. No noticeable surface degradation.
2) Prismo (D1190)	0 Hrs	Dark, shiny smooth black surface.
	304 Hrs	Dull surface with large scales. Sample moved 0.5 inches from original position.
	608 Hrs	Same as above except sample has moved 1.25 inches from original position. Movement seems due to sample not being able to withstand the operating temperature of the weathering machine
3) Sealtight Sofseal LM Sealant (D1190)	0 Hrs	Smooth, shiny black surface, soft compound.
	304 Hrs	Dull, large scales on surface.
	608 Hrs	Same as above with large scales. Sample held up well. Hardness probably decreased as in #12
4) Trumbull (D3405)	0 Hrs	Black, rough, shiny surface.
	304 Hrs	Surface is now dull.
	608 Hrs	Same as above except large scales formed (1/4 inch or larger).
5) Crafco (D3405)	0 Hrs	Black, shiny, slightly rough surface.
	304 Hrs	Dull color, small scales present, sample moving slightly.
	608 Hrs	Same as above except sample has moved 1/8 inch from original position. Pinholes are present on the surface. Pinholes seem to be from the surface bubbling from the heat of the weathering machine.
6) Koch (D3405)	0 Hrs	Smooth, dark, black surface.
	304 Hrs	Dull color, small scales present (less than 1/4 inch), sample starting to run on one corner, wavy (bumpy).
	608 Hrs	Same as above except sample has moved 1/4 inch from original position. Sample shows moderate flexibility. A few pinholes present.

SAMPLE NAME	VISUAL OBSERVATIONS	
7) Neyra Spec Plus (D3405)	0 Hrs	Shiny, rough, black surface.
	304 Hrs	Dull, very bumpy surface, sample has moved 3/4 inch from original position (no scales).
	608 Hrs	Sample has moved 1.5 inches from original position. Slight scaling of surface with many little pin holes. Surface seems to have separated away from the lower part of the sample due to heat which caused the large lumps and scaling. Around each scale is shiny material.
8) Neyra PLS (D3405)	0 Hrs	Rough, dull, black surface.
	304 Hrs	Same as above with a bumpy surface which could be scaling.
	608 Hrs	Same as above, maybe a slight shift in sample position. Best looking of the NEYRA sealants.
9) Sealtight Hi-Spec Polymeric Compound (D3405)	0 Hrs	Smooth, shiny black surface, soft compound.
	304 Hrs	Small round holes on surface (blotches), dull color.
	608 Hrs	Same as above. Sample looks good.
10)Neyra Hi Temp (D3405)	0 Hrs	Shiny, black, rough surface.
	304 Hrs	Sample moved slightly from original position. Dull color, rough surface.
	608 Hrs	Sample moved 1/4 inch from original position. Pinholes in surface. Surface very lumpy with same characteristics as NEYRA Spec Plus.
11)Superseal (D3406)	0 Hrs	Smooth, shiny black surface.
	304 Hrs	Same as above with a faded surface.
	608 Hrs	Same as above, small amount of scale. Sample held up well to testing.
12)Sealtight Sofseal XLM (No Spec.)	0 Hrs	Smooth, shiny black surface, very soft compound.
	304 Hrs	Dull, signs of large scale formation.
	608 Hrs	Same as above with large scales. There are breaks in the surface of one of the samples. Hardness of sample actually decreased from baseline of 15.
None of the samples were tacky after 608 Hrs.		

Figure 2. Examples of Sealants at 0 and 304 hours

the project concluded before the samples could be tested longer. Two studies were reported recently wherein QUV weathering tests were carried out for a much longer duration. Fedor and Brennan[5] studied the change in characteristics of sealants due to natural and artificial weathering in a QUV weathering unit. Comparisons were based on visible surface degradation, changes in surface hardness and, 180^0 bend of the back of the specimen. Based on their results they suggested that, for most sealants the minimum accelerated aging time should be 2000 to 3000 hours and 5000 hours for silicones. Similar studies were reported by Sandberg[6] where silicone and urethane samples were tested in a QUV weathering tester upto 5000 hours. He concluded that, 5000 hours of QUV exposure is not excessive in evaluating the performance of a sealant whose average life is about 10 years. Any evidence of failure prior to 5000 hours could be a strong warning of an early failure. Referring to these studies, a question would arise as to why the present study limited QUV weathering tests to only 608 hours. The purpose of the study was to investigate the feasibility of specifying such a test for acceptance of sealants for pavements and not to establish the needed exposure hours. An exposure period of 2000 hours is certainly not excessive for pavement sealants, but the actual time needs further investigation so as to relate the number of exposure hours in the QUV tester to actual failure in the field. From a study of 170 pavement test sites in Ohio, it was found that while some sealants may maintain a good condition for five years or more, others fail during their first winter. The average effective life of a crack seal is only two to three years[1].

The best looking of the sealants from the point of view of surface degradation was Dow Corning Silicone. There was no noticeable difference in its appearance after 608 hrs. and it remained flexible. Superseal also exhibited good results. Neyra PLS was by far the best sealant of the Neyra products and so was Sealtight Hi-Spec Polymeric sealant of the Sealtight group. These samples withstood the weathering test well. The performance of Prismo 001 was highly unsatisfactory. Not only did it have very large scales but it ran into a large clump at the bottom of the test panel. The placement of the samples in the weathering tester did not seem to have any effect on the results. This may be due to the short duration of study.

Hardness readings were made using a Shore pencil durometer. The samples that had hardness readings taken at 0 hrs. are: Neyra Spec Plus 001 (25), Dow Corning Silicone (25), Meadows SofSeal XLM 001 (15) and Superseal (30). Because of the limited number of samples, hardness comparisons in this test are also limited. The Meadows SofSeal XLM sealant dropped in hardness from a baseline of 15 to below 5 (unreadable on a Shore pencil durometer). The rest of the samples had a constant hardness throughout the test.

CONCLUSIONS AND RECOMMENDATIONS

Proper identification and selection of suitable materials for sealing joints and cracks in pavements is of primary concern. The materials thus chosen should be capable of withstanding the adverse climatic and environmental conditions in the field. With these conditions widely varying from region to region, characterization of the material properties before their application plays a significant role in setting standards for acceptance. In order to characterize the materials, it is essential to determine the various factors that affect their behavior and to establish proper testing techniques. The test procedures should be capable of explaining the short term as well as the long term feasibility of the materials. In general, the test procedures should be capable of estimating material properties with respect to: (i)adhesion, (ii)elasticity, (iii)strength and (iv)longevity. The standard ASTM tests for pavement sealants are not adequate to characterize the sealants with regard to all these properties.

A perfect adhesion of the sealant material to the pavement surface is an essential property which could be determined by the bond tests. Most of the thermal movements in pavements take place at joints and cracks. The sealants should be flexible enough to absorb these movements without experiencing a permanent set or rupture. These properties can, to some extent, be determined by the flow, resilience, modulus and stress relaxation tests. Presently, selection is based mainly on bond, flow and resilience tests. These tests are not enough to characterize the materials. Other tests such as artificial weathering, cyclic shear/tension, stress relaxation, creep and modulus of elasticity should be added to the bank of tests for sealant selection. Continued investigation should be carried out to determine the limits to be set for these tests.

Artificial weathering tests are designed to estimate the long term performance of materials exposed to the outdoor environment. In the present study, weathering tests have been carried out on specimens in the laboratory up to 608 hours. The findings do not provide a satisfactory conclusion with regard to the number of exposure hours required to test the samples. On the other hand, the results of the study do emphasize the need for carrying out the weathering tests. Long term studies should be performed to explicitly determine the proper exposure time for the test. Thus it is strongly recommended that artificial weathering tests along with the tests mentioned above be included in the standard bank of ASTM tests for pavement crack and joint sealants.

REFERENCES

[1] John P. Cook, Frank E. Weisgerber, and Issam Minkarah, "Development of a Rational Approach to the Evaluation of Pavement Joint and Crack Sealing Materials", Final Report, State of Ohio Department of Transportation, Contract No. 5162, August 1991.

[2] Julian R. Panek, and John P. Cook, "Construction Sealants and Adhesives", Second Edition, John Wiley and Sons, 1983.

[3] Michael I. Darter, "Maintenance of Portland Cement Concrete Pavement", Concrete Pavements, Edited by A.F.Stock, Elsevier Applied Science, London and New York, 1988.

[4] Carroll, L.J., "Concrete Pavement Construction", Concrete Pavements, Edited by A.F.Stock, Elsevier Applied Science, London and New York, 1988.

[5] Fedor, G., and Brennan, P, "Correlation of Accelerated and Natural Weathering of Sealants", Adhesive Age, 33(5). 1990.

[6] Sandberg L. Bogue, "Comparisons of Silicone and Urethane Sealant Durabilities", Journal of Materials in Civil Engineering, ASCE Vol 3 No. 4, November 1991.

Frank E. Weisgerber[1] and Chi-Ping Wang[2]

STRESS DISTRIBUTIONS IN HIGHWAY JOINT SEALS

REFERENCE: Weisgerber, F. E., and Wang, C. P., **"Stress Distributions in Highway Joint Seals"**, Science and Technology of Building Seals, Sealants, Glazing, and Waterproofing: Second Volume, ASTM STP 1200, Jerome M. Klosowski, Ed., American Society for Testing and Materials, Philadelphia, 1992.

ABSTRACT: Highway joint seals, and some other butt seals using elastomeric materials, have displayed a high incidence of adhesion failure. An incremental finite element analysis method has been used to determine the magnitudes and distributions of interfacial stresses, treating the material as a low-modulus viscoelastic material in a state of plane strain. The results show that the magnitudes of both normal and shear adhesive stresses are very high at specific points on the contact surface, much higher than the average normal stress predicted by simpler analysis methods. The depth to width ratio for the seal, the rate at which strain is applied, and the relaxation rate of the material are all shown to have significant effect upon the stress magnitudes.

KEYWORDS: stress analysis, viscoelasticity, elastomeric materials, adhesion, stress relaxation, joints, seals

---

INTRODUCTION

The joints in concrete highway pavements are usually sealed in order to prevent water from passing through the joint and acting to degrade the pavement substructure. Fig. 1 shows typical cross sections. Two types of materials are used most frequently: polymer modified asphalt and silicone. A bond-breaking tape or cord is often used at the bottom surface to prevent bond to the bottom of the reservoir.

As the temperature decreases, the concrete pavement contracts, the joint opens and the seal is put in tension. The average tensile strain in the highway pavement joint seal often reaches 10% and it may reach

---

[1]Associate Professor of Civil Engineering, University of Cincinnati, Mail Location 71, Cincinnati, OH 45221.

[2]Former graduate student, Civil and Environmental Engineering Department, University of Cincinnati, Cincinnati, OH 45221.

25% or more. This creates an adhesive stress at the interface between the sealant and the substrate.

Recent studies have shown that adhesion failure is the most troubling obstacle to obtaining effective highway joint seals. Interest in this problem has grown in recent years and there has been much effort directed toward improving long-term adhesion. Research efforts have followed three main routes: field studies, laboratory studies and materials research. The field studies consist mostly of trying many alternate materials and seal configurations in the field, hoping to find a better design. A few laboratory studies have attempted to investigate the behavior of the sealants under more controlled conditions and thus to gain reliable information which may be used to design more successful seals. Some of the materials research has been aimed at developing materials which more tenaciously adhere to the pavement substrate. Additional research has investigated the use of primers to enhance the adhesion.

This paper presents the results of research following a fourth route: stress analysis of the seal system. The magnitudes and distributions of adhesive stresses shown here help to explain why adhesive failure has been a continuing problem and may contribute to resolving the difficulty.

DESCRIPTION OF THE PROBLEM

This problem includes several features which make it unique and also make accurate stress analysis quite challenging. The material is viscoelastic and therefore stresses vary with time even when the displacements are held constant. The materials used for seals have a high rate of stress relaxation so that the stresses reduce significantly even while the displacements are still being imposed. Furthermore, the material is in a state of plane strain and is nearly incompressible with a Poisson ratio which is nearly 0.50. The stresses occur in the seals because of imposed displacements, rather than applied forces. Since the seals are intended to function at large levels of strain, it is necessary to account for the geometric non-linearity.

There have been very few previous stress analyses of seal systems, and none which have accounted for all of the significant features of seal behavior. However, Catsiff, et al [1] and Myers [2] have made outstanding contributions in this area. Because the existing commercial computer programs do not offer options to solve this particular problem, a special purpose program was developed for the present work. Also, the material properties required for such an analysis are not available in the literature, so that a testing program was required prior to performing the numerical analysis.

It should also be noted that the properties of many of the sealants are strongly influenced by temperature. The results shown here neglect temperature changes during the addition of strain but nevertheless provide information on sealant response which is useful to understanding the adhesion problem.

ANALYSIS METHOD

The finite element method was used for the current analyses. Due to the geometric non-linearity and the time-dependent character of the material response, an incremental solution technique was implemented. A Newton-Raphson iteration method is employed within each increment in order to maintain equilibrium of internal forces.

A complete description of the solution scheme is available in Reference [3]. In summary, during the "loading" phase, displacements are applied incrementally, in accord with the selected strain rate. For each increment of displacement, instantaneous elastic stresses are first added, then reduced as stress relaxation occurs over the time increment. For each subsequent time increment, the process of adding instantaneous elastic stresses and reducing them due to stress relaxation is repeated. When the full displacement has been imposed, time may continue to increment, causing additional stress relaxation. However, no new displacements are added and there will be no additional elastic stress increments accumulated. For each increment of added displacement, the current geometry is employed, rather than the original geometry. Consequently, iteration is required to maintain equilibrium at all nodes in the finite element mesh. In essence, these iterations consist of determining unbalanced nodal loads and reapplying them, while time stands still, until equilibrium is achieved.

MATERIAL MODEL

A silicone highway joint sealant was tested in our laboratories to determine its mechanical properties. Based on these tests, the analyses performed here use 1310 kPa for an instantaneous elastic modulus and 0.49 for the Poisson ratio in order to determine instantaneous elastic stress increments.

Modeling the stress relaxation is the most challenging part of representing the material behavior. The sealant is treated here as a linearly viscoelastic material, for which the generalized Maxwell model shown in Figure 2 is appropriate. In the general case, the stress at any point in time, $\sigma(t)$, is given by

$$\sigma(t) = E_0\epsilon(t) + \int_0^t E_i \exp(-(t-t')/\tau_i)\, d\epsilon(t') \qquad (1)$$

in which E_i represents the elastic spring constant, t is the current time, t' is the time since a strain was induced, ϵ represents the strain and τ_i is the relaxation time. The relaxation time is related to the viscosity coefficient of the dashpot, η_i, in the mechanical model as

$$\tau_i = \eta_i / E_i \qquad (2)$$

The values of the coefficients representing viscoelastic behavior, used to obtain the numerical results reported in this paper, are based on regression analysis of the data obtained from our laboratory tests.

With E_i in kPa and τ_i in minutes:

E_0= 322.3, E_1= 697.8, E_2= 169.1, E_3= 67.1, E_4= 55.3

τ_1= 0.05, τ_2= 1.00, τ_3= 10.0, τ_4= 100.0

The expression for stress given in Eqn. 1 is useful in classical viscoelastic analyses, but it is not compatible with the incremental solution scheme described in the preceding section. Therefore it was necessary to derive an incremental form which predicts the decrement of stress $\Delta\sigma_R$ which occurs over an increment of time, Δt_μ:

$$\Delta\sigma_R(t_\mu) = \sum_{i=1}^{n} (\exp(-\Delta t_\mu/\tau_i) - 1)\, Z_i(t_{\mu-1}) \qquad (3)$$

in which t_μ is present time, $t_{\mu-1}$ is the time at the beginning of the preceding time increment and Δt_μ is $t_\mu - t_{\mu-1}$. $Z_i(t)$ is the current value of an internal state variable for component i of the generalized Maxwell model. This internal state variable is continuously updated as time progresses, according to the expression:

$$Z_i(t_\mu) = E_i\Delta\epsilon(t_\mu) + (\exp(-\Delta t_\mu/\tau_i))Z_i(t_{\mu-1}) \qquad (4)$$

in which $\Delta\epsilon(t_\mu)$ is the elastic increment of strain which occurs at the beginning of the time step. This form is similar to that used by Z. P. Bazant for analysis of Portland cement concrete systems [4].

RESULTS OF ANALYSIS

Adhesive Stress Distributions

The distribution of the adhesive stress is found to vary considerably with the depth to width ratio, D/W, and the magnitude of the imposed strain. Fig. 3 shows the distribution of the normal adhesive stresses for seals with three different width to depth ratios: D/W = 1/1, D/W = 1/2 and D/W = 1/4. For each of these cases, the stress distributions are shown for four strain levels ranging from 1% to 10%. It is important to note that the stresses are plotted at different scales for the different D/W ratios.

For all width to depth ratios, at lower strains, say 1% and 2.5%, a maximum stress occurs on the bonded face near the free edge of the seal. This maximum stress is approximately twice the magnitude of the average normal adhesive stress. On the other hand, at higher strains for seals with high D/W ratios, the maximum normal stress occurs at mid-depth of the contact region. These analyses also confirm an often stated maxim: for any given strain level, the seal with lower D/W ratio incurs significantly lower normal adhesive stress than a seal with higher D/W ratio.

Fig. 4 presents the distribution of adhesive shear stress, again for seals with D/W = 1/1, D/W = 1/2 and D/W = 1/4. For seals at low strains there is almost no shear stress developed in the central region of the contact surface, but there are very high shear stresses developed near the top and bottom of the contact surface for all cases. Of particular note is that seals with low D/W ratios incur only slightly lower shear stress magnitudes than seals with high D/W ratios. Furthermore, the shear stress magnitudes, though localized, are higher than the magnitudes of the normal stresses.

Variation of Stress with Strain Rate

All of the stress distributions shown in Figs. 3 and 4 result from a uniformly applied strain rate of 25 percent per minute. The influence of strain rate on the magnitude of stresses incurred is clearly shown by considering the average normal adhesive stress on the contact surface versus time. In Fig. 5, results are plotted for two strain rates, 25% per minute and 25% per hour. In both cases, exactly the same maximum strain, 10%, is applied. For the strain rate of 25% per minute, the maximum strain is reached in 0.4 minutes and subsequently the sealant material continues to relax. For the strain rate of 25% per hour, the maximum strain is reached in 24 minutes. The analysis shows that the peak stress caused by the high strain rate far exceeds the peak stress achieved if a slower strain rate is applied.

Variation of Stress with Relaxation Rate

The rate at which the material relaxes has a strong influence on the magnitude of the stresses incurred. In order to show this, the analysis of the seal with D/W = 1/1 was repeated for a hypothetical material which was assumed to have a higher rate of stress relaxation. (The values of E_i were modified in a manner to cause a higher relaxation rate, but the instantaneous elastic response of the hypothetical material is the same as the original material). Fig. 6 illustrates the two relaxation rates used and Fig. 7 shows the variation of average normal adhesive stress with time for the two materials. In both cases, the strain rate is 25% per minute and a maximum strain of 10% is applied. These curves clearly show that the peak stress is significantly reduced if the relaxation rate is increased.

CONCLUSIONS

The analytical results presented here indicate that the magnitudes of both normal and shear adhesive stresses are very high at certain points in the contact surface, much higher than the average stress predicted by simpler analysis methods. It is also shown that the maximum adhesive shear stresses are very high in comparison with the normal stresses: a circumstance which has not been emphasized in previous literature.

The distribution patterns of the normal stresses suggest that for seals with larger D/W ratios, adhesion failures might originate at the center of the contact region. Since the ASTM bond test for sealants uses specimens somewhat similar to the seal with large D/W, it may be expected that some specimens partially release at the center, but pass the test anyway because a partial release at the center is not visible.

The analyses clearly show the advantage of using seals with low D/W ratios. This supports conclusions from previous laboratory testing and field observations.

Although stress relaxation may lead to certain difficulties in seal performance (particularly if the relaxed material takes a permanent set in the extended position), it is shown here that the materials with a higher rate of stress relaxation will incur lower stress magnitudes for a given joint distortion. Therefore, if the stress relaxation rate could be increased for a material, without decreasing its potential for adhesion, an adhesion failure would be less likely.

Since there is a very strong influence of the strain rate on the magnitude of stresses incurred, there is a great need to determine the actual strain rates occurring in the field. With this information, the analyses and laboratory studies will become more meaningful.

REFERENCES

[1] Catsiff, E.H., Hoffman, R.F. and Kowalewski, R.T., "Predicting Joint Sealant Performance of Elastomers by Computer Simulations - I. Results in Simple Extension and Compression," Journal of Applied Polymer Science, 14, 1970, pp. 1143-1158.

[2] Myers, J.C., "Behavior of Fillet Sealant Joints," Building Sealants: Material, Properties and Performance, ASTM STP 1069, T. F. O'Connor, Ed., American Society for Testing and Materials, Philadelphia, 1990.

[3] Wang, C.P., Development of Stress Analysis Procedure for Low-Modulus Elastomeric Sealants, PhD Dissertation, University of Cincinnati, Cincinnati, Ohio, 1992.

[4] Bazant, Z.P. and Wu, S.T., "Rate-Type Creep Law of Aging Concrete Based on Maxwell Chain," Material and Structures, 7, 1974.

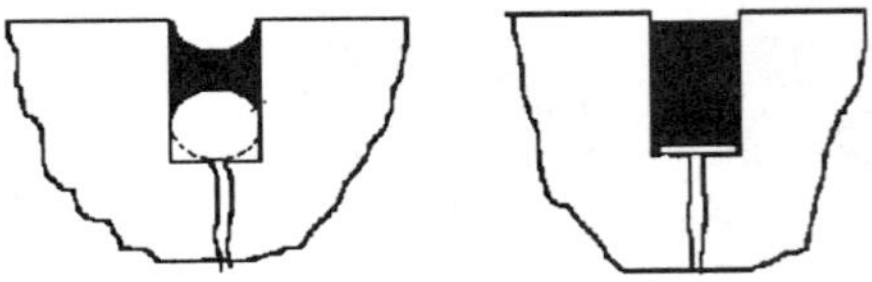

FIG. 1 -- Typical joints.

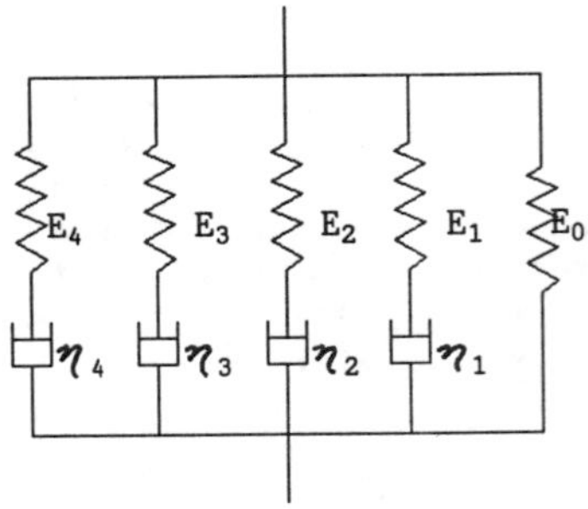

FIG. 2 -- Generalized Maxwell model.

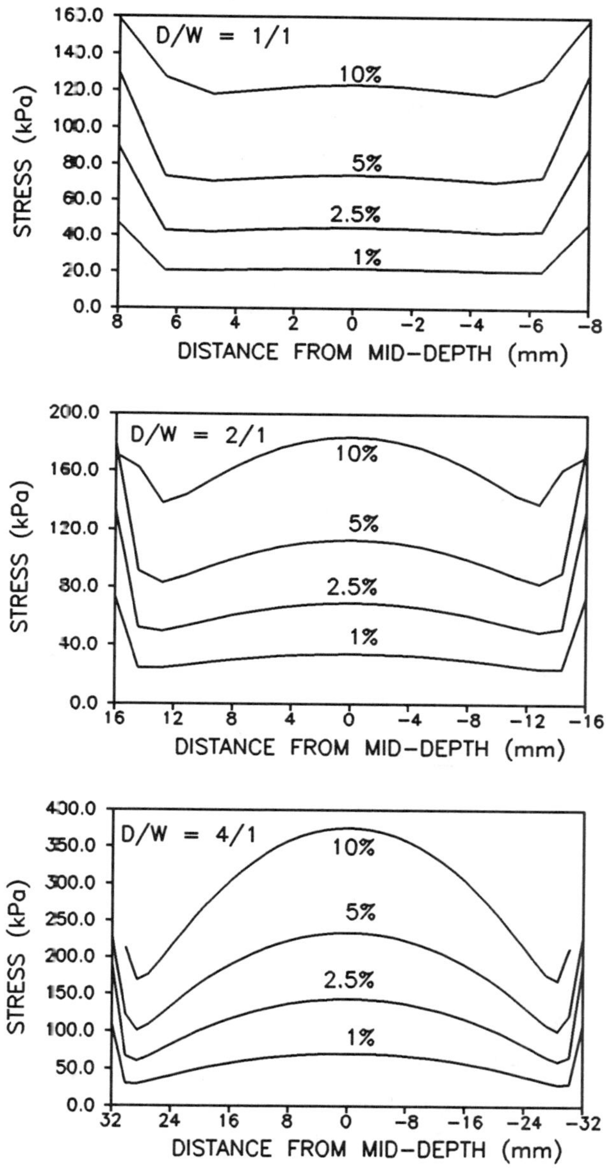

FIG. 3 -- Normal stress distributions at interface.

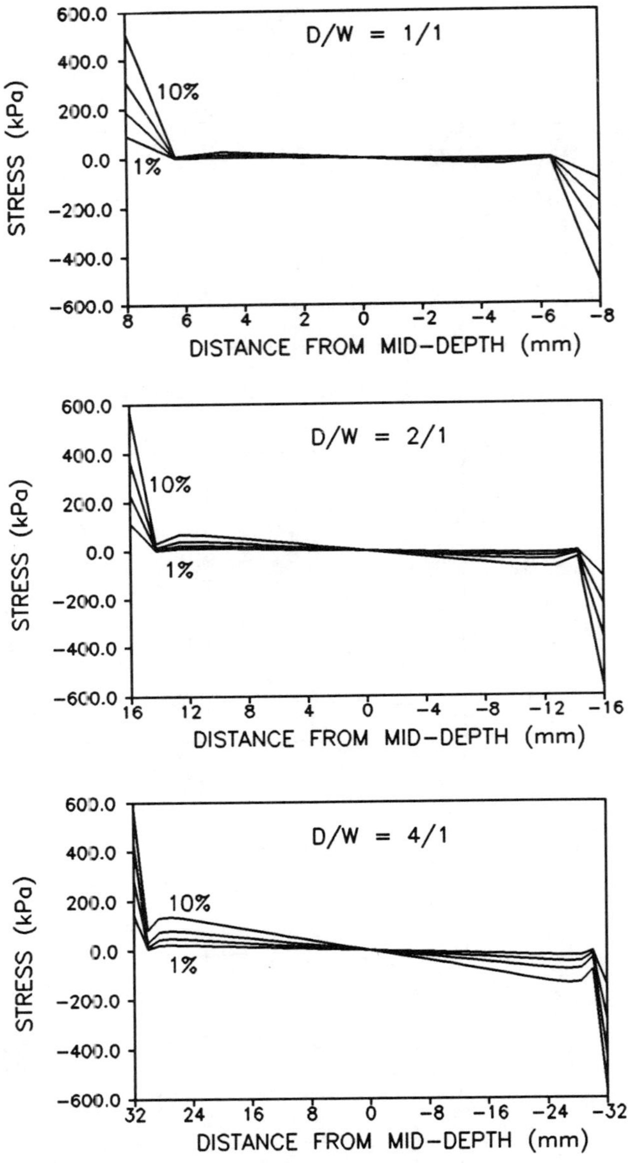

FIG. 4 -- Shear stress distributions at interface.

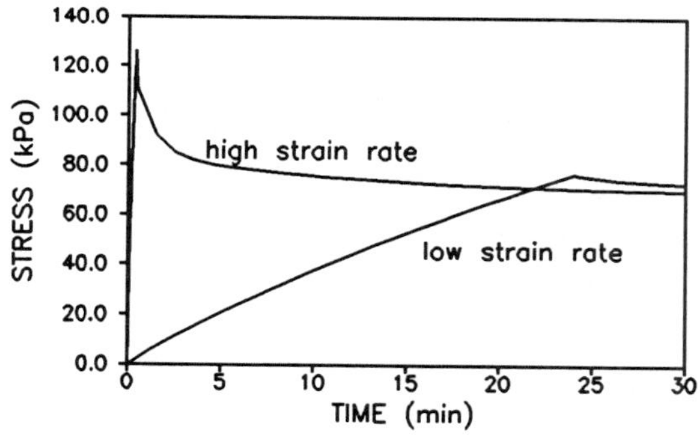

FIG. 5 -- Effect of strain rate on average normal stress.

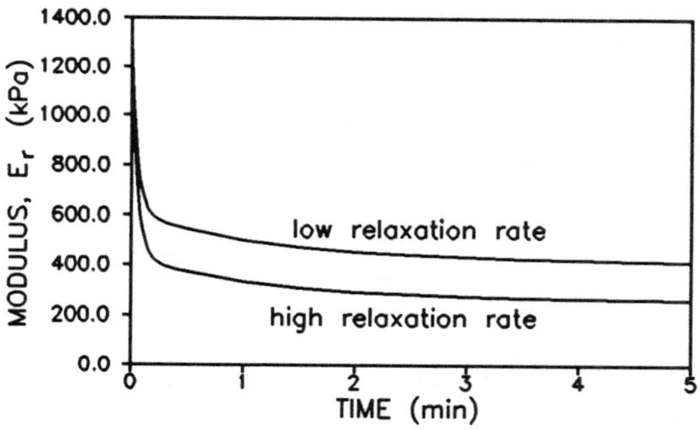

FIG. 6 -- Two relaxation curves for comparison.

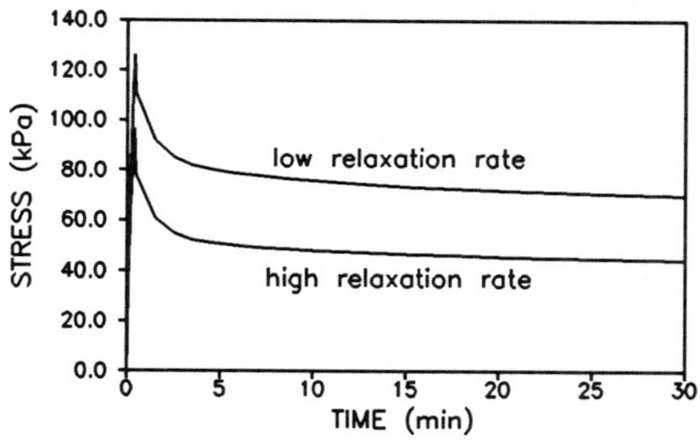

FIG. 7 -- Variation of average normal stress with relaxation rate.

Andreas T. Wolf and Leslie J. Waters

FACTORS GOVERNING THE LIFE-EXPECTANCY OF DUAL-SEALED INSULATING GLASS UNITS

REFERENCE: Wolf, A. T., and Waters, L. J., "**Factors Governing the Life-Expectancy of Dual-Sealed Insulating Glass Units**", Science and Technology of Building Seals, Sealants, Glazing, and Waterproofing: Second Volume, ASTM STP 1200, Jerome M. Klosowski, Ed., American Society for Testing and Materials, Philadelphia, 1992.

ABSTRACT: The water-vapor permeability of the secondary insulating glass sealant plays only a subordinate role in the life-expectancy of a dual-sealed insulating glass unit, since the resistance of the edge-seal to water-vapor diffusion is determined almost exclusively by the low water-vapor permeability of the polyisobutene (PIB) primary seal. On the other hand, great importance must be attached to the viscoelastic properties of the secondary insulating glass sealants, particularly to their tensile stress behavior and their elastic recovery, as these properties affect the ability of the primary seal to function.

KEYWORDS: insulating glass, life-expectancy, durability, sealants

EFFECT OF ENVIRONMENTAL FACTORS ON INSULATING GLASS UNITS

Insulating glass units are exposed to a variety of environmental factors that exert a negative influence on their life-expectancy. Literature on this subject often differentiates between physical and chemical stresses, although certain environmental influences, such as sunlight and water, exert both kinds of stresses.

Dr. A.T. Wolf is Global Technology Manager, L.J. Waters is Technical Service Engineer, Fenestration Industry, both of Dow Corning Corporation, Midland, Michigan 48686, U.S.A.

Temperature

Temperature fluctuations produce differences in pressure within the insulating glass unit. These exert a strong mechanical stress on the edge-seal, especially in the case of small units and unfavorable form factors [1]. Furthermore, differences in the thermal expansion coefficients of glass and spacer bar material result in shearing and peeling forces in the edge-seal. Figure 1 shows the frequency with which certain temperatures occur at the edge-seal of a regularly glazed insulating glass unit over the course of one year [2].

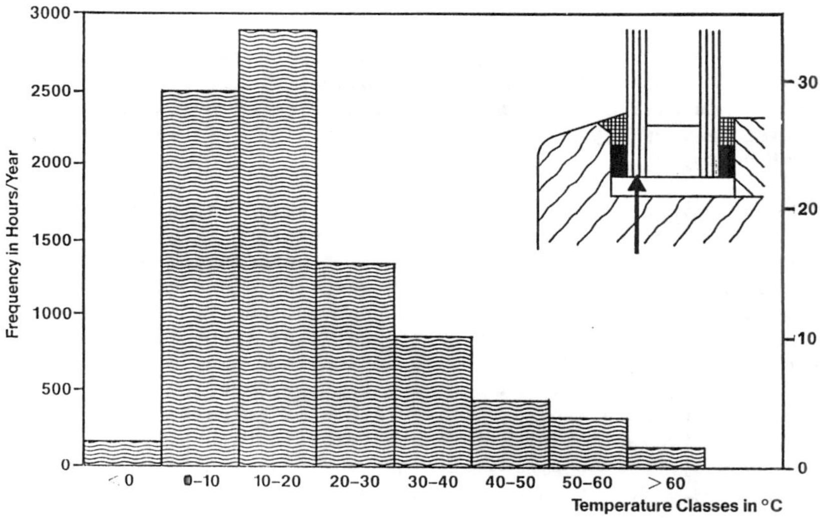

FIG. 1 -- Histogram of glass-edge temperatures measured over the course of one year

Low temperatures may exert a negative effect on the flexibility of the edge-seal, whereas high temperatures accelerate most physical and chemical processes, such as the aging of the insulating glass sealant and the diffusion of water-vapor through the edge-seal.

Atmospheric Pressure/Wind Pressure

Changes in atmospheric pressure produce mechanical stresses on the edge-seal. Added to these are rapid changes in wind loads, such as a sudden build up of pressure caused by a gust of wind or the change from pressure to suction produced with swirling winds. The high frequency with which such pressure loads occur - several hundred thousands times during the service life of an insulating glass unit - and the speed with which the changes in load occur, place extreme demands on the elasticity and the aging resistance of the insulating glass sealant.

Working Loads

Movements resulting from the normal usage of the window generally exert extremely little or no influence on the life-expectancy of the insulating glass unit, presuming that the insulating glass unit is glazed into the window casement in compliance with generally accepted technical guidelines, i.e. a flexible bearing exists between the unit and the window frame. However, repeated abrupt movements, such as occur when the window casement unintentionally slams shut, produce a high mechanical load on the edge-seal.

Sunlight

Sunlight exerts both physical and chemical stresses on the edge-seal. The infrared component of the radiation thermally charges the insulating glass unit. This effect increases, if a tinted solar protection glass is involved, since then a portion of the visible light is also converted into thermal energy.

The short wave component of the solar radiation may also induce photochemical processes in insulating glass sealants. Regular float glass is transparent to ultraviolet light down to a wave-length of about 280 nm [3]. In the case of a regularly glazed insulating glass unit, the unit's edge-seal is protected by the casement, hence no sunlight strikes the edge-seal directly. However, some 3% of the total light falling on the glass area of the window finds its way to the edge-seal due to internal reflection within the glass panes [4]. The glass adhesion of organic insulating glass sealants, such as polysulfides or polyurethanes, can be irreversibly destroyed by the high-energy short-wave spectrum of sunlight [5]. Consequently, the edge-seal of organically sealed insulating glass panes is not to be freely exposed to the sunlight but must be suitably protected, as in the case of sloped glazing, for example, by attaching strips, gaskets or tapes.

Water, Water Vapor

Water and water-vapor may also cause both physical and chemical stresses on the edge-seal. Physical stresses are the water-vapor diffusion through the edge-seal into the inter-pane space as well as the water absorption and the associated swelling of the insulating glass sealant itself. If the edge-seal of an insulating glass unit remains in direct contact with condensed water for a protracted period of time - due to improper glazing, plugging of the drainage holes or for whatever reason - organic insulating glass sealants absorb large quantities of water into the polymer matrix, thereby greatly increasing the volume of the edge-seal [6]. This swelling of the edge-seal inevitably results in an opening of the primary seal and, consequently, a higher rate of water-vapor diffusion into the interior of the insulating glass unit. In their glazing guidelines the manufacturers of insulating glass therefore attach great importance to keeping the glazing rebate free of condensed water.

Water and water-vapor are also capable of triggering chemical reactions. By causing hydrolysis of chemical bonds at the glass surface, water can irreversibly damage the adhesion of the insulating glass sealant [7,8].

Oxygen, Ozone

Oxygen and ozone cause gradual oxidation and embrittlement of organic insulating glass sealants, particularly at elevated temperatures [9].

Aggressive Atmospheric Contaminants

These include gases such as chlorine, hydrogen chloride, sulphur dioxide and nitric oxides, the latter two being major components of acid rain. Depending on the concentration in which these gases occur, they can be expected to exert considerable chemical stresses on the edge-seal. Detailed studies into this aspect are not yet available, however.

Synergetic Effects

The stresses caused by the various environmental factors are not simply cumulative in their effect. Instead, their interaction results in a disproportionately higher stress on the edge-seal, a phenomenon known as synergism. In a study into the aging behavior of insulating glass units conducted by the Institute for Window Technology, Rosenheim (Germany) [2], it was ascertained that the simultaneous action of water, elevated temperatures and sunlight constitutes the greatest stress on the edge-seal of an insulating glass unit.

EFFECT OF THE EDGE-SEAL'S WATER-VAPOR PERMEABILITY ON THE REGULAR SERVICE LIFE OF INSULATING GLASS UNITS

Over the course of time, water-vapor diffusing in via the edge-seal results in a saturation of the desiccant in the spacer, hence water condenses in the inter-pane space of the insulating glass unit, terminating its service life. The service life or the "life-expectancy" of an insulating glass unit depends on the following factors:

- the dimensions of the insulating glass unit,
- the quantity, type and initial loading of the desiccant,
- the air temperature and relative humidity during manufacture of the insulating glass,
- the resistance of the edge-seal to water-vapor diffusion, and
- the effective sealant cross-section through which the water diffusion occurs.

If all conditions are assumed identical and only the influence of the edge-seal itself is considered, the life-expectancy of an insulating glass unit should be the longer, the lower the amount of water-vapor that diffuses through the effective cross-section of the edge-seal. The life-expectancy of a single-sealed insulating glass unit is, thus, determined by the water-vapor diffusion resistance of the insulating glass sealant used to prepare the edge-seal. However, in order to estimate the life-expectancy of such an insulating glass unit, the temperature dependency of the water-vapor permeability of the insulating glass sealant needs to be considered, since - as was already illustrated in Figure 1 - the edge-seal is subject to a broad temperature distribution. The temperature dependency of an insulating glass sealant's water-vapor diffusion rate or, more precisely, of its permeability P is described by the Arrhenius' equation:

$$\frac{P(T_2)}{P(T_1)} = \exp\left(-\frac{E_a}{R}\left(\frac{1}{T_2} - \frac{1}{T_1}\right)\right) \qquad (1)$$

where E_a is the activation energy of diffusion, R the ideal gas constant and T the absolute temperature.

Insulating glass sealants therefore exhibit a considerably higher water-vapor permeability at elevated temperatures than at room temperature. Table 1 shows the water-vapor permeabilities of 15 commercially available insulating glass sealants measured in accordance with DIN 53122, Part 1 [10] at a temperature of 20, 40, 60 and 80°C [11,12]. As can be seen, the permeability at 60°C is an average of six to eight times higher than at 20°C.

TABLE 1 -- Water vapor permeability of insulating glass sealants as a function of temperature

Sealant	Type[a]	Water vapor permeability (10^{-3} kg m^{-2} d^{-1}) for 3 mm membrane thickness			
		T = 20°C	T = 40°C	T = 60°C	T = 80°C
1	Si,Al,1p	13.1	35.7	96.9	241.4
2	Si,Al,2p	11.5	27.5	74.9	197.4
3	Si,Al,2p	7.2	17.4	49.6	.....[b]
4	Si,Al,2p	12.5	28.4	67.2	155.8
5	Si,Ac,1p	19.7	47.7	141.1	.....[b]
6	Si,Oc,1p	16.0	37.7	85.7	209.9
7	PS,2p	4.7	15.8	43.7	.....[b]
8	PS,2p	5.8	18.4	52.6	186.8
9	PS,2p	6.3	18.9	51.9	153.9
10	PU	4.0	12.3	34.3	89.4
11	PS,2p	7.8	24.2	64.7	151.4
12	PS,2p	7.9	24.9	66.0	155.1
13	PS,2p	8.5	27.1	70.4	162.0
14	PS,2p	6.8	28.7	72.3	161.2
15	PS/E,2p	1.9	13.9	40.9	110.9

[a] Type notation: polymer type: Si: silicone, PS: polysulfide, PU: polyurethane, PS/E: epoxi-modified polysulfide
cure system: Al: alkoxy, Ac: acetoxy, Oc: octoate
components: 1p: one-part, 2p: two-parts

[b] For these sealants no meaningful data could be obtained due to deformation of the sealant membrane

By plotting the results of some of these permeation tests on an Arrhenius' graph (Figure 2), it becomes apparent that the activation energy of diffusion depends mainly on the polymer type utilized in the insulating glass sealant, while type and method of cure exhibit only secondary influences. Consequently the water-vapor diffusion rate of silicone sealants varies less as a function of temperature than is the case for organic sealants with a polysulfide base.

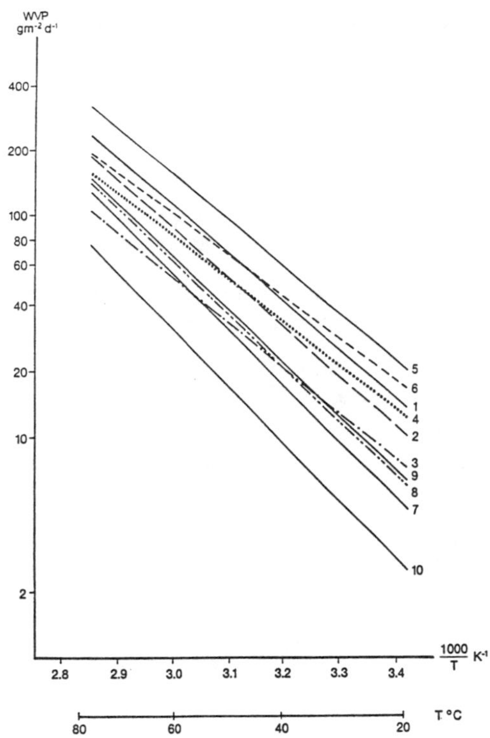

FIGURE 2 -- Arrhenius plot for the water-vapor permeabilities of insulating glass sealants (WVP in 10^{-3} kg m^{-2} d^{-1})

The practical consequences of these results must be examined: Although in Central Europe and in Northern America the edge-seal of an insulating glass unit is subjected to temperatures over 30°C for only about

20% of the year, the insulating glass unit nevertheless sustains about twice as much damage by diffusing water during this brief summer period than during the remainder of the year, when temperatures under 30°C prevail. Higher temperatures in association with high humidity therefore drastically reduce the life-expectancy of an insulating glass unit. Since the diffusion rate of silicone sealants is less dependent on temperature than in the case of organic sealants, the water-vapor permeabilities of silicone and polysulfide insulating glass sealants tend to approach each other at elevated temperatures.

Once the temperature dependency of an insulating glass sealant's water-vapor diffusion rate is known, the life-expectancy of a single-sealed insulating glass unit can easily be mathematically estimated, as has been demonstrated by Backman and others [13-15].

Since the majority of the insulating glass units manufactured are dual sealed, however, the water-vapor permeability of the dual insulating glass seal and its dependency on temperature must be taken into account. The water-vapor diffusion resistance of a plane-sheet laminate is the sum of the individual resistances of the various layers:

$$\frac{d}{P} = \sum_{i=1}^{n} \frac{d_i}{P_i} \tag{2}$$

The water-vapor permeability of the polyisobutene (PIB) primary seal is by far lower than that of the secondary seal, irrespective of whether the secondary seal is made of a silicone, polysulfide, or polyurethane sealant. The water-vapor permeability of the edge-seal is, therefore, determined almost exclusively by the permeability of the PIB sealant:

$$P_1 \ll P_2 \longrightarrow \frac{d}{P} \simeq \frac{d_1}{P_1} \tag{3}$$

Experimental studies into the water-vapor permeability of dual-layer insulating glass sealant membranes [12] confirm this theory, hence it could be assumed that the life-expectancy of a dual-sealed insulating glass unit in the field would depend solely on the PIB primary seal and that it is therefore irrelevant which sealant might be used for the secondary seal. This is not the case, however. It is not unusual for an insulating glass unit not to even realize its regular service life, which is limited by the inward diffusion of water-vapor.

PREMATURE FAILURE OF INSULATING GLASS UNITS CAUSED BY LOSS OF ADHESION OF THE EDGE SEAL

Premature failure occurs, for example, when the insulating glass sealant loses its adhesion to the glass. Both the initial adhesion to the glass surface as well as the aging behavior of the sealant/substrate bond are determined by the type of sealant utilized. Consequently, use of different insulating glass sealants will lead to different probabilities of premature failure of the insulating glass unit.

The varying propensity of insulating glass sealants to premature failure can be illustrated by exposing them to humidity, heat and UV in a Q-UV weathering tester (ASTM Standard Practice G-53) [16]. Table 2 shows the time to failure obtained for 8 commercially available insulating glass sealants based on polysulfide (PS), polyurethane (PU) and silicone (Si) polymers according to the following test procedure: Three ISO 8339 [17] test specimens (12 x 12 x 50 mm^3 tensile/adhesion joints) with glass as substrate were prepared for each sealant and allowed to cure for 6 weeks under standard climate conditions (23°C, 50% r.h.). All sealants were applied without the use of primers, as recommended by their manufacturers. The specimens were then stored in a Q-UV weathering tester (Q-Panel Company, Cleveland, Ohio 44145) for 1000, 2000, 5000, 10000, and 20000 hours. UVA fluorescent lamps (340 nm) were used as UV light source. The specimens were exposed to a weathering cycle consisting of 4 hours of ultraviolet exposure at 65°C followed by 4 hours of condensation at 50°C. After completion of the Q-UV weathering test, the specimens were elongated by 25% and this extension was maintained over 24 hours. The sealant was said to have failed, if two out of the three specimens lost more than 10% of their adhesion in the maintained extension test.

TABLE 2 -- Time to failure of ISO 8339 specimens exposed to Q-UV accelerated aging

Sealant	Type[a]	Exposure Time (hours) 1,000	2,000	5,000	10,000	20,000
1	PS,2p	passed	failed	failed	failed	failed
2	PS,2p	passed	passed	passed	failed	failed
3	PU,2p	passed	passed	failed	failed	failed
4	PU,2p	passed	passed	passed	passed	failed
5	Si,Ac,1p	passed	passed	passed	passed	failed
6	Si,Al,1p	passed	passed	passed	passed	passed
7	Si,Al,2p	passed	passed	passed	passed	passed
8	Si,Al,2p	passed	passed	passed	passed	passed

[a] Type notation: polymer type: Si: silicone, PS: polysulfide, PU: polyurethane
cure system: Al: alkoxy, Ac: acetoxy
components: 1p: one-part, 2p: two-parts

As is shown in Table 2, polysulfide sealants exhibit the poorest resistance to Q-UV, while alkoxy cure one- or two-part silicone sealants did not fail at all within the maximum time duration of the test. The Q-UV resistance of polyurethane sealants obviously is strongly dependent on the sealant formulation itself, possibly due to the use of UV stabilizing additives. The failure of the acetoxy one-part silicone sealant is more likely caused by hydrolysis than a UV degradation mechanism.

EFFECT OF THE EDGE-SEAL'S VISCOELASTIC PROPERTIES ON THE REGULAR SERVICE LIFE OF INSULATING GLASS UNITS

The hypothesis that the performance of a dual-sealed insulating unit is essentially independent of the type of secondary sealant selected, is not valid for the unit's regular life-expectancy either. While the overall water-vapor permeability of a dual edge-seal is almost exclusively governed by the low permeability of the PIB primary seal, the secondary seal still has an influence on the resistance of the edge-seal to the ingress of water-vapor. The decisive factor is the effective cross-section of the PIB primary seal through which the water-vapor diffuses.

This cross-section is determined by

- the workmanship during manufacture of the unit,
- the pressure differential between the inter-pane space and the outside world,
- the dimensions of the panes,
- the tensile stress with which the secondary sealant resists an extension of the edge-seal, and
- the ambient moisture.

Pressure differentials between the inter-pane space and the environment may result from fluctuations in the ambient temperature and pressure or from the unit being exposed to incident sun-light. These pressure differentials are temporary by nature; equilibrium will always be re-established. Permanent pressure differentials may occur for a variety of reasons, such as poor workmanship during unit manufacture, or shipment of the unit to a geographical elevation which is different from that at the place of manufacture. Permanent pressure differentials will exert a highly negative influence on the life-expectancy of an insulating glass unit; however, since they can be avoided through proper quality assurance or equilibration of the unit at the place of installation, this situation will not be further examined here.

The effect that the inwards or outwards bulging of the insulating glass exerts on the edge-seal varies as a function of the dimensions of the insulating glass unit [1]. The smaller the width or the length of the insulating glass unit, or the wider the inter-pane space, the greater the load on the edge-seal is. The degree to which the primary seal opens up when there is a positive pressure differential between the inter-pane space and the environment is determined by the tensile stress with which the secondary sealant resists this applied force. The tensile stress of the secondary sealant is itself a function of the temperature and - although to a much lesser extent - of the equilibrium moisture content in the sealant. In practice, positive pressure differentials occur at low atmospheric pressures or at high temperatures, whereby the temperature is responsible for most positive pressure differentials. The tensile stress behavior of the secondary sealant at high temperatures must therefore be examined.

Three commercially available insulating glass sealants were studied: one two-part polysulfide (PS), one two-part polyurethane (PU), and one two-part alkoxy cure silicone (Si). Three ISO 8339 test specimens (12 x 12 x 50 mm^3 tensile/adhesion joints) with glass as substrate were prepared for each sealant and allowed to cure for 6 weeks under standard climate conditions (23°C, 50% r.h.). Stress/strain curves were measured at temperatures of -20, 0, +20, +40, and +70°C on a ZWICK tensiometer fitted with a climatic conditioning chamber at an extension rate of 6 mm/min. Tensile strength at break, Young's modulus at 25% extension and elongation at break were recorded. Surprisingly, an increase in the Young's modulus with temperature was observed for the polysulfide and polyurethane sealants. It was hypothesized that this increase in the Young's modulus was caused by a post-cure occurring at the elevated temperature. Therefore, the extension test was repeated after exposing all ISO 8339 sealant specimens to the following conditioning: 6 weeks under standard climate conditions (23°C, 50% r.h.) followed by 1 week at 50°C in a convection type oven.

Tables 3 and 4 shows the temperature dependency of the Young's modulus and of the elongation at break for the three insulating glass sealants after the 1 week 50°C heat treatment. It is readily apparent that the tensile stress/strain behavior of the polyurethane sealant exhibits the greatest dependency on temperature, followed by the polysulfide sealant, whereas the tensile properties of the silicone sealant display the smallest dependency on temperature.

TABLE 3 -- Dependency of the Young's modulus measured at 25% elongation on temperature

Temperature	Young's modulus (MPa) of the sealants		
	PU	PS	Si
-20°C	1.36	2.66	2.20
0°C	1.24	2.28	1.92
+20°C	1.12	1.96	1.73
+40°C	0.91	1.84	1.68
+70°C	0.56	1.62	1.48

A further factor affecting the service life of the insulating glass unit is the time during which the opening in the primary seal exists. Certainly this is the case for all of the secondary sealants used - irrespective of their material base - for the duration of the positive pressure difference. As just indicated, the extent of the opening is a function of the tensile stress still present at elevated temperatures.

TABLE 4 -- Dependency of the elongation at break on temperature

Temperature	Elongation at break (%)		
	PU	PS	Si
-20°C	240	100	100
0°C	230	90	160
+20°C	190	70	150
+40°C	160	50	120
+70°C	110	50	100

Once the positive pressure differential decreases and an equilibrium is reached between inside and outside pressure, the length of time, required by the edge-seal to close the primary seal, varies, depending on the kind of secondary sealant utilized.

The cause for this lies in the differing elastic recovery of the secondary insulating glass sealants. Sealants with a low elastic recovery exhibit plastic deformation components in their stress strain behavior; due to relaxation mechanisms (cold flow) their tensile stress decreases during maintained extension. As a result, when the applied force is eliminated, they are no longer capable of quickly closing the primary seal to its original width. The effect, the elastic recovery behavior of secondary insulating glass sealants has on the opening and closing of the primary seal, is presented schematically in Figure 3.

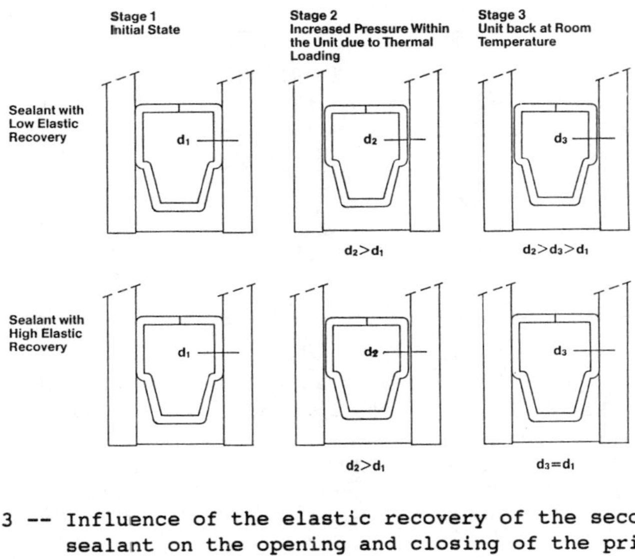

FIGURE 3 -- Influence of the elastic recovery of the secondary sealant on the opening and closing of the primary seal

The elastic recovery of the three insulating glass sealants mentioned above was measured in accordance with ISO 7389 [18] at temperatures of 20, 40, 60 and 80°C. For this purpose, three test specimens (12 x 12 x 50 mm^3 tensile/adhesion joints) with glass as substrate were prepared for each sealant and allowed to cure for 4 weeks under standard climate conditions (23°C, 50% r.h.) and then stored at a temperature of 50°C in a convention type oven. After conditioning, the specimens were extended by 50% for 24 hours and stored at the test temperatures. The stress on the test specimens was then relieved and, one hour later, the extent to which they had recovered their original width was determined.

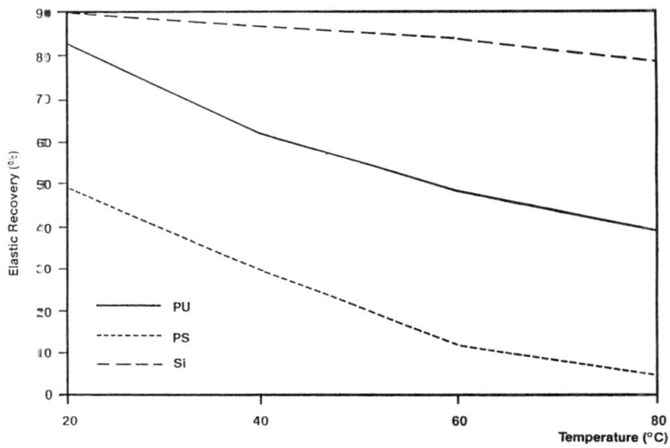

FIGURE 4 -- Dependency of the elastic recovery of insulating glass sealants on temperature

As can be seen from Figure 4, polyurethane and silicone sealants are characterized by a very high elastic recovery, whereas polysulfide sealants exhibit distinct signs of a cold flow even at room temperature. The differences are particularly noticeable at 60°C; a temperature at which fast recovery is especially important for the performance of the insulating glass unit. If the secondary seal does not recover completely, the primary seal remains permanently deformed. As a result, a higher diffusion area is available for the ingress of water-vapor over a certain period of time, until a negative pressure differential arises which again closes the primary seal. Low elastic recovery of secondary sealants at elevated temperatures, thus, markedly shortens the service life of insulating glass units.

How do the various material properties of the secondary insulating glass sealants interact? Is the water-vapor diffusion rate of the secondary insulating glass sealant or its mechanical properties decisive for the life-expectancy of an insulating glass unit? There is no universally valid answer to this question. The interaction of the various material properties differs for every sealant. The ultimate effect of the interaction between the material properties can be examined through experiment, however, by determining the amount of water adsorbed by the desiccant after subjecting the insulating glass units to a humidity/ temperature accelerated aging test cycling. The difference between the initial weight of the desiccant and the weight after concluding the accelerated aging test equals the quantity of water-vapor that has diffused into the insulating glass unit.

Table 5 shows the water pick-up measured on the desiccants of polysulfide, polyurethane or silicone dual-sealed insulating glass units following a humidity/temperature cycling test in accordance with DIN 1286, Part 1 [19]. The lowest water pick-up is observed for the silicone dual-sealed insulating glass units examined.

TABLE 5 -- Water pick-up of desiccants in insulating glass units after temperature/humidity cycling in accordance with DIN 1286, Part 1

Sealant	Type[a]	Desiccant Loading (weight %)		Water Pick-Up (weight %)
		Initial	Final	
1	PS,2p	1.8	2.3	0.5
2	PS,2p	2.0	2.8	0.8
3	PU,2p	1.6	2.1	0.5
4	PU,2p	1.7	2.3	0.6
5	Si,Al,1p	1.2	1.6	0.4
6	Si,Al,2p	1.6	2.0	0.4

[a] Type notation: polymer type: Si: silicone, PS: polysulfide, PU: polyurethane
cure system: Al: alkoxy, Ac: acetoxy, Oc: octoate
components: 1p: one-part, 2p: two-parts

The difference in the behaviors of the secondary insulating glass sealants becomes even more pronounced, when the units are exposed to constant high temperature and high humidity. To this end, insulating glass units made in accordance with DIN 1286 were stored for three months at 55°C and 100% relative humidity.

Table 6 shows the water pick-up measured on the desiccants of the various dual-sealed insulating glass units after the three months high temperature/high humidity storage. The water pick-up of a polysulfide sealed insulating glass unit is almost twice as high as that of a silicone sealed unit. Since the desiccant's water pick-up is inversely proportional to the life-expectancy of an insulating glass unit - the more water-vapor diffuses into the unit, the shorter its life-expectancy -, it can be assumed that dual-sealed insulating glass units manufactured with silicone sealants have a far higher life-expectancy than those sealed with polysulfide or polyurethane sealants.

TABLE 6 -- Water pick-up of desiccants in insulating glass units after three months storage at 55°C and 100% relative humidity

Sealant	Type[a]	Desiccant Loading (weight %)		Water Pick-Up (weight %)
		Initial	Final	
1	PS,2p	2.0	7.4	5.4
2	PS,2p	2.0	7.6	5.6
3	PU,2p	2.2	6.4	4.2
4	PU,2p	2.2	6.7	4.5
5	Si,Al,1p	2.5	4.0	1.5
6	Si,Al,2p	2.5	4.1	1.6

[a] Type notation: as in Table 5

SUMMARY

This study shows that the water-vapor permeability of secondary insulating glass sealants plays only a subordinate role in the life-expectancy of dual-sealed insulating glass units, since the resistance of the edge-seal to water-vapor diffusion is determined almost exclusively by the low water-vapor permeability of the polyisobutene (PIB) primary seal. However, marked differences in the performance of silicone, polysulfide, and polyurethane dual-sealed insulating glass units are experimentally observed, which have to be attributed to differences in the viscoelastic properties of the secondary sealants. Based on the experimental evidence of this study, particular importance needs to be attached to the tensile stress behavior and elastic recovery of secondary insulating glass sealants at elevated temperatures, since these properties affect the opening and closing of the primary seal. It should be noted that the degree to which the viscoelastic properties of cured insulating glass sealants depend on temperature is primarily influenced by their polymer type and cure system. However, other formulation parameters, such as filler type and filler loading, also may affect the viscoelastic properties and their dependency on temperature. Based on the experimental data generated in this study, it can be assumed that dual-sealed insulating glass units manufactured with silicone sealants have a far higher life-expectancy than those sealed with polysulfide or polyurethane sealants.

ACKNOWLEDGEMENTS

The authors would like to thank Mrs. K. Kopp, Mr. M. Schlensog, and Mr. D. Culot for conducting the laboratory measurements.

REFERENCES

[1] Feldmeier, F., "Stresses on Insulating Glass Units due to Climatic Variations" (In German), Fenster und Fassade, 1984, pp. 41-52.

[2] Feldmeier, F., Heinrich, R., Hepp, B., Schmid, J., and Stiell, W., "The Aging Behavior of Insulating Glass" (In German), Institut fuer Fenstertechnik, Rosenheim (FRG), October 1984.

[3] Spauszus, S., Material Science of Glass (In German), Deutscher Verlag fuer Grundstoffindustrie, Leipzig, 1975.

[4] Van Santen, N., "The Aging Behavior of Insulating Glass - Laboratory Tests to Predict the Long-Term Performance of Insulating Glass Units" (In German), Glaswelt, March 1986, pp. 12-20.

[5] Gjelsvik, T., "The Effect of the Outdoor Climate upon Materials and Constructions in Facades" (In Norwegian), The Norwegian Building Research Institute, Offprint No. 234, 1975.

[6] Ludwig, B. and Wolf, A.T., "Insulating Glass Sealants - Test and Evaluation Criteria", Kautschuk + Gummi, Kunststoffe, Vol. 32, 1986, pp. 922-928.

[7] Gan, L.M., Ong, H.W.K., and Tan, T.L., "Durability of Acrylic, Polysulfide, and Silicone Sealants Bonded to Aluminum", Durability of Building Materials, Vol. 4, 1987, pp. 42-52.

[8] Gan, L.M., Ong, H.W.K., Tan, T.L., Chen, C.H., and Lai, F.N., "Durability of Silicone Sealant Bonded on Metal or Glass", Durability of Building Materials, Vol. 2, 1985, pp. 379-387.

[9] Burstrom, P.G., "Aging and Deformation Properties of Building Joint Sealants", University of Lund, Sweden, Division of Building Materials, Report TVBM-1002, 1979.

[10] DIN 53122, Deutsches Institut fuer Normung, Berlin, 1981

[11] Wolf, A.T., "The Temperature Dependency of Water Vapor Diffusion through Insulating Glass Sealants", Kautschuk + Gummi, Kunststoffe, Vol. 38, 1985, pp. 805-807

[12] Massoth, A. and Wolf, A.T., "Studies into the Water-Vapor Permeability of Single-Phase and Dual-Phase Insulating Glass Sealant Systems", Kautschuk + Gummi, Kunststoffe, Vol. 41, 1988, pp. 882-887

[13] Backman, H.E., "The Relationship of Moisture Vapor Transmission to the Durability of Sealed Insulating Glass Units", presented at the Sealed Insulating Glass Manufacturers' Association (SIGMA) Meeting, June 1971

[14] Van Santen, N., "Testing of Insulating Glass" (In Dutch), Mikronik, May 1978, pp. 114-117

[15] Bergmann, G., "Calculation of the Time-Dependency of the Dew Point of Organic Sealed Insulating Glass Units" (In German), Silikattechnik, Vol. 25, 1974, pp. 13-15

[16] ASTM G-53, "Standard Practice for Operating Light and Water Exposure Apparatus (Fluorescent UV - Condensation Type) for Exposure of Nonmetallic Materials", American Society for Testing and Materials, Philadelphia

[17] ISO 8339, "Building Construction - Jointing Products - Sealants - Determination of Tensile Properties", International Organization for Standardization, 1984

[18] ISO 7389, "Building Construction - Jointing Products - Determination of Elastic Recovery", International Organization for Standardization, 1987

[19] DIN 1286, "Multiple Pane Insulating Glass", Deutsches Institut fuer Normung, Berlin (1984)

Philippe J. Hubin-Eschger [1]

COMPARISON OF AN INSULATING GLASS POLYURETHANE SEALANT BASED ON HYDROXYLATED POLYBUTADIENE LIQUID POLYMER AND A POYSULFIDE BASED ONE

REFERENCE: Hubin-Eschger, P. J., "**Comparison of an Insulating Glass Polyurethane Sealant Based on Hydroxylated Polybutadiene Liquid Polymer and a Poysulfide Based One**", Science and Technology of Building Seals, Sealants, Glazing, and Waterproofing: Second Volume, ASTM STP 1200, Jerome M. Klosowski, Ed., American Society for Testing and Materials, Philadelphia, 1992.

ABSTRACT: Most performances of a two-part commercial sealant for insulating glass based on hydroxylated polybutadiene are reviewed and compared with those shown by a conventionnal two-part sealant based on polysulfide. First are examined the handling properties. Then are discussed the adhesion and mechanical characteristics. The ageing effect due to both actions of water and ultra-violet radiations and the permeability to argon are studied on standard insulating glass specimens.

KEYWORDS: Insulating glass, sealant, hydroxylated polybutadiene, polysulfide.

The main target of this work is to draw a technical comparison as complete as possible between the commercial two-component elastomeric sealants for conventional double glazing available today on the european market.We did not include in this study the one-component products more specifically used in some other type of applications such as the direct glazing for instance. To make things as clear as possible in this presentation,we are going to compare only the two main classes of two-component sealants used in this industry and representing more than 80% of the total european market (Approximately 50 million square meters): Those based on polysulfides and those which are polyurethanes based on hydroxytelechelic polybutadiene liquid resins.

So we shall talk neither of the two-component sealants based on silicones which have a very special behavior and thus are not directly comparable nor of those based on other sulfur containing polymers,sometimes called hybrid polyurethanes, whose european development is small. To make things even clearer the results that are discussed in this paper concern only two sealants, one in each class, that used to be in 1990 among the better known in the european elastomeric edge sealant category. All the following results refer exclusively to those two products.

[1] Technical Manager for Speciality Elastomers at Elf Aquitaine, Groupement de Recherches de Lacq, 64170 Artix, FRANCE.

BRIEF DESCRIPTION OF AN INSULATING GLASS UNIT

On the figure below is shematically shown how a double glazing is built to decrease the thermal exchange between both sides.

The whole system is designed to avoid any moisture getting inside what would allow the phenomenon of fogging to occur particularly at low temperature.

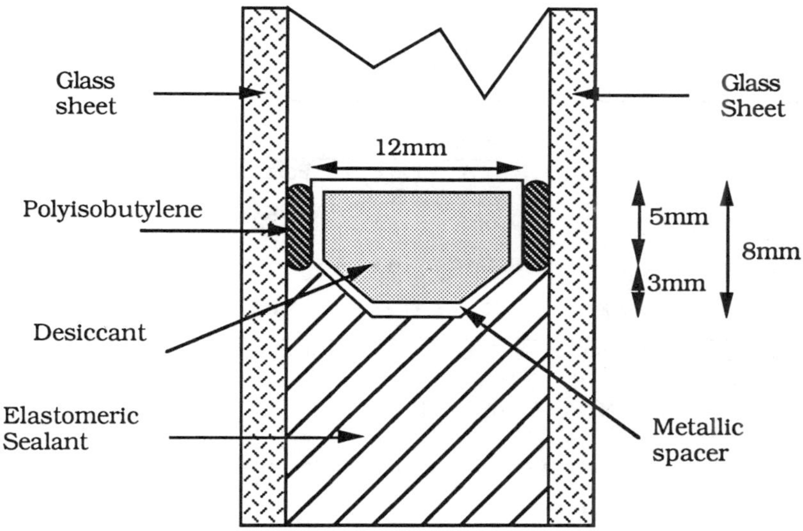

- Two sheets of glass
- In most cases, a metallic spacer maintaining the distance between the sheets, usually 6,8 or 12mm wide. Today this spacer is mainly made of aluminum, and it contains a desiccant to absorb any trace of humidity.
- A tape of polyisobutylene on both side of the spacer, first insures the rigidity of the system during the manufacturing, mostly automatic today.
- Then the elastomeric sealant on top, bonding the entire thing together.

To determine the overall quality of such a double glazing unit, three types of measurements will be successively made on the elastomeric sealant :

1-Handling properties

2-Main mechanical characteristics of the cured sealant on normalised specimens

3-Behavior of the glass unit itself under severe ageing conditions.

HANDLING PROPERTIES OF THE SEALANT

Table 1 shows the physical basic properties at 25°C of the two classes of sealants.

TABLE 1 -- Physical properties

	PBOH based sealant		Polysulfide based sealant	
	Part A	Part B	Part A	Part B
Specific gravity	1.65	1.19	1.72	1.59
Viscosity, Pa.s	800a	1b	700a	200a
Thixotropic Index	2c	N A	1.6c	1.7c

a Brookfield viscosimeter. Model HV8 /Spindle N°11/5 Rpm.
b Same/Spindle N°8/10 Rpm
c Same/Spindle N°11/5 and 10 Rpm

and table 2 the main handling caracteristics of the mixtures.

TABLE 2 -- Handling caracteristics

	PBOH based sealant	Polysulfide based sealant
Mix Ratio by weight	100/7	100/10
by volume	100/10	100/11
Pot life, min	40-45	40-50
Cure time, min	90	70

The pot-life is a very important feature as it represents roughly the time during which you can apply the sealant after the mixing of the two components. It is usually estimated by the weight of material extruded in normalised conditions (DIN 52456) over a certain period of time.

Its evolution in time gives a good idea of the shelf-life.

TABLE 3 -- Pot life

	PBOH based sealant		Polysulfide based sealant	
	Initially	Six month	Initially	Six month
Pot life, min	45	35	45	25

The ultimate Shore A hardness and the way it is reached according to DIN 53505 are shown on table 4

TABLE 4 -- Hardness evolution

	PBOH based sealant	Polysulfide based sealant
Shore A hardness		
1 hour		15
2 hours	10	35
3 hours	25	40
6 hours	40	45
24 hours	45	50
48 hours	45	50

Due to its polybutadiene backbone, hydroxylated polybutadiene based sealants once cured usually keep a slightly lower hardness and a higher flexibility.

MECHANICAL BEHAVIOR AND ADHESION PROPERTIES OF THE CURED SEALANTS

As we said before, an elastomeric sealant for double glazing should have a very low intrinsic moisture vapor transmission (MVT) and an adhesion strength on glass and aluminum sufficient to prevent any leak to occur.

This MVT is first tested on 1mm thick membranes of cured sealants according to DIN 53122B.

TABLE 5 -- Moisture vapor transmission

	PBOH based sealant	Polysulfide based sealant
MVT, g/m2/24h	5.9	8.5

Those results will be cross-checked later on double glazing units.

The method we used to evaluate adhesion-cohesion performances is an adaptation of DIN 52455 where the test-specimen is made of one piece of glass on one side and one piece of anodised aluminum on the other side of the sealing joint.

Table 6 summarizes the results obtained initially and after heat,water and U.V. (applied on the glass side) ageing.

TABLE 6 -- Adhesion performances

	PBOH based sealant		Polysulfide based sealant	
	Initially	After ageing	Initially	After ageing
Tensile strength, MPa	0.784	0.720	1.024	0.448
Elongation at break, %	73.6	86	73.8	28
Type of failure	Cohesive	Cohesive	Cohesive	Cohesive

All failures are cohesive and thus both sealants are considered good relative to this adhesion test . The hydroxylated polybutadiene based sealant has a lower modulus and maintains that lower modulus with ageing . This can be considered a distinct advantage over the polysulfide based material tested .

Compression resistance is also a critical mechanical property of double glazing sealants.

On table 7 are the results we had, following a specification adapted from the rubber industry by one of the big european glass manufacturer: A cylinder of cured sealant is put under a constant stress, then relaxed.
The level of deformation and the elastic recovery are noted.

TABLE 7 -- Compression resistance

	PBOH based sealant	Polysulfide based sealant
COMPRESSION		
Deformation, % 6min	13.4	12.6
1 hour	14.6	13.9
24 hours	14.6	14.9
ELASTIC RECOVERY		
Deformation, % 6min	6.8	10.0
1 hour	6.6	8.9
COMPRESSION SET, %	45	60

The compression set gives more or less an idea of the ratio of total non-recovery, independently of the absolute value of the ultimate deformation itself. The PBOH based sealant did well in the above comparison of this important property.

Compared to the other sealant tested, the tensile modulus in extension of hydroxylated polybutadiene based sealant shows a very small change in a wide range of temperature.

MEASUREMENTS ON THE INSULATING GLASS UNIT ITSELF

All the test-units we used have been made on a semi automatic manufacturing equipment. They are 560*310 mm sized and the aluminum spacer is 12 mm wide.

The former method to evaluate the penetration of moisture inside the glazing consists in measuring the dew point initially and after ageing at 55°C and 95%RH . (NFP 78451 and 78452).

Table 8 shows that, doing this way, both sealant are good and that the results are equivalent.

TABLE 8 -- Results of ageing

	PBOH based sealant	Polysulfide based sealant
Dew point, °C		
initial	<-60	<-60
after ageing	<-60	<-60
Penetration index	0.045	0.060

The latter method is more accurate but more complicated to perform.(NFP 78456). It calculates a moisture penetration index :

$$I = \frac{Tf - Ti}{Tc - Ti}$$

where Tf and Ti are respectively the rate of humidity of the molecular sieves after ageing and initially and Tc the maximum capacity of water absorption of the sieves in precise conditions of temperature and pressure.

On table 8, one can see that in this case, both sealants are considered as good ($I<0.2$). The difference in favor of the hydroxylated polybutadiene material is small but confirms the result of MVT on membranes.

In experiments performed on special units filled with argon as an agent to reduce thermal and acoustic exchanges, both sealants show an equivalent low permeability.

CONCLUSIONS

As usual with that kind of material, each sealant has big strengths and relative weaknesses. Polysulfides sealants have been used satisfactorily for a long time and most of them are still good sealants. The big relatively recent change is that they are not any more ruling alone the market.

Our study shows that hydroxylated polybutadiene based sealants have excellent properties in all the areas needed for the double glazing.

The machinery to dispense two component polyurethanes through automatic manufacturing equipments are now perfectly devised.

By today standards, the two-part polyurethane sealants for insulating glass have a good overall cost quality effeciency and thus all is gathered for their strong and fast development.

William H. Elfring, Jr., and William J. Rosano

CLIMATE AND PERFORMANCE OF ELASTOMERIC WALL COATINGS

REFERENCE: Elfring, W. H., and Rosano, W. J., **"Climate and Performance of Elastomeric Wall Coatings"**, Science and Technology of Building Seals, Sealants, Glazing, and Waterproofing: Second Volume, ASTM STP 1200, Jerome M. Klosowski, Ed., American Society for Testing and Materials, Philadelphia, 1992.

ABSTRACT: Elastomeric wall coatings attempt to protect exterior concrete masonry walls by minimizing water penetration through surface cracks. In contrast to traditional paints, waterborne elastomeric wall coatings used in construction and restoration, are designed as functional crack-bridging membranes. These coatings are highly formulated materials containing pigments, extenders, and a resin binder. The key to the formulation is the binder, which provides the properties that enable the material to elongate and compress repeatedly without flowing apart or embrittling and cracking. A formulation's ability to perform as an elastomeric building material is strongly affected by climatic conditions. An elastomeric coating with a long product life in one geographic location may not hold up as well in another. The coating manufacturer must choose an appropriate binder and tailor his formulation to suit the climate where it will be used in order for the coating to have a long life cycle. Studies of crack width as a function of wall surface temperature have led to a simple model that allows us to predict building movement in different climates. Plotting crack width against wall surface temperature enables us to produce a building movement profile. By factoring in fatigue data for various types of coatings over moving cracks, we can begin to assess how binder type and changes in formulation impact product life cycle. Studies and tests such as these should lead to better understanding on the part of raw material suppliers, formulators, and contractors on how to design, test, and apply high performance elastomeric coatings which are long-lasting and durable.

KEYWORDS: elastomeric, coatings, temperature, climate, fatigue, waterproofing

Drs. William H. Elfring, Jr. and William J. Rosano are research scientists at Rohm and Haas Company, Construction Products Research, 727 Norristown Rd., Spring House, PA 19477, USA.

Weathered and cracked masonry walls have long represented one of the most difficult restoration challenges. The conventional latex paints traditionally employed to protect these substrates simply do not hold up well. As far back as 20 years ago, European experts recognized the need for elastomeric coatings to maintain and restore masonry structures. From the outset, coating manufacturers and suppliers have been involved in the development of products that can supply elastomeric properties. Initially, externally plasticized, hard (high T_g) paint binders (plasticized for flexibility in the final coating) were used as the basis for these elastomeric coatings. These early binders have been replaced by a new generation of internally plasticized, soft (low T_g), dirt resistant polymers. The newer binders offer an unprecedented combination of permanent flexibility and dirt pickup resistance to this market. As a result, during the past five years, recoating with formulations based on these high-technology binders has become one of the more common methods for restoring concrete and concrete masonry buildings in North America.

The increased demand for "elastomeric" wall coatings has had an unfortunate consequence. There are no officially recognized specifications for these coatings. As a result, many of the products introduced for this market have claimed elastomeric performance without possessing the properties inherent in truly dirt resistant, durable, high-performance elastomeric coatings. Consequently, much confusion exists as to what properties are most important when specifying an elastomeric coating and what questions building owners, architects, and specifiers should ask to ensure that the coatings they select perform as expected.

The most important performance properties of a quality elastomeric wall coating are: 1) flexibility over a broad range of temperatures, 2) resistance to dirt pickup, 3) adhesion, 4) resistance to hydrolysis, 5) the ability to allow water vapor to pass through while blocking the transmission of bulk water, and 6) resistance to ultraviolet degradation. Raw material suppliers and formulators share the responsibility for ensuring that formulations possess these properties. Of course, the coating performance also depends on the applicator. For example, even with a high-quality formulation, an applicator must apply ample coating over a sound, properly prepared surface to achieve maximum durability.

Binder suppliers must design polymers that are suited for formulation into elastomeric coatings, while coating manufacturers must produce a coating that is dirt-resistant and has elastomeric properties suited for use in a range of climates. This paper proposes a way to look at building movement and to show how it differs from region to region. Furthermore, it suggests a model that can help support the development of elastomeric coatings closely adapted to the climatic conditions in the regions for which they are intended.

OBSERVATIONS:

Despite the best construction practices, large monolithic structures of concrete, concrete masonry, and stucco exposed to weather eventually develop cracks. Ideally, these are small cracks that form gradually and stabilize at a certain nominal width, as illustrated in Figure 1. Although the crack is stable in the sense that it is not growing, the width of the crack fluctuates as the concrete expands and contracts in response to ambient temperature changes, in much the same manner as a very small expansion joint. Thus the building movement can be estimated by observing the change in the width of a crack and matching it with the daily and seasonal changes in wall temperature.

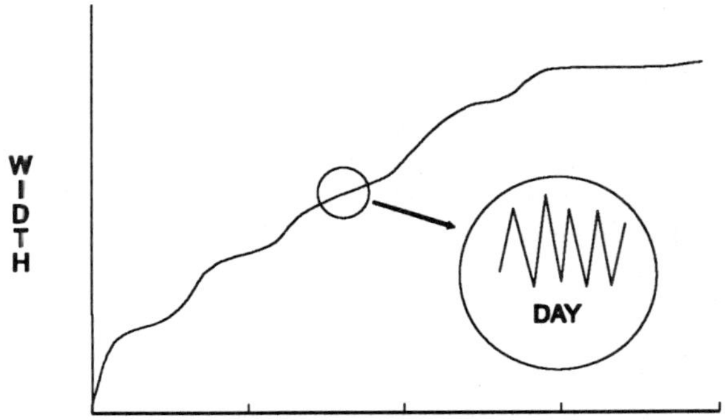

FIG. 1 - -Crack development.

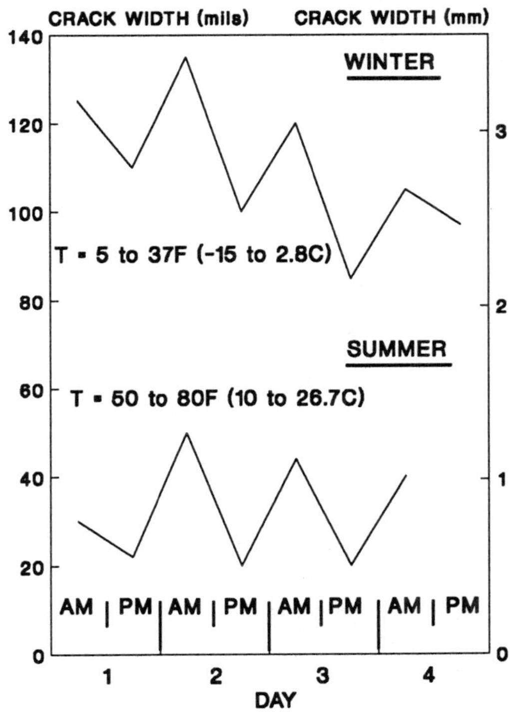

FIG. 2 - - Crack width as a function of temperature.

An example of actual movement is shown in Figure 2. Here the width of a long vertical crack in a sealant expansion joint of a concrete masonry wall was measured as a function of wall surface temperature for a period of days in the summer and winter of 1985.[1] The wall is on the south side of Building 9 at the Rohm and Haas research site in Spring House, Pennsylvania. During the summer, the crack was about 30 mils (0.76 mm) wide as the wall temperature ranged from 50°F to 80°F (10°C to 26.7°C). During the winter, the crack expanded to about 120 mils (3.0 mm) wide and varied daily from temperature changes of 5°F to 37°F (-15°C to 2.8°C). One conclusion that can be made from these results is that the rate of change (e.g., 30 mils per 8 hours) in the crack width averages to 10^{-4} inches per minute, a rate 100 to 1000 times slower than the gauge separation rate of typical mechanical testers used in most laboratories to measure tensile properties. A second conclusion is that it should be possible to model the movement of this building, given the simple wall construction of six 25-foot-wide concrete sections separated by 1/2" expansion joint.

BUILDING MOVEMENT MODEL

This model assumes the total length (L) of a building at two different temperatures (T_1 and T_2) is the sum of the width of each concrete section(w_i) plus the sum of the width of each vertical crack at the expansion joints(c_i):

$$L(T_1) = \Sigma_i w_i(T_1) + \Sigma_i c_i(T_1) \quad \text{and} \quad L(T_2) = \Sigma_i w_i(T_2) + \Sigma_i c_i(T_2).$$

The change in the width of a concrete section is given by:

$$w(T_2) - w(T_1) = \alpha w(T_1)\,[T_2 - T_1],$$ where α is the coefficient of linear expansion of masonry concrete[2]

However, since the change in the total length of the building is very small compared to the total length of the building itself, it can be assumed that the ratio:

$$[L(T_2) - L(T_1)] / L(T_1) = 0.$$ Then: $L(T_1) = L(T_2)$ and

$$c_i(T_1) + w_i(T_1) = c_i(T_2) + w_i(T_2).$$

In the case of Building 9 in Spring House, the concrete sections are about the same width, w, and the cracks are also about the same width, c, then:

$$c(T_1) - c(T_2) = w(T_2) - w(T_1) = \alpha w(T_1)\,[T_2 - T_1]$$ or simply,

---

[1]G.C. Dick and W. H. Elfring, "Binders for Elastomeric Wall Coatings," *Resin Review* XXXIX(3):1989, *passim.*

[2]G. Shortley and W. Dudley, *Elements of Physics,* Englewood Cliffs:Prentice-Hall Company, 1971, p 71.

$$c(T_1) - c(T_2) = \alpha w(T_1)\,[T_2 - T_1]\,.$$

Thus, the change in crack width is proportional to the change in temperature which is in agreement with other studies of temperature-driven building movement.[3] A crack width of 30 mils was assumed at 100°F and a coefficient of expansion of 5.56 x 10^{-6}/F degree in English units. In the metric system, the appropriate values are a crack width of 0.762 mm at 37.8°C and a coefficient of expansion of 1.0 x 10^{-5}/C degree.[2] In both instances, the width of each concrete section was 25 feet (7.62 meters). Thus, to calculate the crack width in mils (1/1000 inch) from the wall temperature in Fahrenheit degrees use:

$$c(\text{mils}) = 197 - [1.67 \times T(°F)].$$

Or to calculate the crack width in millimeters from the wall temperature in Centigrade degrees use:

$$c(\text{mm}) = 3.64 - [0.0762 \times T(°C)].$$

A comparison of the predicted movement with the observed movement for the winter of 1985 is shown in Figure 3. The agreement is close enough to qualify the model as a basis for predicting building movement.

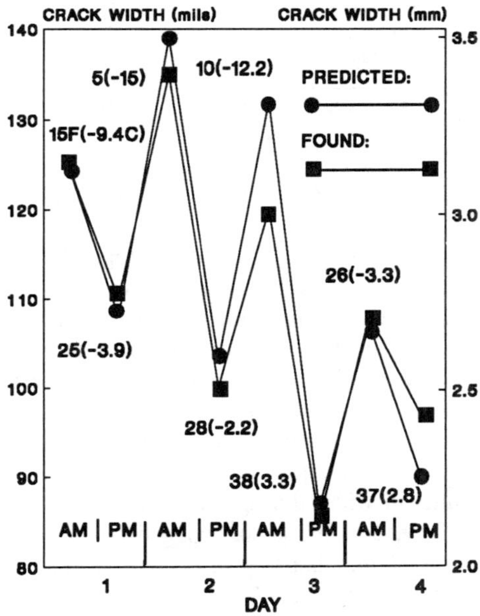

FIG. 3 - - The agreement between observed and predicted crack movement.

[3]R. Solomon, *Double Liasion-Chimie Des Peintures, Vol. 36, No. 360,* 1985, p. 307.

RESULTS

The following is an example of how to use the model. Figure 4 shows a diagram of the monthly extremes in ***air*** temperature for the Philadelphia region in 1985 (obtained from the National Climatic Data Center) and the calculated monthly extremes in the width of a wall crack from the same geographic region. [Note: Normally, only air temperatures are available on a National or International basis. Thus, the movement predicted by the model will agree with observed movement to the extent that wall and air temperature agree. However, regardless of how close the agreement between wall and air temperatures, the pattern of change will not be affected.]

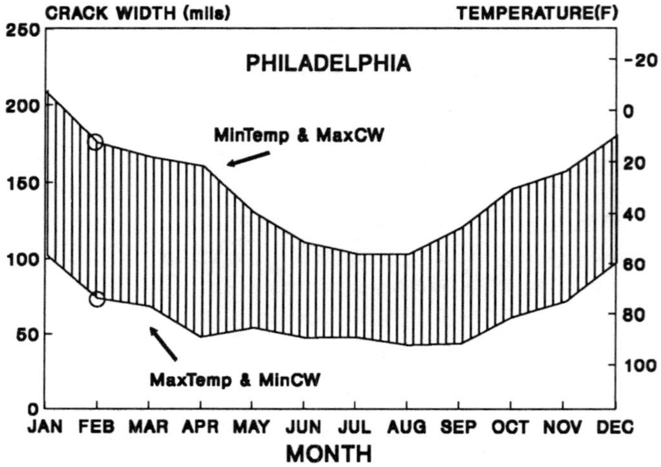

FIG. 4 - - Crack width profile for Philadelphia, PA, USA, for 1985.

The top line in the figure shows the minimum temperature and the maximum crack width while the lower line shows the maximum temperature and the minimum crack width. In this example, the diagram shows the crack width in mils and the temperature in Fahrenheit degrees. For the month of February (see circled area in the figure), the maximum temperature is about 75°F (24°C); this temperature results in a minimum crack width of 73 mils (1.9 mm). Also, at some time during February, the minimum temperature was 13°F (-11°C); and this gave a maximum crack width of 175 mils (4.4 mm). During the restoration season from April to October, the graph shows that the test crack was approximately 50 mils in width during the daylight hours. However at night, the surface shrank substantially, and the crack opened to more than twice its daytime width(!). Over the course of the changing seasons from summer to winter, the mean width of the crack grew by approximately 300%.

This kind of analysis can be done for any location in the world. Figure 5 shows the location of seven cities investigated in the United States and Puerto Rico. For each of these cities, the published maximum and minimum monthly temperatures were used to generate building movement profiles for 1985.

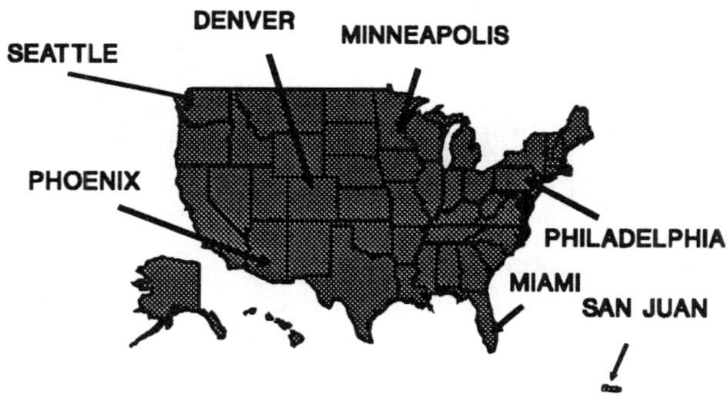

FIG. 5 - - Geographical representation of the United States.

These profiles are shown in Centigrade degrees and in mm of movement (see Figures 6 - 12). They are arranged in order from highest building movement (Minneapolis, Minnesota) to the lowest (San Juan, Puerto Rico).

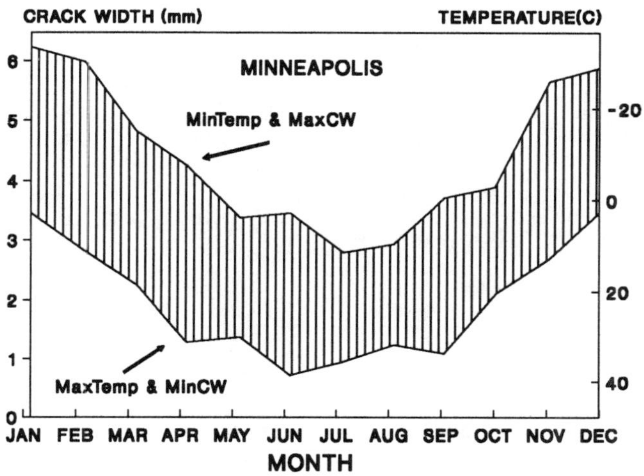

FIG. 6 - - Crack width profile for Minneapolis, MN, USA, for 1985.

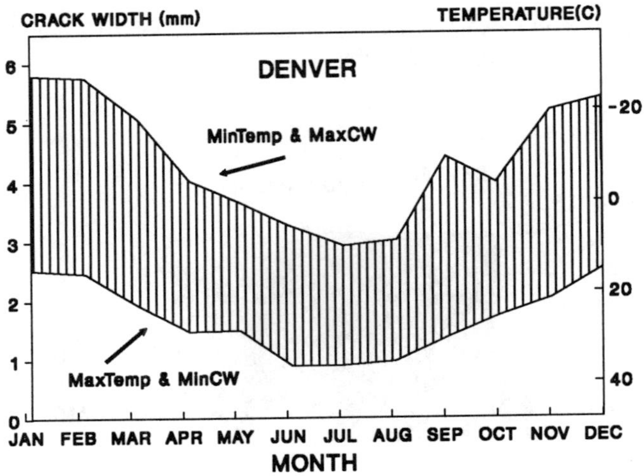

FIG. 7 - - Crack width profile for Denver, CO, USA, for 1985.

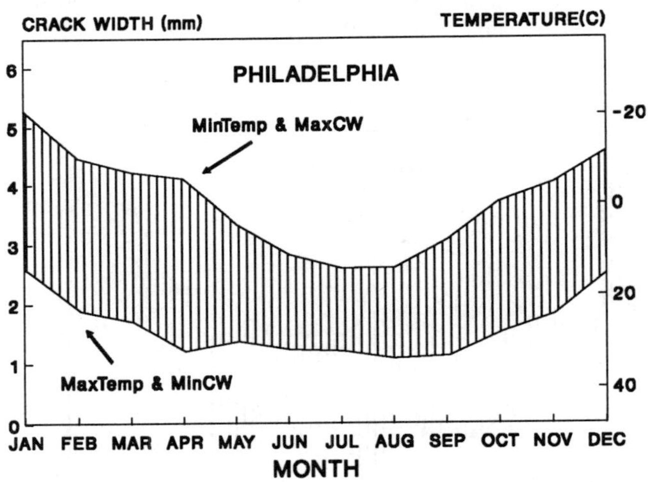

FIG. 8 - - Crack width profile for Philadelphia, PA, USA, for 1985.

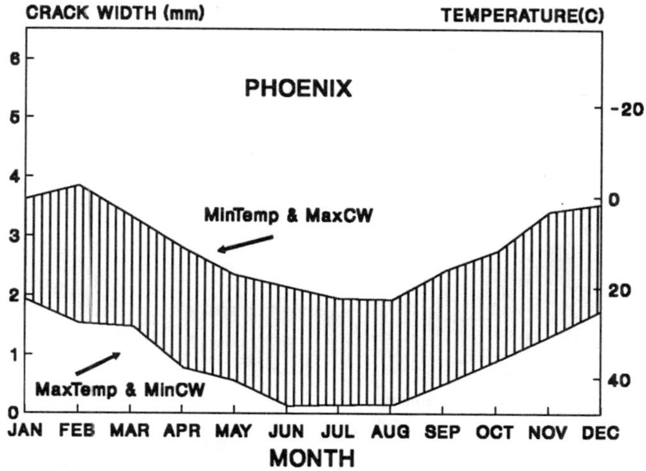

FIG. 9 - - Crack width profile for Phoenix, AZ, USA, for 1985.

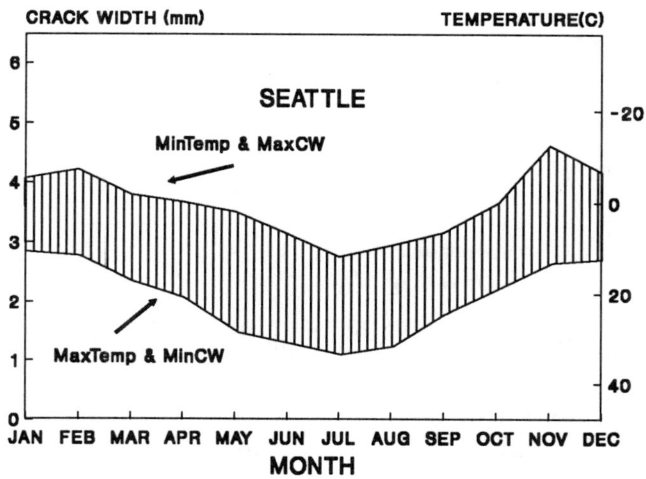

FIG. 10 - - Crack width profile for Seattle, WA, USA, for 1985.

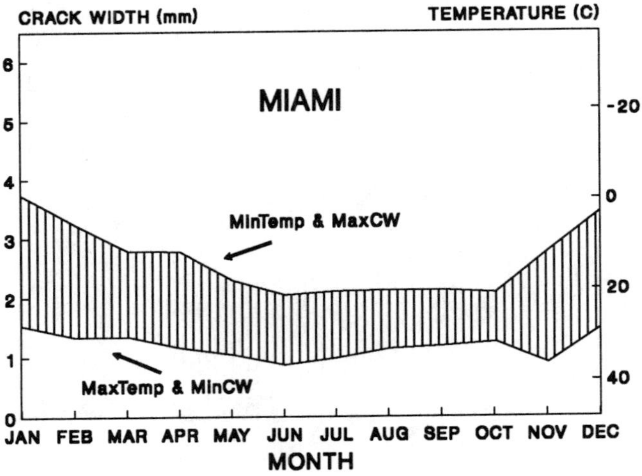

FIG. 11 - - Crack width profile for Miami, FL, USA, for 1985.

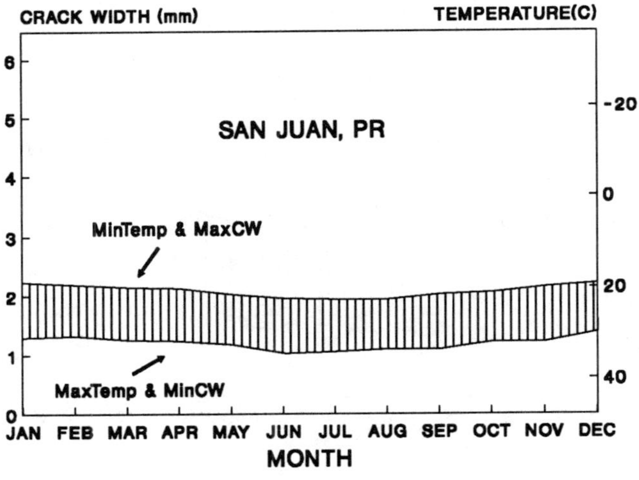

FIG. 12 - - Crack width profile for San Juan, PR, USA, for 1985.

What these building movement profiles suggest is that to protect a building in Minnesota requires an elastomeric coating with a high degree of flexibility and recovery in the range of temperatures between about -30°C and 40°C, as a minimum. On the other hand, a building in Puerto Rico would require an elastomeric coating with flexibility and recovery properties in a narrower and higher temperature range of 10°C to 40°C. In either case, the elastomeric wall coating must have excellent dirt pickup resistance and color retention as well as functional properties appropriate to the climate. To elaborate on this point, the building in Minneapolis would require an elastomeric wall coating that would bridge cracks and survive elongation and compression cycles of 400% to 500%. In San Juan, however, the ability to tolerate 100% to 200% elongation across a much narrower temperature range of 20°C to 30°C may be sufficient. Nevertheless, in both cases, the elastomeric coating must be able to elongate and recover a minimum of 365 cycles per year; over a 10-year span, this means the ability to survive a minimum of nearly 4,000 cycles.

Although the comparison in the previous two paragraphs involves North American locations, building movement profiles could be used to analyze elastomeric coating needs virtually anywhere in the world. In fact, these profiles have already been very useful to the Rohm and Haas Company in establishing the relative amount of building movement in other geographic regions.[4]

COMPARING BASE POLYMERS

There are several types of polymers that can be formulated into coatings that are labeled as "elastomeric": certain 100% acrylic emulsions sold under the Rhoplex® trademark, various generic acrylic polymers, styrenated acrylic copolymers, and vinyl acetate acrylic copolymers. These types of binders differ significantly in their ability to provide elasticity, dirt resistance, hydrolysis resistance, and resistance to ultraviolet radiation.[1,5,6] Only the specially designed 100% acrylic elastomeric emulsion polymers[7] have demonstrated the ability to deliver a sufficiently high level of performance in all four areas. In general, styrene acrylic copolymers, generic all acrylic polymers and modified acrylic copolymers can not form coatings with the same outstanding low-temperature flexibility as the specially designed 100% acrylic polymers. Figure 13 demonstrates the relative merits of the different binder types in a 45% PVC formulation in the Climate-Driven Durability Tester (CDDT).[8]

---

[4]J. Waldie, "Expand Your Business with Elastomeric Wall Coatings," *Journal of the Surface Coatings Association (Australia)*, Vol. 28 (11 - 12):1991.

[5]G. C. Dick and W. H. Elfring, "Specifying Elastomeric Wall Coatings," *Walls & Ceilings*, (February) 1991, p. 20.

[6]L. R. Freimiller and M. B. Freedman, "All-Acrylic Latex Paints for Top Performance," *Resin Review* XXXI(1):1981, pp. 9-20.

[7]The Rhoplex polymers designed for elastomeric coatings are chemically different from other all-acrylic polymers marketed by other companies for the same purpose. The distinctions between binder polymers will be discussed later in this article.

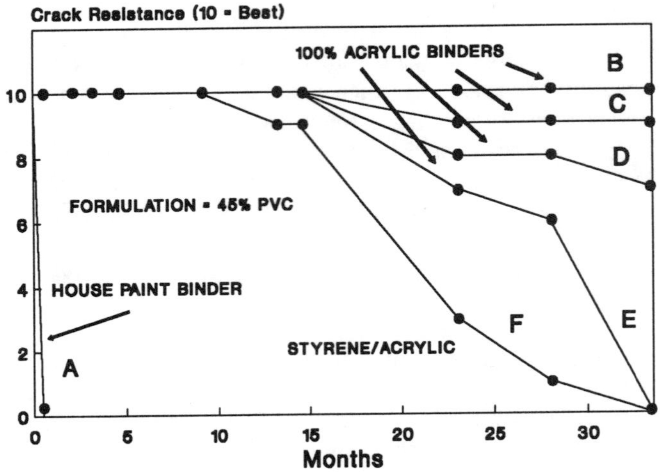

FIG. 13 - - Functional exterior durability of a selection of 45% PVC coatings

This experiment compared versions of the same formulation that differed only in their binder. For a point of reference, a conventional high-quality latex paint binder was included in this comparison of products designed specifically for elastomeric wall coatings. The coatings employed in the study were cast at a thickness of 20 dry mil (0.5 mm) by conventional techniques over thin precracked fiber-reinforced concrete sheets. The films were then placed on exposure in the CDDT located on the South wall of Building 9 at the Rohm and Haas Research facility in Spring House, Pennsylvania (USA). They were observed for a period of 36 months and then rated for the degree of crack resistance in order to establish their relative functional durability. A value of 10 indicates the film was 100% intact and uncracked. A value of 5 indicated 50% of the film remained. A value of 0 indicated that none of the film was intact over the crack.

As is evident, the formulation based on the quality exterior house paint binder (coating A) failed almost immediately. It simply did not have the flexibility to survive more than a few days. The styrene acrylic (coating F) and the generic all-acrylic formulation (coating E) showed a significant amount of failure between 15 and 23 months; both failed completely between 28 and 33 months. In contrast, the formulations based on the specially designed elastomeric binders (coatings B, C, and D) remained intact after 33 months. In addition to the variations in film integrity, the appearance properties of these coatings differed dramatically. Table I summarizes the difference in dirt resistance of the coatings after one year of exposure and the relative crack resistance performance. The data in Table I indicate that elastomeric coatings based on specially designed 100% acrylic binders have both excellent crack and dirt resistance.

---

[8]J. D. Tanzer, "Climate-Driven Durability Tester," Science and Technology of Building Seals, Sealants, Glazing, and Waterproofing: Second Volume, ASTM STP 1142, Jerome M. Klosowski, editor, American Society for Testing and Materials, Philadelphia, 1992.

TABLE I
COMPARISON OF POLYMER PERFORMANCE PROPERTIES

POLYMER CHARACTERISTICS			**PERFORMANCE IN A 45% PVC TEST FORMULATION**	
Sample ID	**Type of Binder**[a]	**Tg(°C)**[b]	**Crack Resistance**[c]	**Dirt Resistance**
A	Quality House Paint	10	Poor	Excellent
B	100% Acrylic	-43	Excellent	Excellent
C	100% Acrylic	-43	Excellent	Excellent
D	100% Acrylic	-22	Excellent	Excellent
E	100% Acrylic	-10	Good	Poor
F	Styrene/Acrylic	-22	Fair	Poor

a. Samples B, C, and D are Rhoplex 100% acrylic elastomeric binders.
b. Tg(°C) as measured by Differential Scanning Calorimetry.
c. Refer to Figure 13.

Clearly, the nature of the polymer is crucial to the ability to maintain elasticity and dirt resistance. By manipulating composition, Tg(°C), molecular weight and degree of cross-linking, polymer chemists were able to design 100% acrylic binders with a special blend of performance properties. Coatings based on these polymers develop an elastic, dirt-resistant film that lasts far longer than coatings based on other types of binders.

ELASTOMERIC WALL COATING APPLICATION FILM THICKNESS

In addition to the type of polymer employed as a binder, the film thickness of the applied coating plays an important role in the ability of an elastomeric coating to successfully accommodate thermally-driven crack movement. To illustrate the impact of dry film thickness, an elastomeric wall coating was applied over a hairline crack at a known film thickness.[1] The crack was then widened in an Instron Tester until the film ruptured completely. Figure 14 shows the results of this procedure. Plotting the coating thickness against the width of the crack at failure reveals that the coating's ability to span a moving crack is proportional to its film thickness. The slope of the line we call a B-value. In this example, the B-value is 3. This suggests that the coating can bridge a new crack that expands from "hairline" width to a width 3 times the dry film thickness of the coating before breaking. B-values can be used to compare the performance of elastomeric wall coatings and to illustrate the effect of application film thickness on durability.

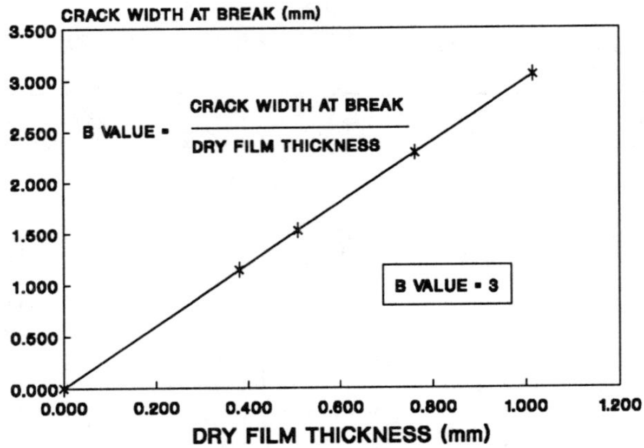

FIG. 14 - - The effect of dry film thickness on coating performance.

FORMULATING SUCCESSFUL ELASTOMERIC COATINGS

The preceding sections suggest that the single most important key to success is the choice of a binder suited to the climate where an elastomeric coating will be used. However, while the choice of binder is paramount, there are several other critical considerations in formulating elastomeric wall coatings.[5] The pigment volume concentration (PVC) and the volume solids of the coating are particularly important.

The PVC level expresses the relative amounts of pigment, extender, and binder present in the formulation. The higher the PVC, the greater the proportion of pigment and extender to the binder in the coating. Increasing the amount of extenders and pigments at the expense of binder results in a loss of elastomeric properties which may threaten the durability of the coating. In general, the more binder a formulation contains, the greater its ability to elongate, bridge thermally driven moving cracks, and recover. It will also adhere better to the substrate, chalk less and have superior flexibility at low temperatures.

Lastly, the volume solids of the coating is significant due to its influence on a formulation's ability to bridge cracks. It is the volume solids of the formulation coupled with the number of application coats and the wet film thickness of each coat, that determine the dry film thickness of the elastomeric coating. A coating with a higher volume solids contains less water. As a result, the formulation will shrink less when the water evaporates, which translates into a thicker film per application. Consequently, fewer applications (and less labor) are required to achieve an adequate dry film thickness (approximately 10 to 20 mils). In practice, these performance standards are ultimately balanced with raw material and labor cost considerations.

SPECIFICATION OF ELASTOMERIC WALL COATINGS

In the United States, where the market in elastomeric wall coatings is relatively new, there are relatively few specifications.[9] In contrast to the United States, France supports a much larger and more mature elastomeric wall coatings market. In that country, elastomeric coatings are regarded as high-technology construction products. As a result, testing agencies there are classifying elastomeric coatings on the basis of performance. Under the proposed French Norms (similar to our ASTM), elastomeric coatings would be assigned to performance categories according to their ability to bridge cracks as a function of temperature.[10] Although this system is not yet official it is already in use. The proposed classifications are being employed as performance guides for building owners, formulators, raw material suppliers, architects, and specifiers. In addition, agencies such as Bureau Veritas test and approve elastomeric wall coatings systems and certify their installation, which allows the building owner to insure the performance of the coating.[11] Together the French Norms and the Veritas approval process provide the building owner with a strong system of warranties to ensure successful installation and performance.

CONCLUSIONS

In summary, climate, binder choice, formulation and application thickness all play a role in the ability of elastomeric wall coatings to protect and waterproof exterior concrete masonry walls. On balance, only specially designed 100% acrylic elastomeric binders can provide both the permanent flexibility and dirt resistance required in elastomeric wall coating formulations.

In the United States, the currently existing specifications for elastomeric wall coatings simply are not adequate enough to ensure that building owners, architects, applicators, and contractors obtain formulations that provide a satisfactory level of performance. There is an obvious need to develop the standards and specifications that recognize the interdependent relationship between temperature and elastomeric properties. An appropriate set of functional performance standards in combination with the dirt pickup resistance, adhesion, alkali resistance, UV resistance, and other durability standards should go a long way to promote the growth of elastomeric wall coatings as high-performance construction products.

RHOPLEX is a trademark of the Rohm and Haas Company or of its subsidiaries or affiliates.

---

[9]See for example, American Institute of Architects (AIA):MASTERSPEC 09830 and Construction Specifications Institute (CSI):SpecGUIDE 09830.

[10]Proposed French Norms NF P 84 403 Version 5 (September 1988).

[11]BUREAU VERITAS, "Service de controle des Constructions et de la Securite," 31, rue Henri Rochefort, 75821 Paris Cedex 17. Tel. (1) 766.51.05.

Harold V. Lefler III

A NEW ELASTOMERIC COATING TECHNOLOGY FOR THE 90's

REFERENCE: Lefler, III, H. V., **"A New Elastomeric Coating Technology for the 90's"**, Science and Technology of Building Seals, Sealants, Glazing, and Waterproofing: Second Volume., ASTM STP 1200, Jerome M. Klosowski, Ed., American Society for Testing and Materials, Philadelphia, 1992.

ABSTRACT: In the 20th century masonry buildings evolved from massive support systems to thin facades attached to structural supports. Minor cracks in this thin facade allowed severe water ingression and damage, and brought about the development of elastomeric coating system for above grade waterproofing. Emulsion progression from butadiene–styrene and vinyl acetate polymer, through acrylics, and now into the silicone emulsion systems, has resulted in increasingly superior waterproofing elastomers. This paper reviews the basic chemistry relevant to silicone emulsion elastomers, their relationship to current acrylic elastomeric coatings for weathering, and physical property profiles. Also discussed are silicone emulsion elastomeric coatings resistance to mildew, dirt, graffiti and staining.

KEYWORDS: elastomeric, silicone, waterproof, coating, graffiti, staining.

Masonry is an ideal building material with good weathering ability and strength, but unfortunately, many times it cracks. In the past, when buildings required massive masonry walls for support, surface cracks, limited porosity, and minor water penetration were of little concern, but with todays thin skinned buildings, a small crack can mean big water problems.

At the same time elastomeric waterproof coatings provide aesthetic beauty, weatherability and waterproofing for the vertical walls and horizontal shelves of masonry clad buildings.

Recently commercialized silicone emulsion polymer technology is being utilized to produce exterior, above grade elastomeric waterproof coatings. These silicone elastomeric coatings maintain a continuous waterproof film on masonry surfaces, by bridging non structural dynamic cracks during thermal and wind cycling.

Harold V. Lefler III is in the Science and Technology Department of the Dow Corning Corporation, Midland, MI, 48686

The 1940's styrene–butadiene and vinyl acetate polymer emulsions became available, and were utilized for commercial architectural coating applications. [1] Emulsion systems progressed slowly at first, continually building new polymer systems to and through the acrylic age, and now into the silicone emulsion polymer age. Silicone emulsion polymer coatings have been commercially available for about 10 years, with specifically formulated silicone emulsion elastomeric waterproof coatings available for three years.

SILICONE EMULSION POLYMER

Preparation

Silicone emulsions are made via emulsion polymerization of cyclic oligomers or polysiloxane polymers, in the presence of a strong acid catalyst, such as the sodium salt of DBSA. [2] [3] Polymerization is carefully controlled to produce a molecular weight between 50,000 and 200,000, depending on the end use desired for the polymer. Particle size distribution is also selectively varied to be very narrow or broad, again depending on the specific end use and filler considerations.

Crosslinking

Polydimethylsiloxane (PDMS) emulsions produced from this processes, if made directly into an elastomeric coating formulation, coalesce into soft gummy films with low physical properties. To gain the strength required for an elastomeric coating the silicone emulsion must be internally crosslinked. Crosslinking or toughening of the emulsion polymer can be accomplished by using a tin catalyst such as a dialkyltindicarboxylate in conjunction with a suitable crosslinker such as silicic acid. [4] When this step is completed, the result is a crosslinked dispersion of silicone rubber. This silicone rubber dispersion can now be formulated to coalesce into a tough, elastomeric film suitable for an exterior waterproof coating.

Plasticizing

Silicone rubber particles are known to be very soft. No plasticizer or coalescing solvent, as used in acrylic systems, is required to achieve film formation. Hence, the silicone emulsion system can not harden over time from the loss of plasticizer. Silicones also have inherent resistance to the debilitating effects of ultra violet light, heat, most chemicals, and water. This excellent silicone property profile has been utilized to produce a new generation of silicone emulsion exterior waterproof elastomeric coatings.

The following portion of this paper compares a silicone emulsion elastomeric system to three acrylic systems which are commercially available. For future reference, they will be referred to as acrylic coating one, two, and three.

COATING PROFILES

Low Temperature Flexibility

The ability to maintain waterproof dynamic joints in cold as well as warm climates, is related to the coatings ability to move and conform in these severe conditions. While the glass transition temperature (Tg) of uncrosslinked or unfilled PDMS is –120°C, the crystallization point of the current filled and crosslinked elastomeric coating, as measured by differential scanning calorimeter (DSC), is –41°C (–41.8°F). DSC conditions for this measurement began at –180°C ramped up 20°C per minute to an end point of +20°C. –40°C is considered the lower temperature at which the coating would remain elastomeric. Dynamic mechanical thermal analysis (DMTA) evaluation also determined the silicone emulsion coating to be useable to –40°C.

Uncured Coating Profile

There is little difference between uncured silicone and acrylic elastomeric emulsion coatings, (Table 1) except for the specific gravity (SpG). This relates to the amount and types of extending filler used, and indicates that the silicone emulsion system is more polymer rich. It also means that the waterproofers roller is lighter when applying silicone based coating, making the days work a little easier.

TABLE 1--Uncured Elastomeric Coating Properties

Property	Avg. Acrylic Range	Silicone
Volume Solids (%)	45 - 55	55
Viscosity (Pa-sec)	42 - 85	80
Voc (g/l minus H2O)	70 - 100	85
Specific gravity	1.3 - 1.5	1.1

Cured Property Profile

Cured physical properties of the silicone and acrylic elastomeric coatings are below. (Table 2) As shown, the silicone based elastomer is at the low end in tensile strength, but has a much higher elongation for superior dynamic joint movement protection. Chloride ion ingression resistance is necessary if the coating is to be used in a high salt environment.

TABLE 2--Cured elastomeric coating properties

Property	Avg. Acrylic Range	Silicone
Tensile Strength (MPa) (ASTM D 2370)	2.1 - 4.4	2.1
Elongation (% at Break) (ASTM D 2370)	90 -300	900
Durometer (Shore A) (ASTM D 2240)	80 - 86	50
Permeance (metric perms) (ASTM D 1653)	4.9 - 11	8.2
Alkali Resistance (FTM TTC555)	poor - fair	excellent
Chloride Ion Ingress (NCHRP 224)	76 -99	93

Water Vapor Permeability

Table 2 shows the relationship of the silicone and acrylic coatings at equal film thickness. Water vapor permeance of the silicone emulsion coating is in the high region of the acrylic counterparts. Good water vapor permeability is important since many applications involve over coating a previously waterproofed or coated substrate. When this happens, it is very important that the final coating thickness be permeable enough to allow moisture vapor through the system to eliminate moisture condensation within the masonry and subsequent spalling during freeze cycles.

Silicone emulsion elastomeric coatings allow greater water vapor permeability when applied at it's recommended dry film thickness. Water vapor permeability, measured at the manufacturers recommended dry film thickness of 0.25 mm (10 mils) is 14.5 metric perms for the silicone. Permeability for the acrylics at 0.38 mm (15 mils) dry mils is 8 for acrylic #1, 5 for acrylic #2 and 3.5 metric perms for acrylic #3.

Alkali Resistance

Silicones excellent resistance to alkali is beneficial in that often the failure mode of organic coatings is blistering caused from internal moisture driving alkalinity to the masonry/coating interface. At this interface the Ph has been measured at 11, which can dissolve acrylic coatings causing adhesion loss and blistering. Because of silicones superior resistance to alkali attack, the silicone elastomer remains unaffected. When the alkaline resistance test was performed, the silicone based film remained intact and elastomeric, while two of the three acrylic based films dissolved.

The exceptional alkali resistance can allow a masonry surface to be coated with silicone based elastomeric after a two days cure rather than the standard 28 day waiting period for stucco, concrete, and mortar.

Other Testing

All coatings passed other important tests such as Low Temperature Flexibility (ASTM C–711), Mold Resistance (ASTM D–3273), Fungus Resistance (ASTM D–3274), and Wind Driven Rain (Fed. TT–C–555B). ASTM C–711 was used because no other paint/coating test involved flexibility at low temperature. However, a problem with ASTM C–711 is that it allows minor cracks and crazing from the mandrel bend test to pass, and the silicones do not crack or craze.

Accelerated Weathering

An elastomeric coating is by definition flexible, and flexibility must be maintained to keep masonry cracks waterproofed. Silicone elastomeric systems need no additives to maintain flexibility and elongation in long term weathering conditions.

After application of an elastomeric coating to a building, the manufacturers data sheet may quickly become irrelevant. Weathering is a combination of ultra violet light, moisture, temperature cycling, and pollutants. Common accelerated methods, xenon–arc, carbon–arc, and florescent ultra violet light use only three, light, moisture, and temperature.

To evaluate the effect of accelerated weathering on a silicone elastomeric coating, it was exposed with three acrylic counterparts to Xenon–arc as prescribed in ASTM Standard Practice for Operating Light–Exposure Apparatus (Xenon–arc Type) With and Without Water for Exposure of Nonmetallic Materials (G–26). The test cycle was 40 minutes of light followed by 20 minutes of light with water spray on the panel front and 60 minutes of light at 70°C. Then one hour of darkness at 38°C with water spray on the panel back.

Films of coating were drawn down to the dry film thickness recommended by the manufacturer and cured for two weeks at 72°F and 50% relative humidity. These films were installed unbacked in the weathering machine. Samples were removed at 500, 1000, and 2500 hours exposure and tested for elongation. Results (Figure 1) show a steep decline in elongation during the first 1000 hours exposure in the silicone coating. This is believed to be caused by removal of excess surfactants and glycols from the color pigmenting system. After 5000 hours xenon–arc exposure, the silicone system retains over 350% elongation while the acrylic coatings all fall below 100% elongation. For the building owner this assures that cracks initially bridged by the elastomeric coating will not reappear after a few years of weathering.

FIGURE 1 – Elongation With Xenon–Arc Exposure

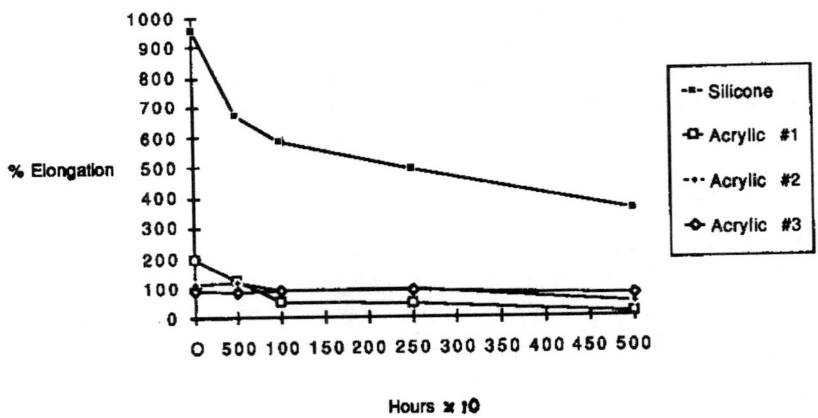

Color Stability

Color retention stability was also evaluated with the spectrophotometer via ASTM G 26 method of weathering (Figure 2). At 500, 1000, and 2500 hours of exposure, Delta E values were measured. As expected, the silicone elastomer exhibited better color retention than the acrylic systems. Surface chalking of the acrylic samples may have affected the color retention measurements.

FIGURE 2 – Color Retention Stability

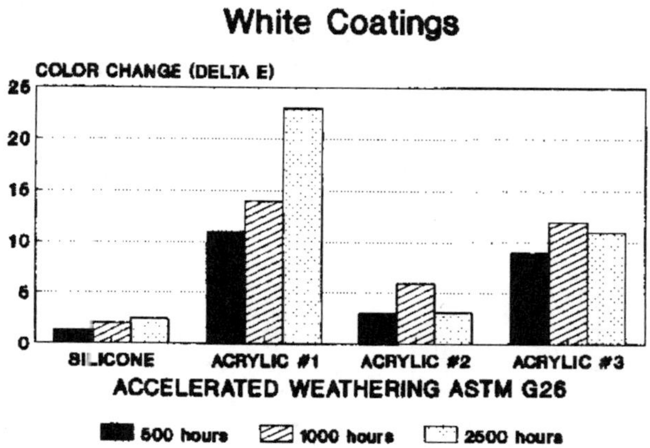

Conventional Weathering

In outdoor weathering, pollutants create conditions that current accelerated weathering devices do not. Airborne pollution from automobile exhaust, acid rain, efflorescence, roof run–off, ozone, mold and mildew compete for their opportunity to destroy the coating. Early elastomeric coatings gained a reputation for attracting contaminants to their surfaces quicker than common latex paints. One reason was the necessary use of either plasticizers or a soft polymer to meet the flexibility requirements for an elastomeric coating. Recent advances with internally plasticized acrylic polymers have greatly cleaned up on the airborne surface dirt problem.

For a comparison of todays cleaner acrylic elastomeric coatings, a silicone elastomeric was evaluated against two acrylic coatings. In this test, samples of cured coating were exposed to real time weathering in Los Angeles, California. At one and five months the samples were measured for color change on a spectrophotometer. The Delta E color change, which incorporates both color fade and dirt pick–up, (it is assumed that at only five months most of the change was due to dirt and not to real color change) was plotted as shown. (Figure 3). Dirt pick–up, in this less than pristine environment, with todays acrylic and silicone elastomers, are acceptable and essentially the same.

Stain Resistance

Silicones posses an inherent resistance to waterborne stains. Possibly the cured silicone surface is so hydrophobic that waterborne stains can't get close enough to tightly bond. Red iron staining by water from irrigation of lawns or shrubbery is a severe problem in many municipalities. To show the silicone emulsion coatings resistance to iron staining, a concrete block was coated with a silicone elastomeric and acrylic elastomeric coatings one and two. After curing two weeks at room temperature, they were exposed to an iron rich water irrigation system in Florida for six months. All of the coatings appeared iron red. However, the iron stain was easily removed from the silicone elastomer by rubbing with a wet cloth, while it was not removable from the acrylic coating. This easy way of cleaning silicone emulsion coatings also relates to most other airborned dirt, mildew, mold, plant and animal contaminates. A low pressure water wash is usually sufficient to remove most common dirt and stains.

FIGURE 3 – Dirt Pick-up in Outdoor Weathering

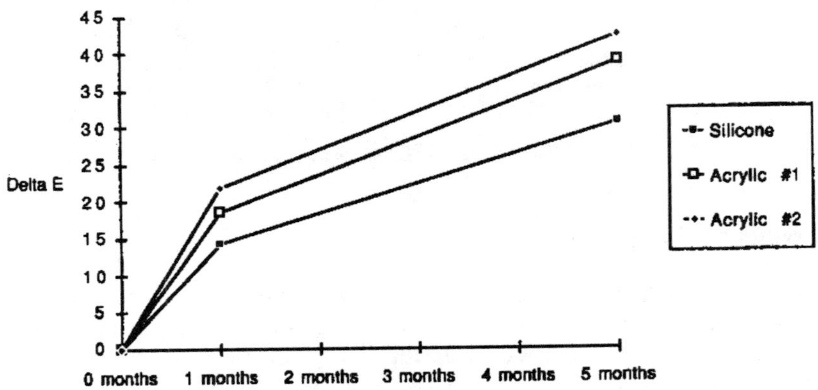

Graffiti Resistance

This cured silicone emulsion coating also exhibits limited silicone release properties in its resistance to attack by graffiti artists. Solvent born epoxies, vinyls and enamels were sprayed onto silicone elastomeric coating. After cure the most of the organic paint was removeable with only water and a brush.

CONCLUSION

As discussed, silicone emulsion elastomeric waterproof coatings have several features that make them the elastomeric waterproof coating of choice. They are formulated to be environmentally friendly and meet the most stringent VOC regulations. Silicones are inherently mold and mildew resistant and require no biocide which can leach out into the environment. Silicone emulsion elastomers posses excellent weathering performance which allows less silicone coating 0.25 mm (10 mils) as compared to 0.38 mm (15 mils) for acrylics, and still keep the building waterproof under dynamic conditions. Silicone emulsion elastomeric coatings are easy to apply, and stay clean and aesthetically pleasing. And last, most common types or graffiti, stains, and dirt are easily removed from the coating using only water wash methods.

REFERENCES

[1] Seymore, R. B. and Mark, H.F., Handbook of Organic Coatings, Elsevier, New York, 1990.

[2] Findlay, Donald E. and Weyenberg, Donald R., U.S. Patent 3,294,725 (1966).

[3] Smith, S. B., "Silicone Latex Review," Proc. Water–Borne and Higher – Solids Symp., 1983, pp 67–81.

[4] Johnson, Robert D., Saam, John C. and Schmidt, Christine M., U.S. Patent 4,221,688 (1980).

Keith B. Potts[1]

REPLACEMENT OF SOLVENT BASED GLAZING COMPOUNDS WITH HIGH PERFORMANCE WATER BASED ACRYLICS

REFERENCE: Potts, K. B., **"Replacement of Solvent Based Glazing Compounds with High Performance Water Based Acrylics"**, Science and Technology of Building Seals, Sealants, Glazing, and Waterproofing: Second Volume, ASTM STP 1200, Jerome M. Klosowski, Ed., American Society for Testing and Materials, Philadelphia, 1992.

ABSTRACT: Historically waterborne systems have had somewhat lower performance than most solventborne products. However, with the increased emphasis on environmental concerns and EPA mandated reduction in volatile organic content, window manufacturers are looking at alternate sealants. New advances in the waterborne acrylic polymers have led to the development of waterborne acrylic sealants with performance characteristics that meet or exceed most solvent-based systems. These sealants show excellent cold temperature flexibility, little age hardening, good U.V. resistance and excellent wet and dry adhesion. Compared to most solvent based compounds, water-based acrylics offer superior compatibility with all types of insulating glass sealants. In addition solvent emissions are eliminated and clean-up is accomplished with water only. These systems exhibit easy gunnability and application properties, and can be used on all existing gunning equipment.

KEYWORDS: Solventborne, Waterborne, Coalesce, Flexibility, Volatile Organic Content

---

Current glazing/backbedding consists of either a wet applied compound or a dry preformed compound. Wet applied compounds commonly consist of a variety of solvent or waterborne acrylics, solvent butyls,

---

[1]Research and Development Group Leader for Insulating Glass Sealants at Morton International, 1275 Lake Avenue, Woostock, Illinois 60098

urethanes or silicones. Dry preformed compounds generally consist of vinyl boots or gaskets, foam tapes or butyl tapes. Wet applied compounds can be used as the primary glazing sealant as in conventional drop-in glazing or can be used in conjunction with dry preformed com-pounds as a toe bead, heel bead or cap bead. Wet applied compounds are less labor intensive than vinyl boots or tapes. They generally provide a tighter, longer lasting air and water seal between the glass and framing members.

Wet applied backbedding compounds account for approximately one third of the total backbedding market (the remaining two thirds consist of tapes or boot glazing). Solvent borne acrylics account for 50-60% of the wet applied compounds, the remainder being divided between water borne acrylics, solvent butyls, silicones, urethanes and polysulfides.

Historically, waterborne systems have had somewhat lower performance than most solvent borne products. However with the increased emphasis on environmental concerns and Environmental Protection Agency mandated reduction in volatile organic content, window manufacturers are looking for alternative sealants. New advances in waterborne acrylic technology have led to the development of waterborne acrylic sealants with performance characteristics that meet or exceed most solvent borne products.

CHEMISTRY/FORMULATING

Water borne acrylic resins and solvent borne acrylic resins are produced by two distinctly different processes. The waterborne acrylics are produced by batch emulsion polymerization. In this process, monomer, water, surfactants and an initiator are charged into a reactor and polymerization proceeds until the acrylic monomer has been depleted. Extremely high molecular weight polymers are formed with low viscosity. Maximum percent solids are approximately 60-65%. Above this level viscosity increases sharply and latex stability is compromised due to loss of free volume between the individual latex spheres or particles. Solution polymerization, on the other hand, replaces the water and surfactants with solvent, in addition to the monomers and initiator. Molecular weights of these solution polymers are limited due to high solution viscosities. Percent solids, however, are typically higher for solution acrylics and range between 80-83%. Table 1 compares various properties of the two types of polymers. [1]

TABLE 1 -- Polymer Characteristics

Characteristic	Latex Polymer	Solution Polymer
Appearance	opaque	Clear
Typical molecular wt.	1,000,000	100,000
Viscosity @ 25°C	100-500 cps	50,000-200,000 cps
% wt. solids	60-65%	80-83%

Solvent borne acrylics film form by polymer concentration through solvent evaporation. Water borne acrylics film form due to water evaporation and subsequent particle fusion (coalescing). Due to chain entanglement and secondary bonding, solvent acrylics tend to have a high degree of wet or "green" cohesive strength. Waterborne acrylics will have lower "green" cohesive strength due to their individual polymer particles being separated by surfactant. This cohesive strength develops rapidly upon removal of water and particle fusion. Waterborne polymers also possess a characteristic known as "cold flow" or improved wetting/adhesion upon aging.

Formulation of waterborne sealants is accomplished in much the same manner as solvent borne sealants. Typically, a high speed mixer/disperser is used, normally having a sweep blade and vacuum capabilities. A typical ingredient list is given in table 2.

TABLE 2 -- Raw materials

Ingredient	Function
Acrylic latex	Binder for sealant
Surfactant	Pigment wetting/stability
Dispersant	Pigment dispersion/stability
Plasticizer	Improved flexibility
Pigment/Fillers	Color/rheology/cost reduction/performance properties/reduce volume shrinkage
Adhesion promoter	Improved adhesion
Freeze/thaw stabilizer	Prevent gellation upon freezing
Thickener	Slump control
Defoamer	Foam reduction
Biocide/mildewcide	Microbial protection

The sealants may be clear or pigmented. The clear sealants are typically lower solids than the filled/pigmented sealants which may contain as much as 40% filler. Depending on the resin type used, plasticizer may or may not be required. Ammonia or other highly alkaline materials may be used to adjust the pH of the finished sealant. This is usually required to improve storage stability.

SEALANT COMPARISON

Sealants formulated with high performance water borne acrylics compare very favorably in performance to many solvent borne sealants, especially when compared to other non-chemical cure products such as solvent borne acrylics, solvent borne butyls, and solvent borne block copolymer systems. These water borne products generally contain a much lower volatile organic content than their solvent borne counterparts (see Table 3).

As mentioned earlier this area has become a critical concern for many window manufacturers as new E.P.A. mandated V.O.C. reduction comes into effect. In addition, clean-up can be accomplished with water only,

TABLE 3 -- Volatile organic content

Product	V.O.C. (grams/liter)
Water borne acrylic backbedding	0-40
Water borne acrylic seam sealer	0-40
Solvent borne acrylic backbedding	200-260
Solvent borne acrylic seam sealer	450-600
Solvent borne butyl backbedding	250-400
Solvent borne block co-polymer	250-350

thus eliminating the need for open solvent containers in the workplace. Worker exposure to solvent is thus reduced dramatically. In addition, materials with extremely high flash points (>200°F (93.3°C)) pose much less of a storage risk than lower flash point products.

Compatibility with other window components is an extremely important aspect of backbedding/glazing compounds. Window components include insulating glass sealants (which include polysulfide, polyurethane, silicone and butyl based sealants), laminated glass resins (normally polyvinyl butyral or polyurethane), setting blocks or stops (usually vinyl, neoprene, EPDM), and the framing member itself (normally wood, vinyl, or aluminum). An incompatability between glazing compounds and any of these components can lead to unsatisfactory window performance. Due to their lack of solvent and plasticizer, these water borne acrylics have proven to be extremely compatible with virtually all window components.

Basic wet physical properties are quite similar between water borne and solvent borne sealants. Table 4 compares these properties.

TABLE 4 -- Sealant properties

Product	wt/gallon	% solids	Viscosity[a]	Slump[b]
Waterborne clear acrylic sealant	8.65	60-65%	1,000,000 cps	<.1"
Waterborne filled acrylic sealant	11.5-12	80%	1,500,000 cps	<.1"
Solvent borne clear/translucent acrylic sealant	10.0	70-75%	>2,000,000 cps	<.1"
Solvent borne filled acrylic sealant	12.0	80%	>2,000,000 cps	<.1"

[a] - Brookfield cps spindle TE
[b] - Modified boeing slump ASTM D-2202

Weight solids are typically slightly higher in the clear/translucent solvent acrylics when compared to clear waterborne acrylics. This difference is generally negligible when the products contain filler. Viscosities of water borne acrylics are usually significantly lower than their solvent borne counterparts. This difference is readily apparent in window plants where manufacturers find that when using water borne acrylics pump pressures can be drastically reduced.

Base physical properties of various sealants are compared in table 5. Solvent borne acrylics generally vary considerably between harder high modulus products to extremely soft low modulus products. Solvent borne butyls are typically very low modulus products and remain extremely low durometer. Solvent borne urethanes usually are higher modulus products with mid-range Shore A durometers. The high performance water borne acrylics will usually range between low to mid range modulus materials with low Shore A durometers.

TABLE 5 -- Sealant physical properties

			Modulus[a]			
Product	Tensile[a]	Elongation%[a]	10%	20%	50%	100%
Water borne acrylic	.41 MPa	650%	.028	.034	.048	.090
Solvent borne acrylic low modulus	.19 MPa	400%	.11	.137	.172	.207
Solvent borne acrylic high modulus	.96 MPa	550%	.296	.372	.448	.551
Solvent borne butyl	.17 MPa	500%	.014	.024	.038	.055
Solvent borne urethane	2.07 MPa	250%	.345	.562	1.05	1.38

[a] ASTM D412-83

In general the ultimate elongation properties are quite similar between all types of materials.

Weathering characteristics and a low degree of age hardening is extremely important in backbedding/glazing compounds. These compounds need to remain soft and pliable throughout the life of the sealant. This is needed to allow for movement in a window without causing undue stress on the glass itself. In addition, soft/pliable sealants allow for much easier field reglazing in the event of broken glass or other types of field failures. Q.U.V. accelerated weather was run on a variety of sealant types and compared in Table 6. These sealants were run for a total of 500 hours and compared to non-exposed controls for durometer change and visual differences.

The ability to remain ductile and flexible in cold temperature conditions is also an extremely important property of a backbedding compound. The ability for a window to move and flex without glass stress in cold temperatures is critical to avoiding glass breakage. In addition the ability to retain movement capability in cold temperatures puts less stress on bedding/substrate bond lines. Loss of adhesion in cold weather leads to seal failure and ultimately water and air infiltration. Table 7 compares shore A durometers at room temperature and 0°F (-17.8°C) as well as 180° 1" (25.4 mm) mandrel bends run at 0°F (-17.8°C) of various sealants. Graph 1 charts change in durometer over a temperature range of 0°F (-17.8°C) → 100°F (37.8°C) for various sealants, after 1 month ambient cure.

TABLE 6 -- Weathering properties

System	Initial / Final Shore A	Shore A	Visual Effects
Waterborne Acrylic	23A /	26A	Slight yellowing, no chalking or surface cracks
Solvent borne Acrylic Low modulus	30A /	62A	Slight yellowing, no chalking or surface cracks
Solvent borne Urethane	55A /	76A	Slight yellowing, slight chalking, no surface cracks

TABLE 7 -- Cold Temperature Flexibility

System	R.T. Durometer	0°F(-17.8°C)Duro	Mandrel Bend[a]
Water borne Acrylic	30A	60A	Pass
Solvent borne Acrylic low modulus	36A	85A	Fail
Solvent borne Acrylic high modulus	60A	94A	Fail
Solvent borne Butyl	32A	65A	Pass
Solvent borne Urethane	51A	83A	Pass

[a] - 0°F (-17.8°C), 1" (25.4 mm) bend, 180°

High performance water borne acrylics remain relatively soft and flexible, even in extremely cold weather conditions.

Backbedding compounds need to provide service over a wide range of temperature and wind load conditions. These materials act as a shock absorber between the glass and framing member, as well as provide an air and water tight seal between the two. Table 8 compares the lap shear strength of a high performance waterborne acrylic to a solvent borne acrylic over a temperature range of -60°F (-51.1°C) to 220°F (104.4°C), after 1 month ambient cure. In this study glass to wood lap shears were prepared using a 1" (25.4 mm) overlap and approximately .030" of backbedding compound. The pounds of lap shear strength as well as the mode of failure were noted at each test temperature.

As can be noted from the test data, even at temperatures as low as -60°F (-51.1°C) the high performance water borne acrylic remains flexible enough to allow for cohesive sealant failure. This maintains the integrity of the air and water tight seal as well as preventing glass breakage. Water borne acrylics are also able to maintain more high temperature shear strength than their thermoplastic solvent borne acrylic counterparts. The higher degree of branching and chain entanglement dictated by the monomers used in these waterborne acrylics enables their strength to be maintained at high temperatures. This also aids in maintaining the window integrity in high temperature, high windload conditions.

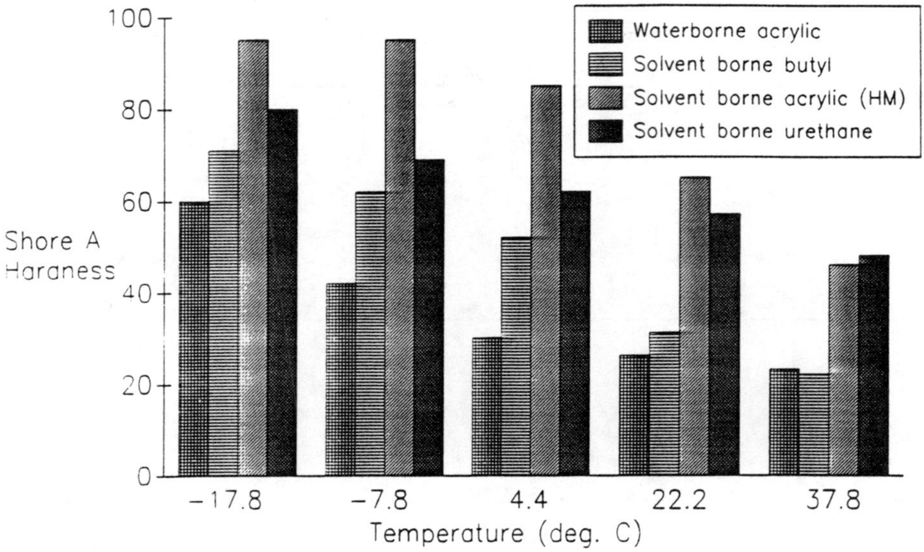

GRAPH 1 -- Hardness versus temperature

TABLE 8 -- Service temperature evaluation

Test Temperature	Waterborne Acrylic	Solventborne Acrylic (H.M.)
-60°F (-51.1°C)	3.74 MPa/Cohesive	3.47 MPa/Glass broke
-30°F (-34.4°C)	2.91 MPa/Cohesive	3.10 MPa/Glass broke
0°F (-17.8°C)	2.36 MPa/Cohesive	3.38 MPa/Glass broke
30°F (-1.1°C)	.65 MPa/Cohesive	2.05 MPa/Cohesive
72°F (22.2°C)	.34 MPa/Cohesive	.44 MPa/Cohesive
150°F (65.6°C)	.11 MPa/Cohesive	.036 MPa/Cohesive
180°F (82.2°C)	.048 MPa/Cohesive	.024 MPa/Cohesive
220°F (104.4°C)	.045 MPa/Cohesive	.009 MPa/Cohesive

APPLICATION

Waterbase Acrylics can be used successfully as a replacement for solvent borne sealants in many applications where solvent borne sealants excel today. These include traditional drop in glazing (both interior and exterior glazed) as well as cap beads, toe beads, and heel beads. Figures

1, 2, 3, 4 show these particular end uses.[2][3] Water borne acrylic sealants can be used on any existing sealant equipment.

Figure 1 Backbedding

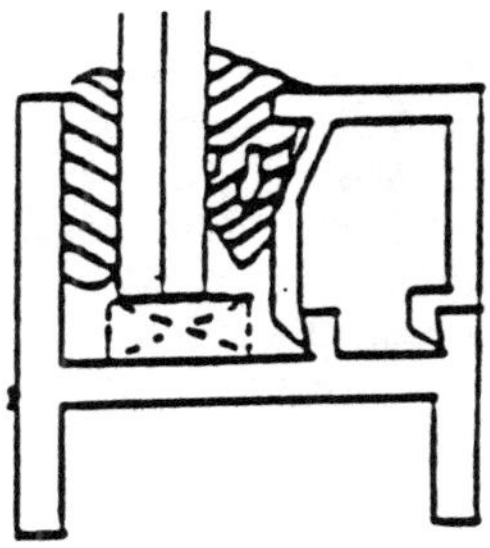

Figure 2 Cap Bead

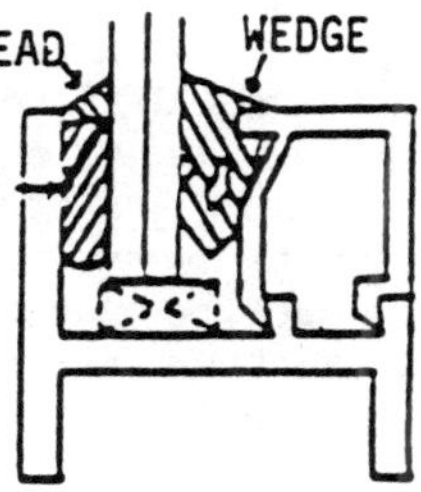

Figure 3 Toe Bead System

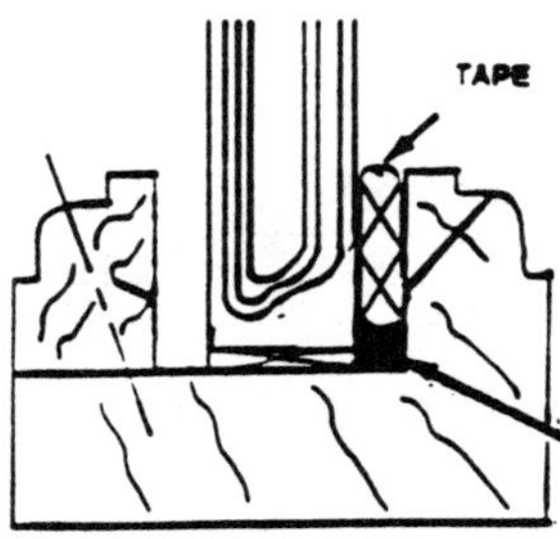

Figure 4 Heel Bead System

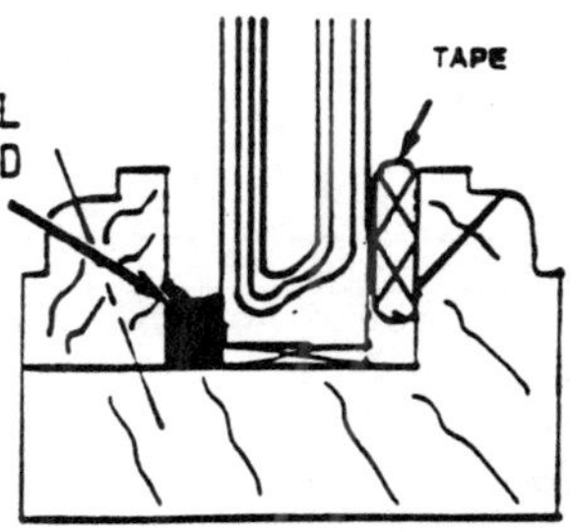

In some instances the switch to stainless steel parts may be required, in place of cold rolled steel. Water borne acrylic products are generally highly thixatropic sealants that break down readily under pressure and flow relatively easily. For this reason pump pressures can be reduced significantly from the pressures used to pump solvent borne products. In many cases, one central pump can accommodate a series of application stations. This not only reduces the number of pumps but can reduce maintenance costs as well. Water borne acrylics tend to show very good snap or break-off when being applied. They do not tend to string out when removing the gun from a finished bead like solvent borne butyls or solvent borne acrylics. This produces a cleaner operation with less need for post application window clean-up.

The same requirements exist for applying water borne sealants that currently exist for solvent borne sealants. Ambient temperatures $\geq$40°F (4.4°C), and a relatively clean substrate devoid of water and oils that may effect adhesion. Damp substrates are acceptable with waterborne sealants, an advantage over many solvnent borne products.

Waterborne acrylics do show several differences from solvent borne sealants when compared in actual window applications. Solvent borne sealants possess higher initial or "green" cohesive strength. This is due in large part to polymer entanglement and is particularly apparent in solvent borne butyls and solvent borne acrylics. Water borne acrylics, being a series of distinctly separate particles, need to begin to fuse or coalesce for cohesive strength to build. This fusion begins as soon as water begins to evaporate from the sealant. Initially a manufacturer may see more window movement with a water borne acrylic. This has usually not presented a problem once the stops have been placed in the window. In extremely large windows setting blocks are often used to prevent window settling. This difference in wet "green" strength is only a temporary condition and is alleviated once water begins to evaporate from the sealant. Table 9 compares lap shear strength development of water and solvent borne acrylics in a glass/aluminum lap shear (1" (25.4 mm) overlap, .030" backbedding). As noted from the data, lap shear development is quite similar between the water borne acrylic and the high modulus solvent acrylic between 3 days and 1 month.

Another difference existing between water borne and solvent borne sealants is the rate of evaporation of the volatile component. Both types of products need evaporation of their volatile component before full physical properties can be achieved.

TABLE 9 -- Lap shear strength development

Product	3 day	1 week	2 week	3 week	1 month
Solvent borne Acrylic Low modulus	.015 MPa	.018 MPa	.043 MPa	.059 MPa	.076 MPa
Solvent borne Acrylic High modulus	.032 MPa	.057 MPa	.081 MPa	.139 MPa	.327 MPa
Water borne Acrylic	.029 MPa	.092 MPa	.181 MPa	.210 MPa	.319 MPa

Waterborne sealants with water as their main volatile component are more sensitive to ambient relative humidity than solvent borne sealants which rely on solvent evaporation. In high humidity conditions water will leave the sealant much slower than in relatively dry/low humidity conditions. These sealants are not affected as much by temperature as they are by relative humidity. Solvent borne systems, on the other hand, show very little difference in solvent evaporation over a large relative humidity range but are more effected by temperature in their rate of solvent evaporation. Graph 2 compares a solvent borne and water borne acrylic in 3 day lap shear strength at a constant temperature with varying relative humidities. Graph 3 compares the same sealants in 3 day lap shear strength at a constant relative humidity with varying temperature.

SPECIFIC TEST REQUIREMENTS

Sealants formulated with high performance waterborne acrylics have been able to meet the requirements of a number of independent specifcations. American Architectural Manufacturers Association (AAMA) "Voluntary Specifications and Test Methods for Sealants" contains three

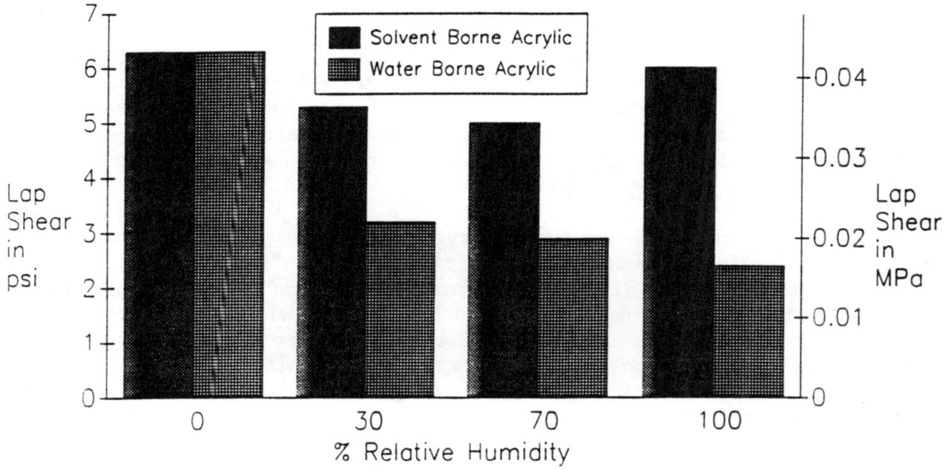

Graph 2 -- Lap shear strength versus relative humidity

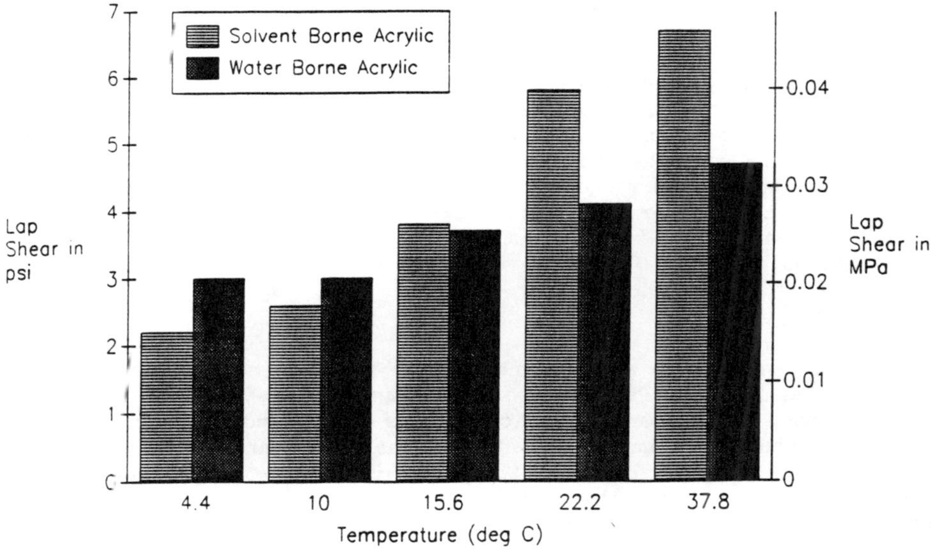

Graph 3 -- Lap shear strength versus temperature

test specifications which have been met with these waterborne acrylics. These include AAMA 802.3-85 for ductile backbedding compounds, AAMA 805.2-85 for bonding backbedding compounds and AAMA 803.3-85 for narrow joint seamsealers. These requirements all specify good peel adhesion to glass and aluminum substrates. Sealants must remain reasonably soft and have good low temperature flexibility. Weathering and heat resistance is also tested as well as vehicle migration and water resistance.

High performance water borne acrylics have also been able to perform as the backbedding compound in meeting AAMA 101-88 "Voluntary Specification for Aluminum Prime Windows & Sliding Glass Doors". In this procedure windows must go through a series of tests inside a windload chamber. The backbedding compound must be able to maintain its air and water tight seal through air infiltration, water resistance and uniform structural load testing. In addition, a deglazing test is run to determine the actual strength required to deglaze the window from the framing member.

Sealants have also been formulated to meet the requirements of ASTM C834-76 "Standard Specification for Latex Sealing Compounds". In this test procedure sealants must meet shrinkage and weathering requirements in addition to having good low temperature flexibility.

Sealants also must be able to pass 25% extension with minimal bond loss between glass and aluminum substrates. In this procedure, percent recovery of the test specimen is also examined. A minimum of 75% recovery is allowed.

Historically, waterborne sealants have had trouble performing to ASTM C-920-86 "Standard Specification for Elastomeric Joint Sealants". With the development of higher performance water borne acrylics most of the performance requirements for this specification can be met. Sealants using these high performance water-borne acrylics can meet both the peel adhesion and cold temperature flexibility requirements. Adhesion/cohesion cycling has also been met for both class 12 1/2 and class 25 with some formulations.

CONCLUSION

High performance water borne acrylic sealants have proven to be a viable alternative to solvent borne sealants in a variety of glazing applications. Water borne acrylic sealants allow window manufacturers a means to reduce or eliminate volatile organic content emissions without sacrificing window performance. Worker exposure to solvent emissions is also reduced in both the application and window clean-up stages. Waterborne sealants usage will continue to grow as E.P.A. mandated reductions become nation-wide.

ACKNOWLEDGEMENTS

The author acknowledges with appreciation the assistance of Terry Rossi in compiling test data for this paper.

REFERENCES

[1] Lomax, Jr., "Acrylic Polymer Caulks and Sealants", Report Caulks and Sealants short course, Adhesives and Sealants Council, Dallas, TX 1990

[2] American Architectural Manufacturers Association, "Fenestration Sealants Guide Manual", 1991

[3] Morton Thiokol, Inc. "Joints Design Digest", 1990

Donald T. Liles and Nick E. Shephard

SILICONE RUBBER LATEX SEALANTS

REFERENCE: Liles, D. T., and Shephard, N. E., **"Silicone Rubber Latex Sealants"**, Science and Technology of Building Seals, Sealants, Glazing, and Waterproofing: Second Volume, ASTM STP 1200, Jerome M. Klosowski, Ed., American Society for Testing and Materials, Philadelphia, 1992.

ABSTRACT: Silicone rubber latex is an aqueous dispersion of silicone rubber particles that forms elastomers upon evaporation of water. Silicone rubber latex is formulated into sealants by incorporating fillers, pigmnets and other additives. Silicone rubber latex is prepared by crosslinking emulsions of poly(dimethylsiloxane) using a tin catalyst. Properties of silicone rubber latex sealants differ in some respects from those of traditional RTV silicone sealants. This paper describes preparation of silicone rubber latex and silicone rubber latex sealants. Properties of both uncured and cured silicone rubber latex sealant are also discussed.

KEYWORDS: silicone, sealant, latex, rubber, emulsion

INTRODUCTION

Silicone rubber latexes are aqueous dispersions of silicone rubber that form elastomers upon evaporation of water. Sealants and coatings can be formulated from silicone rubber latex by incorporating fillers, pigments and other additives. Silicone rubber latex is prepared by crosslinking emulsions of poly(dimethylsiloxane), PDMS. Removal of water causes the latex particles to coagulate irreversibly, which leads to silicone elastomers. Solvents, coalescing aids or plasticizers are not required for elastomer formation, thus products based on silicone rubber latex can be formulated to have low volatile organic content and consequently, low odor.

Advantages of silicone rubber latex sealants include ease of application, water clean-up, paintability, low odor, high elongation, low modulus and low temperature flexibility. Limitations of silicone

Donald T. Liles and Nick E. Shephard are in the Science and Technology Department of the Dow Corning Corporation, Midland, MI 48640.

rubber latex sealants are those that are typical of water-based sealants, which include low solids content (shrinkage), variable cure rate due to humidity and subject to early rain washout. Silicone rubber latex sealant has been available for about the last seven years and steady improvements in the science and technology of this unique material have occurred during this time. This paper describes preparation of silicone rubber latex, silicone rubber latex sealant and its properties.

DISCUSSION

A. Preparation of Silicone Rubber Latex

1. Emulsion Polymers

Emulsion polymers used to prepare silicone rubber latex consist of oil in water emulsions of hydroxyl terminated poly(dimethylsiloxane), PDMS. In order for elastomers derived from silicone rubber latex to have useful mechanical properties, molecular weight (M_w) of the polymer should be above 50,000. Although it is possible to emulsify high molecular weight silicone polymers, it is much more convenient to obtain them via the emulsion polymerization process of Weyenberg and Findlay [1].

2. Crosslinking

Silicone emulsion polymer must be crosslinked in order for the emulsion to produce an elastomer. Without crosslinking, removal of water from silicone emulsion polymers produces tacky silicone gums that flow due to gravity or other applied forces. Removal of water from crosslinked silicone emulsion polymers produces tack-free, non-slump silicone elastomers.

Crosslinking silicone emulsion polymers leads to precured or precrosslinked emulsions, or in other words, aqueous, dispersed silicone rubber. Since silicone rubber latex particles are crosslinked, migration of polymer chains between particles cannot and does not occur. As a result, the original integrity of the particles is maintained throughout the life of the cured (dry) latex, see Figure 1.

There are numerous means for crosslinking silicone emulsion polymers, but only the two most common methods are discussed. These two methods consist of crosslinking via emulsion copolymerization and crosslinking via tin catalysis. Additional methods for crosslinking silicone emulsion polymers have been previously described [2].

Crosslinked silicone emulsion polymer was first patented by Weyenberg and Findlay in 1966 [3]. In this process, a small amount of multifunctional alkoxysilane crosslinker such as methyltrimethoxysilane (MTM) was copolymerized in emulsion with the oligomeric PDMS. Removal of water produced a film of silicone rubber. In recent years, several improvements to this process have been reported in the patent literature [4,5,6]. The method works best when crosslinker is added to the oligomer prior to emulsion compolymerization. The chief advantage of the copolymerization method is that it is a one step process.

In 1980, Johnson, Saam and Schmidt reported that silicone emulsion polymers could be crosslinked by adding a tin curing catalyst and colloidal silica [7]. This latex differed from previous examples in that no crosslinker, the alkoxysilane, was required. Saam, Graiver and Baile proposed that the crosslinker in this case was actually silicic acid, which is always present in aqueous silica systems [8]. They also proposed that the tin catalyst, a dialkyltin dicarboxylate salt, transported crosslinker into the latex particles.

Saam et. al. advanced the concept of a precured silicone emulsion, which until that time, it was not known how silicone rubber latex formed elastomers [7,8]. Saam's concept of a precrosslinked emulsion has been accepted as a realistic mechanism for crosslinking and cure processes in silicone rubber latex.

The crosslinker for silicone rubber latex prepared via tin catalysis can be practically any source of a multifunctional silanol compound. Silicic acid from silica, sodium silicate, silicone resins

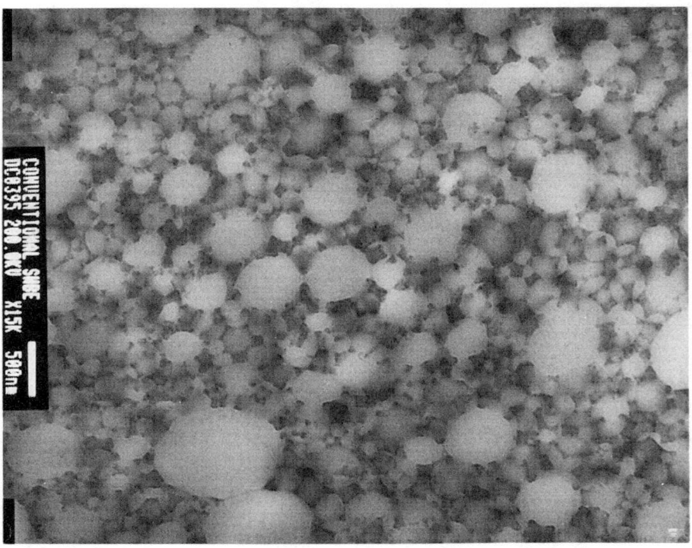

Figure 1. Transmission electron micrograph of cured silicone rubber latex. Latex was Sn catalyzed and filled with 15 pph silica as NALCOAG® 1115. Latex particles are seen as spheres while silica is not resolved. White vertical bar represents 500 nm.

and hydrolyzable precursors such as alkoxysilanes or alkenyloxysilanes can all function as crosslinkers in tin catalyzed silicone rubber latex [7-12]. It would be expected that all multifunctional silicone compounds having hydrolyzable groups on silicon will function as crosslinkers in tin catalyzed silicone rubber latex.

When tin catalyzed silicone rubber latex that contains silica is aged in emulsion form, it produces weaker elastomers with respect to time. The reasons for this are not fully understood, but it is believed that the tin catalyst remains active and crosslinking continues unabated while the latex is in emulsion form. When reinforcing silica is present, there is an unlimited supply of crosslinker, silicic acid. However, upon removal of water, the cured latex becomes much more stable.

The active tin catalyst in silicone rubber latex results in products that have limited shelf lives. One method that is used to overcome this shortcoming is to replace silica with a non-siliceous filler such as calcium carbonate and use a limited quantity of alkoxysilane as crosslinker. This has been the approach used to prepare shelf stable silicone rubber latex sealants [10,11]. Another approach is to prepare silicone rubber latex using the copolymerization method of Weyenberg and Findlay [3]. In formulations that require silica for reinforcement, a sequestering agent to deactivate the tin catalyst can be added after crosslinking is complete [13].

3. Filler

a. Reinforcing Fillers

Silicone rubber requires reinforcement in order for it to have any appreciable mechanical properties and silica is the reinforcement of choice. For example, unreinforced silicone rubber has a tensile strength of only 0.34 MPa, but this strength can be increased by a factor of 40 with proper reinforcement. Silicone rubber latex also requires reinforcement to be useful and silica provides the best reinforcement in terms of strength.

Silica can be added to silicone rubber latex as either a neat solid or as an aqueous dispersion. Aqueous dispersions are the preferred forms of silica for reinforcing silicone rubber latex because they provide excellent reinforcement and they are simply stirred into the latex. However, aqueous dispersions of silica also lower solids content of the sealant. Therefore such dispersions must be used sparingly or solid silica used instead.

Types of silica that provide excellent reinforcement to silicone rubber latex are aqueous, colloidal, precipitated silica and fumed (pyrogenic) silica. Fumed silica is also available as aqueous, colloidal dispersions [14]. The type and amount of silica has a significant effect on mechanical properties of the cured latex. When used to reinforce silicone rubber latex, aqueous, colloidal silica generally leads to high modulus elastomers, while fumed silica (either dry or aqueous) gives elastomers of lower modulus, see Figure 2. However, it is possible to combine these types of silica along with extending filler to obtain elastomers having practically any (reasonable) modulus desired. Levels of silica in silicone rubber latex usually range from 5 to 50 parts silica per 100 parts of emulsion polymer (both on a solids basis).

b. Extending Filler

Amount of shrinkage in silicone rubber latex sealants is directly related to the nonvolatile content of the product. Most silicone emulsion polymers contain from 50 to 60 weight percent solids. Solids content of the latex sealant is further reduced by the addition of other volatile, stabilizing ingredients, see section A-4. Therefore it is desirable to incorporate extending filler into the latex sealant to obtain a solids content of 70-80 percent by weight. Calcium carbonate, diatomaceous earth, sand, titania, ceramics and various clays are typical extending fillers for silicone rubber latex sealants.

Calcium carbonate is the material of choice for silicone rubber latex because of the properties it imparts to the cured latex and its low cost. Either precipitated or ground calcium carbonate can be used as an extending filler for silicone latex sealants. Calcium carbonate is usually used from 10 to 50 percent by weight of the final sealant composition.

4. Other Ingredients

Similar to most latex sealants and coatings, silicone rubber latex sealants contain numerous ingredients other than binder and filler. Examples of these additives include thickeners and rheology modifiers, adhesion promoters, freeze/thaw stabilizers, pigments, dispersants, and additional stabilizers.

B. Properties of Silicone Rubber Latex Sealants

1. Rheological Properties

Fully formulated silicone rubber latex sealants are tacky, putty-like materials that cure by the loss of water. Therefore shrinkage of silicone rubber latex sealants is higher than that of non-latex silicone sealants, but it is similar to other (organic) latex sealants. Shrinkage in silicone rubber latex sealants is typically 19 percent by volume or 17 percent by weight, see Table 1.

Although silicone rubber latex sealants typically have very high viscosities, they also exhibit thixotropic behavior (shear thinning). Silicone rubber latex sealants have a high yield point and a high shear-thinning index. Typically silicone rubber latex sealant has a low shear (0.5 reciprocal seconds) viscosity of 120,000 Pa-sec., a high shear (20 rec. sec.) viscosity of 200 Pa-sec. and a yield point of

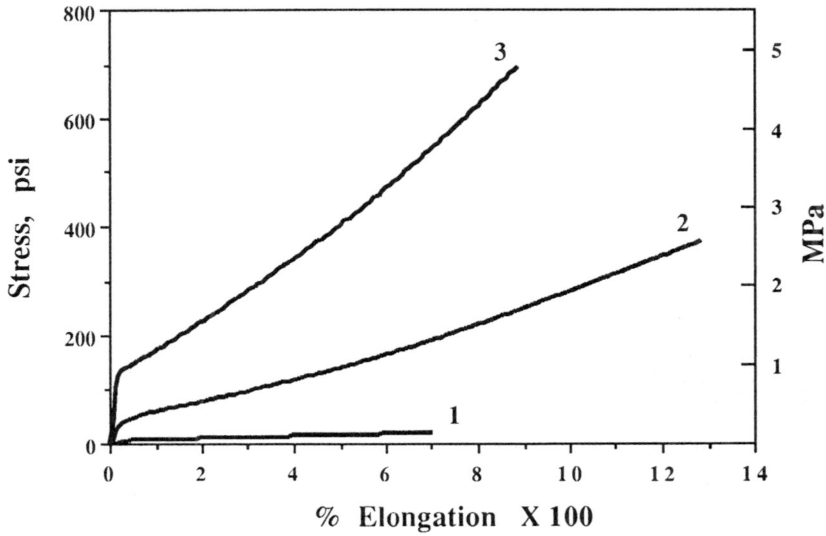

Figure 2. Effect of silica on silicone rubber latex. 1 is unfilled. 2 is filled with 10 pph CABOSPERSE® F (aqueous fumed silica). 3 is filled with 15 pph NALCOAG® 1115 (aqueous colloidal silica). All latexes are Sn catalyzed. pph = parts filler per 100 parts emulsion polymer (both on a solids basis).

approximately 1000 Pa-sec. A typical silicone (non-latex) sealant has a low shear viscosity of 1.5 MPa-sec., a high shear viscosity of 3,000 Pa-sec. and a yield point of approximately 10,000 Pa-sec. This combination of properties is manifested in silicone rubber latex sealant as a non-slump caulk that is easily gunnable from a cartridge (high extrusion rate). In terms of application, silicone rubber latex sealants resemble organic latex sealants more than they do non-latex silicone sealants, see Table 1.

2. Cure Requirements

As silicone rubber latex sealant cures by the loss of water, ambient temperature and humidity have a significant effect on cure rate. At room temperature (RT) and at 50 percent relative humidity (RH), tack free time is 30 minutes to 1 hour. Under these conditions, the sealant will cure in a channel at a rate of about 1mm of depth per day. The sealant, of course, cures from the outside to the inside. At higher RH levels, silicone rubber latex sealant cures much slower and above 95 percent RH under static air, it will not cure at all. It should be realized that organic latex sealants have similar behavior, as they too cure by the loss of water.

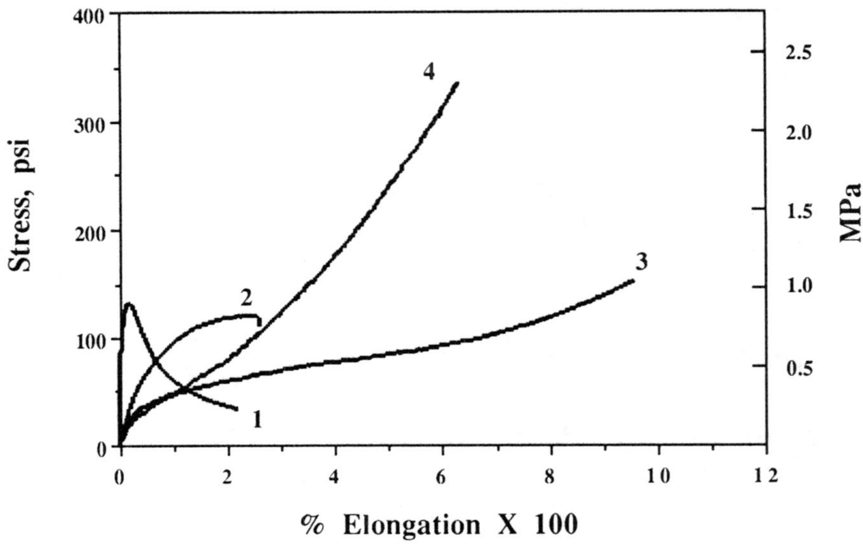

Figure 3. Stress strain behavior of typical sealants. 1. Acrylic latex sealant I. 2. Acrylic latex sealant II. 3. Silicone rubber latex sealant. 4. Silicone (non-latex) sealant.

Silicone rubber latex can coalesce and function as an elastomer at and below freezing temperatures (Tg of polymer is - 120° C). Typical organic latex sealants differ from silicone rubber latex in this property as they require a minimum film forming temperature, which is normally above 5° C. However, at low temperatures, cure rate of all latex sealants is extremely low, due to the low vapor pressure of water.

Since the binder in silicone rubber latex sealant is an oil in water emulsion, that is, water is the continuous phase, the sealant can be diluted to any degree with water. Uncured silicone rubber latex sealant can easily be cleaned from surfaces simply by washing with soap and water. This contrasts with non-latex sealants, which require solvent for removal. It should be emphasized that only uncured or partially cured silicone rubber latex sealant can be removed by water. Once cured, the sealant is impervious to water washing and must be removed by physical abrasion.

3. Tensile Properties

Tensile properties of cured silicone rubber latex sealants

TABLE 1

PROPERTIES OF TYPICAL SEALANTS

Sealant	NVC Wt. %	Duro. A	Tensile MPa	Elong. %	150% M MPa	Ext. Rate g/min
Silicone Latex	81	25	1.38	1000	0.43	700
Acrylic Latex	79	57	1.05	156	1.05	760
Silicone (non-latex)	94	37	2.14	550	0.62	95

NVC = non-volatile content; Durometer as per ASTM C 661; Tensile, elong., 150% modulus as per ASTM D 412 die C; Extrusion rate as per ASTM C 603 (30 psi)

resemble those of traditional (non-latex) silicone rubber sealants more than they do those of acrylic latex sealants, see Table 1. However, modulus of silicone rubber latex sealants is slightly lower and ultimate elongation is significantly higher than that for traditional silicone sealants. Stress-strain behavior of several types of sealants is given in Figure 3. Due to its low modulus and high elongation, silicone rubber latex sealants pass cyclic joint movement test, such as ASTM C-920 Type S, grade NS, Class 25, which means it can withstand a joint movement of +/- 25 percent.

4. Low Temperature Properties

The very low Tg of PDMS (-120° C) is manifested in silicone rubber latex as a material that remains flexible at low temperature. Thus silicone rubber latex sealant is useful at temperatures where most organic latex sealants become inflexible. In general, silicone rubber latex sealants are flexible to temperature to -40° C. PDMS tends to crystallize at approximately -60° C, which leads to more rigid materials near this temperature.

5. Adhesion

Adhesion of silicone rubber latex sealants varies from excellent to poor, depending upon the substrate. Adhesion to mineral surfaces such as concrete, masonry and glass is excellent, with adhesion to glass and porous masonry being especially good. Cohesive failure is almost always observed using these substrates. Adhesion to metals is generally

TABLE 2

ADHESION OF TYPICAL SEALANTS

180° Peel Adhesion (KN/m) ASTM C 794

Sealant	Glass		Anodized Aluminum		Pine	
	dry	wet	dry	wet	dry	wet
Silicone latex	1.2	0.7	0.6	0.3	1.2	0.4
Acrylic latex	0.3	0.2	0.4	0.2	0.5	0.1
Silicone (non-latex)	0.6	0.7	0.7	0.3	0.5	0.2

Dry: Sealant cured 14 days @ room temp. (RT), 50% RH.
Wet: Dry cured sealant submerged 7 days in water, @ RT.

TABLE 3

QUV ACCELERATED WEATHER TESTING OF TYPICAL SEALANTS

Sealant	Time Hr.	Duro A	Break Tens. MPa	Elong. %	150%M MPA
Silicone latex	0	25	1.38	1000	0.43
Silicone latex	1000	23	1.38	980	0.48
Acrylic latex	0	57	1.05	156	1.06
Acrylic latex	1000	46	1.24	195	1.24
Silicone (non-latex)	0	24	2.21	490	0.63
Silicone (non-latex)	1000	24	2.21	490	0.63

QUV: Testing as per ASTM G 23 Type E Cycle; Tensile properties: ASTM D 412 die C; Durometer: ASTM C 661

good. Adhesion to wood is good to fair, depending upon the type of wood. Adhesion to plastics such as polystyrene, vinyl polymers and polyethylene is fair to poor. Adhesion results for silicone rubber latex are shown in Table 2.

6. Weatherability

Silicone rubber is well known for its outstanding durability toward exterior exposure [15]. Silicone rubber latex retains this characteristic as it also has extremely good resistance to exterior exposure and accelerated weathering conditions [16]. Silicone rubber latex sealants can also be formulated to have good weatherability properties. Table 3 gives results of QUV accelerated weather testing as per ASTM G-23 type E cycle.

7. Miscellaneous Properties

Silicone rubber is well known to have excellent dielectric properties, good high temperature stability, excellent low temperature properties and superb resistance to weathering. Silicone rubber latex sealants retain some, but not all of these characteristics. Generally, silicone rubber latex has poor dielectric properties, moderate high temperature stability, good low temperature properties and excellent resistance to weathering [16]. It would be expected that the same is true for silicone rubber latex sealants.

8. Paintability

Paintability of sealants can be divided into two categories: ability of the sealant to accept paint, or in other words, ability of the paint to form a film on the sealant, and ability of the paint to adhere to the sealant. Silicone sealants are well known for being unpaintable [15]. Paints, both water-based and oil-based, do not form films on nor do they adhere to silicone (non-latex) sealants. The extremely low surface tension of silicone rubber (20 dynes/cm) hampers film formation and polymer incompatibility prevents paint adhesion to traditional silicone sealants.

Silicone rubber latex sealants however, are paintable. Oil-based paints perform better on silicone rubber latex sealants than do water-based paints. Sealant must be fully cured before applying oil-based paint, otherwise the coating will retard or even prevent cure of the sealant. Water-based paints can be applied as soon as possible, which means as soon as the sealant becomes tack free. Water from the sealant can easily migrate through a water-based paint film, but not so with an oil-based paint film.

Optimum adhesion of paint to silicone rubber latex sealant occurs after about two weeks at room temperature. Adhesion of paints to silicone rubber latex sealants varies with the paint formulation and even from color to color within the same brand. Adhesion of paint to silicone rubber latex sealant is conveniently measured by the cross-hatched/tape method as per ASTM D 3359.

Although it has not been proven, it is believed that silicone rubber latex sealant is paintable because of its unique morphology. That is, silicone rubber is the discontinuous phase while filler is the continuous phase [8]. This is the opposite of traditional silicone (non-latex) sealants, which constitute a dispersed phase of filler in a continuum of silicone rubber.

C. Availability of Silicone Rubber Latex

Although silicone rubber latex had its origins in the early 1960's, it was not until the late 1970's and early 1980's that commercial products became available. Silicone rubber latex coatings were developed first followed by silicone rubber latex sealants. Silicone rubber latex has been commercialized by the Dow Corning Corporation as additives and both fully formulated sealants and coatings.

SUMMARY

Silicone rubber latex and its preparation have been described. Silicone rubber latex sealant and its properties have also been described. Silicone rubber latex is prepared by crosslinking silicone emulsion polymer. Silicone rubber latex sealants consist of silicone rubber latex, filler and other additives. Rheology of silicone rubber latex sealants resembles that of organic latex sealants while tensile properties resemble those of traditional (non-latex) silicone sealants. Silicone rubber latex sealants have good adhesion to most surfaces, good weatherability characteristics, good low temperature flexibility and they are paintable.

ACKNOWLEDGMENT

H. A. Freeman and S. A. Wackerle of the Analytical Research Laboratories, Dow Corning Corporation are sincerely thanked for obtaining the transmission electron micrograph of Figure 1.

LITERATURE CITED

1. Weyenberg, D. R., Findlay, D. E., Cekada, J., Jr. and Bey, A. E. (1969), Journal of Polymer Science, Part C., Vol. 27, pp. 27-34.

2. Liles, D. T. and Lefler, H. V. III, "Silicone Rubber Latex Technology," Proceedings of the 18th Water-Borne, Higher-Solids and Powder Coatings Symposium, Story, R., F. and Thames, S. F., Editors, 1991, pp. 161-173.

3. Findlay, D. E. and Weyenberg, D. R., US Patent 3,294,725; 1959.

4. Huebner, D. J. and Saam, J. C., US Patent 4,567,231; 1986.

5. Huebner, D. J., and Landis, J. K., US Patent 4,661,556; 1987.

6. Bauman, T.M. and Huebner, D. J., US Patent 4,618,645; 1986.

7. Johnson, R. D., Saam, J. C. and Schmidt, C.M., US Patent 4,221,683; 1980.

8. Saam, J. C., Graiver, D. and Baile, M., (1981), Rubber Chemistry and Technology, Vol. 54, p. 976.

9. Saam, J. C., US Patent 4,244,849; 1981.

10. Schoenherr, C. J., US Patent 4,618,642; 1986.

11. Freiberg, A. L., US Patent 4,608,412; 1987.

12. Feder, M., (1989) European Patent Application EP 0,364,375.

13. Freiberg, A. L., Graiver, D., Lim, T. F. and Wright, A. P., US Patert 4,609,486; 1986.

14. Burns, S. A., Liles, D. T., Schoenherr, C. J., US Patent 4,957,963; 1990.

15. Klosowski, J. M., (1989), Sealants in Construction, Marcell Dekker: New York, 1989.

16. Smith, S. B., "Silicone Latex Technical Review," Proceedings of the Water-Borne and Higher-Solids Symposium, 1983, pp. 67-81.

Anthony R. Fiorillo[1]

ADVANCES IN ONE PART POLYSULFIDE SEALANT TECHNOLOGY

REFERENCE: Fiorillo, A. R., **"Advances in One Part Polysulfide Sealant Technology,"** Science and Technology of Building Seals, Sealants, Glazing, and Waterproofing: Second Volume, ASTM STP 1200, Jerome M. Klosowski, Ed., American Society for Testing and Materials, Philadelphia, 1992.

Abstract: One of the major drawbacks of one part polysulfide sealants is their extremely slow cure rates compared to other types of sealants.

Major studies have been conducted in an effort to find ways of accelerating the cure of one part polysulfide sealants. New formulating techniques and new developments in curative technology have shown that one part polysulfide sealants can be formulated that have cure rates similar to other generic types of sealants while maintaining good shelf life stability.

Sealants will be discussed that have been formulated based on these studies. These sealants have been tested and found to meet all the requirements of ASTM C-920.

Keywords: polysulfide, lead dioxide, manganese dioxide, bairium peroxide, calcium peroxide, phthalate.

---

[1]Group Leader, LP Technical Service and Applications, Morton International, 1275 Lake Avenue, Woodstock, IL 60098

Liquid polysulfide polymers have been in use for almost fifty years. Traditionally the curing agents used with these polymers have been metallic peroxides such as lead dioxide, manganese dioxide, barium peroxide, calcium peroxide and zinc peroxide. These curing agents work fine in two part systems and are still being used.

In one part sealants, formulators have only been able to get stable systems based on barium, calcium and zinc peroxides. The cure rates of these sealants are very dependent on the ambient temperature and relative humidity. In order to be activated, these peroxides all need water to be drawn into the sealant. Because of the low water vapor transmission properties of the polysulfide polymer, once a skin is formed on the sealant it is difficult for water to penetrate the skin and, depending on the relative humidity, it may take a month or longer to obtain full cure of a sealant bead.

Based on the knowledge that water is an accelerator for curing polysulfide polymers, a study was done on trying to develop a one part, manganese dioxide cured polysulfide polymer, drying all the ingredients, including the manganese dioxide, before they were compounded into the formulation.

A typical starting formulation for a one part, manganese dioxide cured polysulfide sealant would be:

INGREDIENT	**PARTS BY WEIGHT**
Polysulfide Polymer	100.0
Molecular Sieve	2.0
Phthalate Plasticizer	80.0
Calcium Carbonate (coated)	95.0
Titanium Dioxide	25.0
Stearic Acid	0.5
Epoxy Adhesion Promoter	5.0
Fumed silica	2.0
Manganese Dioxide (active)	6.0
Barium Oxide Paste (75% active)	5.0

A sealant based on this formulation was evaluated for tack free time, cure rate and tube stability with the following results.

Tack Free Time	3 hours
Cure Rate at 22°C/50% RH (1/4" bead)	
1 day	0.04" skin
3 days	0.07" skin
6 days	0.12" skin
12 days	full cure
Tube Stability	
at:	
50°C	63 days
60°C	49 days
70°C	29 days

This same formulation was evaluated against ASTM C-920 and found to meet all the requirements except for hardness. A slight adjustment in the levels of plasticizer and filler should correct this problem without significantly affecting other properties.

ASTM C-920 SPECIFICATION TEST

SAG	**SPEC LIMIT**	**TEST RESULTS**
Horizontal, 122°F (50°C)	No Change	No Change
40°F (4.4°C)	No Change	No Change
Vertical, 122°F (50°C)	3/16 Max. Sag	No Change
40°F (4.4°C)	3/16 Max. Sag	No Change
Shore A hardness	15-50	12
Weight loss, %	10 Max.	7.15
Tack free time	72 hrs. Max.	3 hours
Stain		
Unprimed concrete	No visible stain air or UV	0
Primed concrete	No visible stain air or UV	0

Accelerated Weathering (ASTM G-23; Type D)		
250 hrs exposure	2	<1 (slight hairline cracks)
250 hrs exposure plus bending 180 degrees while frozen at -15°F(-26°C)	2	<1 (slight hairline cracks)
Durability	**Maximum Failure**	
Concrete	1.5 in. sq. (9.7 cm. sq.)	<0.25 in. sq. (<1.6 cm. sq.)
Aluminum	1.5 in. sq. (9.7 cm. sq.)	<0.25 in. sq. (<1.6 cm. sq.)
Glass	1.5 in. sq. (9.7 cm. sq.)	<0.25 in. sq. (1.6 cm. sq.)
Peel Adhesion, lbs./in. (kg./2.54 cm.)		
Concrete	5 lbs. (2.27 kg.) min. 75% cohesive min.	12.1 lbs.(5.5 kg.) 100% cohesive
Aluminum	5 lbs. (2.27 kg.) min. 75% cohesive min.	15.6 lbs.(7.1 kg.) 100% cohesive
Glass	5 lbs. (2.27 kg.) min. 75% cohesive min.	15.0 lbs.(6.8 kg.) 100% cohesive
Glass, UV exposure	5 lbs. (2.27 kg.) min. 75% cohesive min.	13.0 lbs.(5.9 kg.) 100% cohesive

Another study was done on evaluating other potential curing agents for polysulfide polymers. Among the various curatives that were evaluated, sodium perborate gave the most promising results.

In developing a formulation based on sodium perborate it was seen that the level of perborate was critical and for best results should be between three and five parts per hundred polysulfide polymer. Below three parts did not give an adequate cure; levels higher that five parts caused blistering in the sealant.

A typical starting formulation for a one part sodium perborate cured polysulfide sealant would be:

INGREDIENT	PARTS BY WEIGHT
Polysulfide Polymer	100.0
Molecular Sieve	3.0
Epoxy Adhesion Promotor	5.0
Phthalate Plasticizer	50.0
Calcium Carbonate, Coated	60.0
Calcium Carbonate, Uncoated	30.0
Titanium Dioxide	20.0
Amine Accelerator	2.5
Fumed silica	3.0
Barium Oxide Paste (75% active)	5.0
Sodium Perborate Paste (40% active)	8.0

A sealant based on this formulation was evaluated for tack free time, cure time and tube stability, with the following results.

Tack Free Time	4 hours
Cure Rate (1/4" bead)	
at 38°C/95% RH	Full cure in 24 hours
at 22°C/50% RH	Full cure in 72 hours
Tube Stability	
at 50°C	>90 days
at 60°C	>90 days
at 70°C	49 days

This same sealant was tested against ASTM C-920 and found to meet all the requirements of the specification.

ASTM C-920 SPECIFICATION

SAG	**SPEC LIMIT**	**TEST RESULTS**
Horizontal, 122°F (50°C)	No change	No change
40°F(4.4°C)	No change	No change
Vertical, 122°F(50°C)	3/16" (0.475 cm.) Max. sag	No change
40°F(4.4°C)	3/16" (0.475 cm.) Max. sag	No change
Shore A hardness	15-50	18
Weight loss, %	10 Max.	3.32
Tack free time	72 hours	4 hours
Stain		
Unprimed concrete	No visible stain air or UV	0
Primed concrete	No visible stain air or UV	0
Accelerated Weathering (ASTM G-23; Type D)		
250 hrs exposure	2	<1 (slight hairline cracks)
250 hrs exposure plus bending 180 degrees while frozen at -15°F(-26°C)	2	<1 (slight hairline cracks)

Durability	**Maximum Failure**	
Concrete	1.5 in. sq. (9.7 cm. sq.)	0.38 in. sq. (2.47 cm. sq.)
Aluminum	1.5 in. sq. (9.7 in. sq.)	0.60 in. sq. (3.9 in. sq.)
Glass	1.5 in. sq. (9.7 cm. sq.)	<0.42 in. sq. (2.73 cm. sq.)

Peel Adhesion, lbs./in. (kg./2.54 cm.)

Concrete	5 lbs. (2.27 kg.) min. 75% cohesive min.	14.0 lbs.(6.4 kg.) 80% cohesive
Aluminum	5 lbs. (2.27 kg.) min. 75% cohesive min.	26 lbs. (11.8 kg.) 90% cohesive
Glass	5 lbs. (2.27 kg.) min. 75% cohesive min.	22 lbs. (10.0 kg.) 90% cohesive
Glass, UV exposure	5 lbs. (2.27 kg.) min. 75% cohesive min.	28 lbs. (12.7 kg.) 90% cohesive

Traditionally one part polysulfide sealants have a reputation for being extremely slow to cure through. These studies show that now stable, one part polysulfide sealants can be formulated that cure at very fast rates and, when properly formulated, meet all the requirements of ASTM C-920 as a class 25 sealant.

Dale R. Flackett

EFFECT OF EARLY JOINT MOVEMENT ON OXIME SILICONE SEALANTS AND COMPARISON TO OTHER ONE-PART SILICONE SYSTEMS

REFERENCE: Flackett, D. R., "**Effect of Early Joint Movement on Oxime Silicone Sealants and Comparison to Other One-Part Silicone Systems**", Science and Technology of Building Seals Sealants, Glazing, and Waterproofing: Second Volume, ASTM STP 1200, Jerome M. Klosowski, Ed., American Society for Testing and Materials, Philadelphia, 1992.

ABSTRACT: Two separate investigations are discussed concerning effect of early joint movements.

1. Continuous Cycling While Curing

   The overall physical properties of sealants cured in a joint undergoing continuous +/- 10% deformation for one week are compared to those of sealants cured under static conditions. Acetoxy, benzamide, alkoxy and oxime systems were studied.

2. "Crack Test"

   Freshly prepared joints of different formulations were elongated to 25% of their original width starting 15 minutes after preparation. The effect on the surface and body of the sealants as well as the displacement from substrate was observed.

   This test, the so called "crack test" was performed on acetoxy, benzamide, alkoxy, amine, and oxime.

KEYWORDS: Early joint movement, cracking, curing systems, oxime cure.

INTRODUCTION:

Concern has existed over the problems caused by early joint movement in expansion joints as this has resulted in various kinds of failures; cohesive failure, adhesive failure, cracking of the surface and body of the sealant and a general weakening or lowering of the sealant performance characteristics.

At the 1991 ASTM C24 Symposium on Science and Technology of Building Seals, Sealants, Glazing and Waterproofing, two papers on this topic were given by John Margeson[1] and Yoichi Matsumoto[2] with a third paper by Jim Brower[3] being an alternate. ASTM also has a Task Group headed by Jim Brower looking at this issue, to determine a "standard test method for effect of movement on sealants during cure".

It seems interesting and useful at this time to look at some different cure chemistries and formulations employed in silicone sealants to see if there were any significant differences in regard to early joint movement effects.

Dale R. Flackett is the Market Applications Manager at Allied-Signal, 101 Columbia Road, P.O. Box 1053, Morristown, NJ 07962-1053

Two tests were chosen: 1) Continuous cycling (+/- 10%) during cure for one week, 2) 25% elongation during the very early stage of cure. The former representing relatively small movements continuously occuring throughout the complete cure of the sealant, and the latter a large movement which occurs early after the sealant is applied.

The first test was designed to look at the effect on the physical property buildup of the sealant and the second to observe the surface effects.

1. CONTINUOUS CYCLING WHILE CURING

Sealants: For this test six silicone systems were tested.

	Skin Over Time
1. Alkoxy - tin catalyzed	3- 5 minutes
2. Acetoxy - translucent	5- 6 minutes
3. Alkoxy - titanium catalyzed	20-40 minutes
4. Oxime - translucent (medium cure rate)*	12-15 minutes
5. Oxime - filled (medium cure rate)*	14-18 minutes
6. Benzamide	20-30 minutes

*Allied Lab formulations

Test Procedure

Before starting such a test many criteria had to be considered. The following were decided upon as appropriate in representing a typical joint and situation. The width 12.5 mm is slightly less than a typical expansion joint however this was utilized so that at the given speed of movement 25% joint extension could be reached in approximately 1 hour (important in the crack test), and also so that almost complete through cure could be attained in one week.

a) Joint size - 50mm (length) x 12.5mm (width) x 5.5mm (depth at center)
b) Joint shape - flat surface, concave underside
c) Closed-cell polyethylene backer rod
d) Extent of deformation +/- 10%
e) Begin in extension
f) Speed of movement = 3.175 mm/hr (1/8 inch/hour)
g) Substrate - aluminum
h) Temperature - 23°C
i) Humidity 40-60%
j) Movement to commence 2 hours after sealant applied
k) Duration of cycling = 1 week or approximately 105 cycles
l) Speed of physical property determination = 0.5mm/min to 20% and 12.5mm/min to break

All joint movement during cure was performed using an Applied Test Systems model 520 sealant tester, with subsequent physical property analysis by a Lloyd L500 tensile tester.

The following six specimens were prepared for each system:

Specimen 1a- Static Joint Control - cured one week under static conditions

Specimen 1b- Static Joint Control - cured one week under static conditions

Specimen 2a- Cycled - cured one week under continuous +/- 10% deformation

Specimen 2b- Cycled - cured one week under continuous +/- 10% deformation

Specimen 3 - 5 x 10% - cured one week under static conditions and then extended to 10% elongation 5 times at 2.5mm/min

Specimen 4 - Cured 2 weeks under static conditions

Specimens 1 and 2 were selected to directly compare the difference in physicals of sealants cured under static conditions to those cured under continuous motion for one week. Specimen 3 was utilized to observe possible hysteresis effects, and Specimen 4 to give an indication of completeness of cure after one week since this was an identical specimen to 1a and 1b cured an additional week.

For the cycling test the joints were prepared using foam spacers below the joint and between the grips, and release tape around the grip surface; the sealant being applied, tooled and then left for 2 hours before turning on the cycling machine. Neither the foam spacer or release tape restricted the movement of the sealant, however the sealant in extension would move while the spacer did not and in compresion the spacer would slightly push up into the sealant. No cracking or indenting was observed at the interface between the sealant and the release tape or foam. After one week the specimen was removed from the machine and allowed 2 hours to recover before tensile testing. The static joints were prepared in a plastic jig.

Tensile testing was carried out in two stages, to 20% at a slow speed and then to break at a faster speed.

All pieces were first extended to 20% elongation at a rate of 0.5mm/min. They were then allowed to recover 2 hours before being pulled to break at a rate of 12.5mm/min. The physical properties were recorded.

TABLE 1 - - Physical Properties of Alkoxy-tin System

Modulus at: (MPa)	5%	10%	15%	20%	100%	Elong. % at break
Specimen 1a/Control	0.019	0.037	0.058	0.070	0.173	436
Specimen 1b/Control	0.014	0.033	0.058	0.071	0.181	427
Specimen 2a/Cycled	0.017	0.028	0.040	0.050	0.146	292
Specimen 2b/Cycled	0.013	0.026	0.034	0.045	0.143	332
Specimen 3 /5 x 10%	0.022	0.042	0.060	0.071	0.163	389
Specimen 4 /2 weeks	0.044	0.079	0.115	0.136	0.302	275

TABLE 2 - - Physical Properties of Acetoxy System

Modulus at: (MPa)	5%	10%	15%	20%	100%	Elong. % at break
Specimen 1a/Control	0.051	0.107	0.159	0.186	0.481	145
Specimen 1b/Control	0.047	0.110	0.164	0.195	0.522	169
Specimen 2a/Cycled	0.046	0.089	0.122	0.149	0.392	144
Specimen 2b/Cycled	0.042	0.085	0.113	0.138	-	65
Specimen 3 /5 x 10%	0.031	0.094	0.145	0.173	0.476	115
Specimen 4 /2 weeks	0.025	0.089	0.136	0.177	0.504	226

TABLE 3 - - Physical Properties of Alkoxy-titanium System

Modulus at: (MPa)	5%	10%	15%	20%	100%	Elong. % at break
Specimen 1a/Control	0.062	0.116	0.163	0.193	0.585	332
Specimen 1b/Control	0.039	0.113	0.164	0.198	0.634	370
Specimen 2a/Cycled	0.018	0.051	0.087	0.117	0.509	211
Specimen 2b/Cycled	0.021	0.064	0.102	0.135	0.543	203
Specimen 3 /5 x 10%	0.041	0.104	0.152	0.178	0.562	347
Specimen 4 /2 weeks	0.029	0.097	0.146	0.186	0.695	282

TABLE 4 - - Physical Properties of Oxime Translucent System

Modulus at: (MPa)	5%	10%	15%	20%	100%	Elong. % at break
Specimen 1a/Control	0.045	0.077	0.105	0.128	0.345	137
Specimen 1b/Control	0.034	0.066	0.099	0.127	0.345	166
Specimen 2a/Cycled	0.024	0.045	0.067	0.087	-	87
Specimen 2b/Cycled	0.026	0.046	0.073	0.094	0.258	109
Specimen 3 /5 x 10%	0.046	0.088	0.112	0.137	0.355	142
Specimen 4 /2 weeks	0.050	0.084	0.115	0.140	0.363	124

TABLE 5 - - Physical Properties of Oxime-filled System

Modulus at: (MPa)	5%	10%	15%	20%	100%	Elong. % at break
Specimen 1a/Control	0.072	0.127	0.171	0.207	0.576	138
Specimen 1b/Control	0.064	0.118	0.162	0.203	0.569	139
Specimen 2a/Cycled	0.042	0.078	0.111	0.142	(0.420)*	98
Specimen 2b/Cycled	0.040	0.074	0.108	0.136	(0.410)*	94
Specimen 3 /5 x 10%	0.061	0.116	0.163	0.199	0.549	140
Specimen 4 /2 weeks	0.071	0.133	0.181	0.221	0.615	102

TABLE 6 - - Physical Properties of Benzamide System

Modulus at: (MPa)	5%	10%	15%	20%	100%	Elong. % at break
Specimen 1a/Control	0.008	0.016	0.021	0.026	0.097	409
Specimen 1b/Control	0.008	0.016	0.021	0.026	0.103	406
Specimen 2a/Cycled	0.006	0.011	0.016	0.018	0.081	377
Specimen 2b/Cycled	0.004	0.010	0.012	0.015	0.076	295
Specimen 3 /5 x 10%	0.009	0.016	0.022	0.026	0.089	455
Specimen 4 /2 weeks	0.007	0.015	0.024	0.032	0.127	306

*extrapolated

Figure 1 - Modulus comparison of static and cycled joint.
● = Static ■ = Cycled
(average of values from tables 1-6 of specimens 1a,1b and 2a,2b)

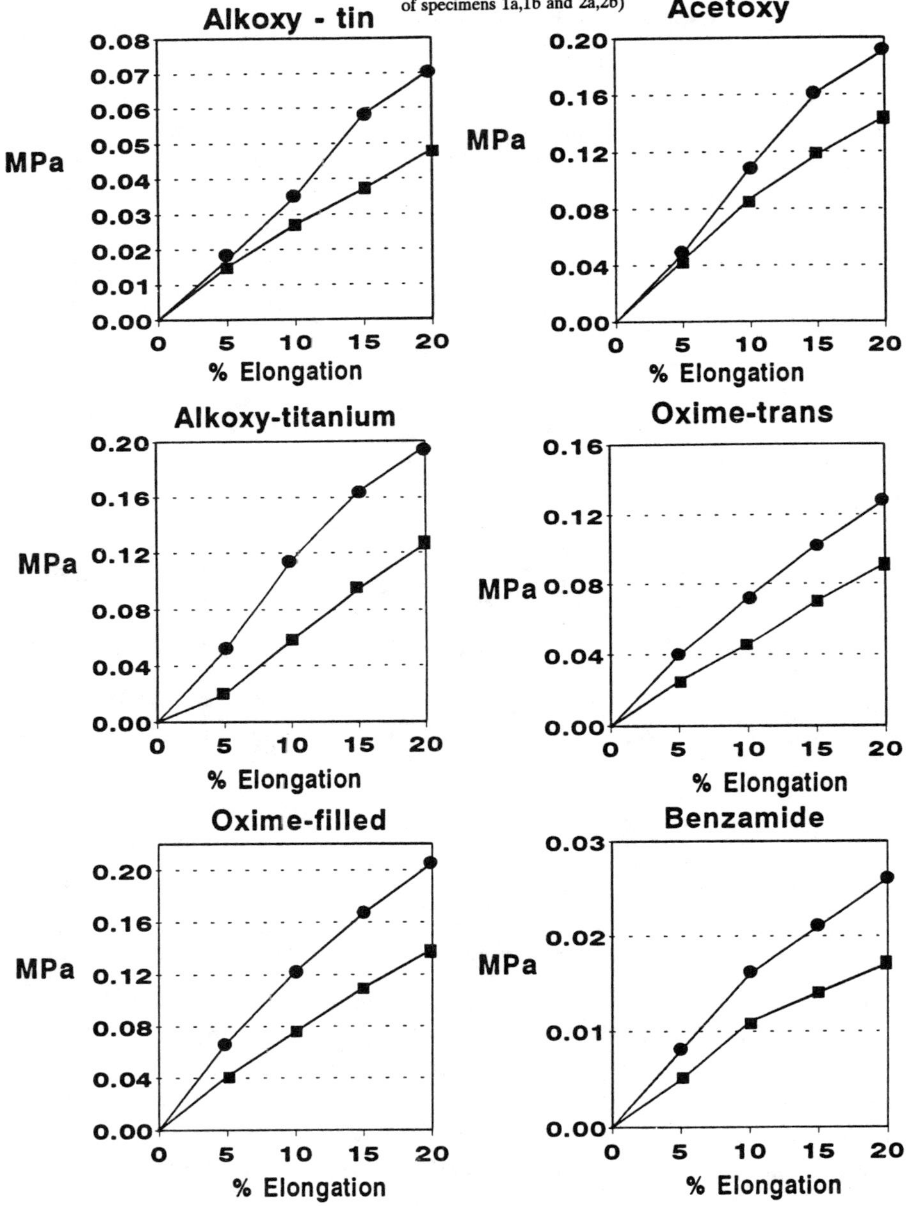

Table 7 -- Change in Properties of Cycled Joint Compared to Static Joint (Control)

	20% Modulus	100% Modulus	Elongation at Break
Alkoxy-tin	-32%	-18%	-28%
Acetoxy	-25%	-21%	-33%
Alkoxy-titanium	-36%	-14%	-41%
Oxime-translucent	-29%	-25%	-35%
Oxime-filled	-32%	-27%	-31%
Benzamide	-35%	-21%	-18%

Figure 2- Percent Reduction in 20 and 100% Modulus of Cycled Joint Compared to Static Joint.

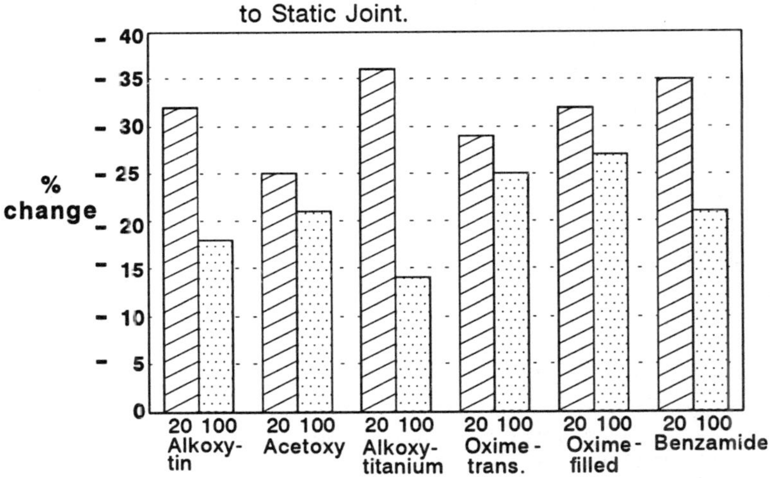

Figure 3- Percent Reduction in Elongation of Cycled Joint Compared to Static Joint.

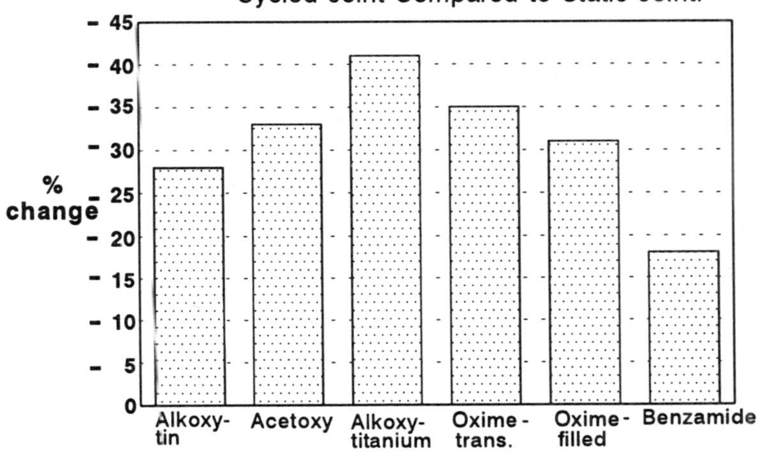

Table 8 -- Change in Properties of
5x10% Joint Compared to Static Joint (Control)

	20% Modulus	100% Modulus	Elongation at Break
Alkoxy-tin	0%	- 8%	-10%
Acetoxy	- 9%	- 5%	-27%
Alkoxy-titanium	- 9%	- 8%	- 1%
Oxime-translucent	+ 7%	+ 3%	- 6%
Oxime-filled	- 3%	- 4%	+ 1%
Benzamide	0%	-11%	+12%

Table 9 -- Change in Properties of
2 Week Cure Joint Compared to Static Joint (Control)

	20% Modulus	100% Modulus	Elongation at Break
Alkoxy-tin	+94%	+71%	-37%
Acetoxy	- 7%	+ 1%	+44%
Alkoxy-titanium	- 5%	+14%	-20%
Oxime-translucent	+ 9%	+ 5%	-18%
Oxime-filled	+ 8%	+ 7%	-27%
Benzamide	+23%	+27%	-25%

Discussion:

As can be seen the results have been documented quantitatively, with the data reproducibility appearing ver good, however due to the small number of samples tested the overall result should be taken qualitatively namely:

- All systems that were cycled during cure showed reduced modulus and elongation.
  On examination of these cycled test pieces, after destruction, it was apparent that the top 60-70% of the joint was intact as normal, however the lower portion and underside had many voids and cracks which would result in a lower modulus and lower elongation.
  The texture of the underside had definitely been effected by the curing of the sealant in contact to the backer rod while under motion.

- The difference in 20% modulus of the cycled joint compared to the control joint is always greater than at 100% modulus, perhaps because of some lack of recovery caused by the continual motion to +/-10%. The Benzamide and Alkoxy systems show the largest difference here.
  However, the specimens that underwent the 5 x 10% extensions after cure showed no real significant change when compared to the control. All changes being around 10% or less.

- It should also be pointed out that all systems failed 100% cohesively except the acetoxy which failed 100% adhesively and the alkoxy -titanium system which failed 50% adhesively (although this system failed 100% cohesively after 2 weeks). This could account for the larger fluctuations in elongation at break values.

- Finally, as mentioned the size of joint utilized in this test, was selected so that almost full cure could be attained within one week. It can be seen from the 2 week cure joint comparison that this was essentially true for the two oxime, acetoxy and alkoxy-titanium systems which increased in modulus by only about 10% or less during the second week, whereas the benzamide system increased 25% and the alkoxy-tin system by 70-90%. It is difficult to say whether this lack of complete cure at the time of measuring would effect the test.

2. CRACK TESTING

Sealant

Many different commercial sealants from several manufacturers were tested.

	SEALANT 1		SEALANT 2		SEALANT 3	
	S.O.T.	SCR	S.O.T.	SCR	S.O.T.	SCR
3 Acetoxies	5	25	11	20	6	23
3 Alkoxy-titaniums	25	40	5 hrs	5½ hrs	32	50
1 Benzamide	25	40	-	-	-	-
1 Alkoxy-tin	3	12	-	-	-	-
3 Oximes	20	55	20	45	6	12
1 Amine	4	13	-	-	-	-

(minutes unless otherwise stated)

In addition 5 other oxime formulations were made in the lab:

	SEALANT 1 (Translucent)		SEALANT 2 (Filled)	
	S.O.T.	SCR	S.O.T.	SCR
2 Methyl tri-oximino silane cured	15	2 ½ hrs	14	1 hr 45 mins
2 Medium cure rate	10	30	12	46
1 Fast cure rate	3	5	-	-

(minutes unless otherwise stated)

S.O.T. = Skin over time - time taken to form a skin
SCR = Scratch time - time until skin has some elastic properties (able to be scratched lightly)

Test Procedure

In this test all the testing conditions were identical to those used in the cycling tests except:

a) The magnitude of deformation was +25%

b) The standard time before commencing movement was 15 minutes - although other times were used.

c) The samples were prepared in a special plastic jig which facilitated not only easy and quick preparation but also immediate removal of the test piece at the end of the test without making a mess of the machine with uncured sealant.

Standard Test:

The sealants were tooled into the joint left for 15 minutes and then extended to 25% of their original width.

This took approximately 1 hour, at 3.175 mm/hour.

Each test involved 4 identical samples.

The effect on surface and interface with the aluminum substrate was observed, continuously through the 25% extension.

Results/Discussion

None of the commercial formulations showed any significant cracking.

The benzamide developed a very small indentation/crack about 1/3 the length of the joint in the center at around 18% elongation.

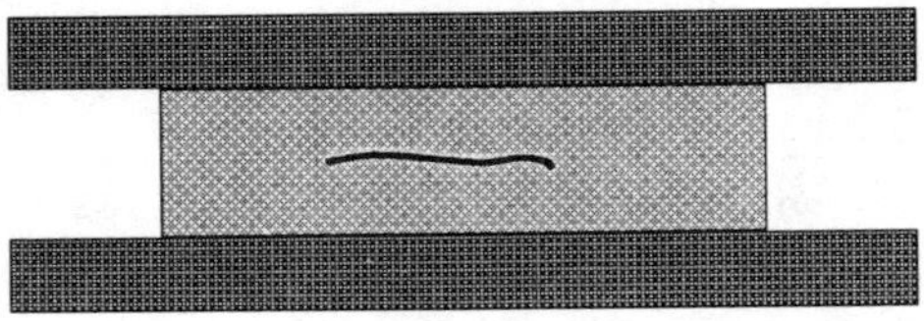

The alkoxy-tin system developed enlarged bubbles.

Some of the highly filled systems separated slightly from the substrate at around 20% extension.

All the sealants moved away from the substrate behind the surface.

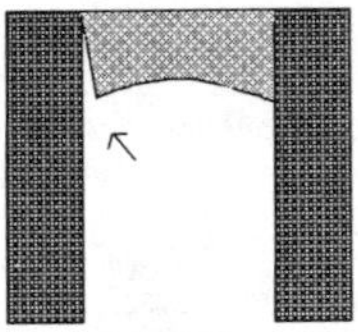

This indicates that 25% is indeed a stringent test for a completely uncured sealant.

However, in the methyl tri-oximino cured sealants made in the lab, severe cracking occurred.
Cracks started to appear after only 0.9mm movement or 7% elongation. Usually starting at a corner or edge the cracks enlarged as the joint continued to widen.

This was in fact a very difficult test for such a sealant since the skin over time is around 10-20 minutes and the "scratch time" or time when sealant surface attains some rubber like properties is around 1 ¾ hours for the filled system and 2 ½ hours for the translucent system. Thus a weak skin is formed in the 15 minutes after the sealant is applied and then it is immediately elongated causing cracking.

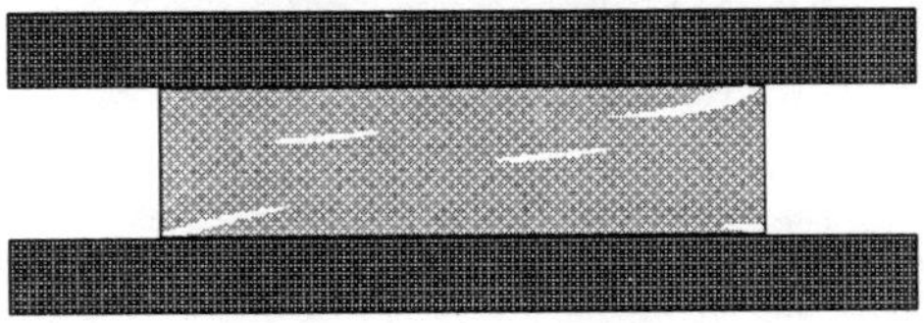

The test was re-run in the filled oxime sealant case after pre-cure times of 1 ¼, 1 ½, and 1 ¾ hours. The samples pre-cured for 1 1/4 hours showed only slight cracking at edge while the others showed no cracking even up to the full 25%.

In the translucent case a pre-cure time of 3 hours was selected and no cracking was observed.

This may explain the idea from the field that filled oxime systems do not tend to crack while some translucent formulations do, as the window of opportunity for cracking to occur is larger in the translucent case than it is in the filled. However it is clear that if conditions are right, a filled system will crack as well, and therefore pure methyl tri-oximino silane crosslinked sealants should not be recommended for use in expansion joints.

The medium and fast cure rate oxime systems did not show any cracking. In fact the translucent fast cure oxime was placed in the machine and immediately extended to 30% elongation without any cracking or substrate separation.

It is known in the field that other systems do also crack - so it is important to mention that two important factors in addition to many other diverse factors that occur in the outside world, have been omitted from the previous criteria. These factors are shelf-life and temperature variation at time of cure.

All the commercial samples received were fresh samples and all the oxime sealants made in our lab had controlled cure rates. Shelf-life of the various systems leading to a slowing of the cure rate will undoubtedly change the behavior concerning cracking as will lower cure temperatures which severely retard the cure rate of some systems.

Therefore the key to cracking seems to be the time difference between the S.O.T. and the scratch time, the smaller this difference the less chance exists for cracking to occur. It appears that in this test that a difference of 30-40 minutes was low enough to prevent cracking where as 1 ½ hours was too long.

Finally, it is important to emphasize that the initial cure time before the movement starts, is critical for the material being tested; for it is apparent that a sealant with a greater than 4 hours skin over time will not crack when extended 1 hour after application as no skin has yet formed to crack.

Further, work needs to be carried out to look more comprehensively at different pre-cure conditions, faster-more sudden smaller movements and the effect of shelf-life and cure temperature on the overall cracking situation.

Conclusion

In terms of cure chemistries in the continuous cycling test there seems little difference, and in the crack test the difference seems more related to the rate of cure of the various formulations, in so much as how fast they develop an elastic skin, than to the actual chemistry.

It appears that large early joint movements (within the first 12 hours) result in visible surface effects and substrate failure where as smaller continual movement during cure creates voids and cracks on the underside of the sealant.

Both are areas for concern, however it is clear that the surface effects are more detrimental than the underside effects where the top 60-70% of the sealant is still intact and functioning properly. In the case of silicone, with its excellent physical and aging properties together with design safety factors, a joint with the top 70% functioning normally will surely perform well.

However in the real world both factors (along with others) can come into play so it would be prudent to develop a sealant that can, at least, survive the early cure stress conditions.

Acknowledgements

The author wishes to thank: Jeff Knepper for meticulously carrying out the cycling tests and subsequent tensile measurements, AET for providing some of the backing foam and Dow Corning, General Electric, Mobay, Rhone-Poulenc and Wacker for providing samples of commercial silicone sealants.

REFERENCES

1. Margeson, J., "Effect of Joint Movement During Cure on Sealant Strength Development" Science and Technology of Building Seals, sealants, glazing and waterproofing. First Volume, ASTM 1168, Charles J. Parise, editor, American Socity for Testing and Materials, Philadelphia, 1992.

2. Matsumoto, Y., "The Effect of Building Joint Movements on Outdoor Performance of Sealants During their Cure" Science and Technology of Building Seals, sealants, glazing and waterproofing. First Volume, ASTM 1168, Charles J. Parise, editor; American Socity for Testing and Materials, Philadelphia, 1992.

3. Brower, J.R. "Deformation of Building Sealants Due To Movement During Cure" Science and Technology of Building Seals, sealants, glazing and waterproofing. First Volume, ASTM 1168, Charles J. Parise, editor, American Socity for Testing and Materials, Philadelphia, 1992.

077.drf
gg/jm

Udo C. Pernisz, Jerome M. Klosowski

Fraction of the Global Normal Irradiance Incident from the Northern Part of a Clear Sky

REFERENCE: Pernisz, U. C., and Klosowski, J. M., **"Fraction of the Gloval Normal Irradiance Incident from the Northern Part of a Clear Sky"**, Science and Technology of Building Seals, Sealants, Glazing, and Waterproofing: Second Volume, ASTM STP 1200, Jerome M. Klosowski, Ed., American Society for Testing and Materials, Philadelphia, 1992.

ABSTRACT: The luminous flux received from a clear blue sky at the north side of a building was determined relative to the global normal illuminance with a commercial photometer using a hemi-spherical diffusor. Measurements with narrow band interference filters were also made to obtain information on the ratio of these quantities in the violet and blue parts of the visible spectrum. It was found that the blue sky provides 6.3% of direct sunlight at a wavelength of 370 nm. In comparison, a value of 4.7% was found for the luminous flux (not spectrally resolved).

It is concluded that materials which degrade principally by absorption of UV light will last approximately 15 times longer on the north side of a building than when fully deployed on a southern exposure.

KEY WORDS: Sunlight, blue sky radiation, UV absorption, sealant degradation, Northern and Southern exposure

Dr. Pernisz is a Senior Research Specialist in the Electronic Materials group and Jerome Klosowski is a Scientist in the Sealants department of the Dow Corning Corporation, Midland, MI 48686-0994.

INTRODUCTION

Building materials such as sealants, adhesives, and surface coatings degrade by weathering. A major driving force for this process is the blue and ultraviolet (UV) radiation received from the sun. The south side of a building (on the northern hemisphere) is exposed to the direct global radiation while the north side receives only the scattered light from the blue sky.

We are interested in determining how much slower photodegradation can be expected to proceed under diffuse irradiation than it does in direct sunlight. With the underlying assumption that the degradation rate is proportional to the irradiance in the active (blue and near UV) part of the spectrum this question can be resolved by a measurement of the relative luminous flux for the two exposure conditions.

In addition to a measurement of the total luminous flux, that is, the full visible spectrum, spectrally resolved data were also obtained in order to determine whether the relatively higher content of short wavelengths in the blue light of the sky, compared with direct sunlight, could be significant for degradation effects. The change in the spectral distribution of the diffuse light received from the blue sky is caused by Rayleigh scattering in the atmosphere; the intensity of the scattered radiation in this process increases as the fourth power of the inverse wavelength.

EXPERIMENTAL DETAILS

Relative values of luminous fluxes were obtained with a commercial photographic exposure meter equipped with a diffusor.[1] The diffusor has a nominal transparency of 0.17 for visible light and a 2π (hemispherical) field of view. The taken were taken in front of a square building with concrete walls, four storeys high (20 m), its faces oriented at 45° to the N-S compass direction, see Figure 1, left panel. Two series of measurements were taken: (I) with the sun approximately in the SW (Nov.1, 14:00 h EDT); and (II) in the SE (Jun.6, 11:00 h EST). In both cases the vertical plane of incidence was approximately normal to one of the building walls ("South" side). Readings of the illuminance were taken at a distance of 7 m from the building wall. The ground in front of the walls was covered with green grass; there was no other building within 150 m. Both days were essentially cloudless.

For each of the two series of measurements, a reading of the global and direct luminous flux was obtained by pointing the entrance aperture of

---

[1] Sekonic multi-lumi Model L-248, by Sekonic Co., Ltd

the meter directly at the sun with the diffusor in place. The meter was raised 2 m above ground, see Figure 1. Spectrally resolved readings were taken with the same geometry by replacing the diffusor with narrow band interference filters in front of the aperture. The filter characteristics are listed in Table 1.

Readings of the illuminance from the sky (and the foreground) were taken in front of the opposite ("North") side of the building within approximately 15 min of the time the sunlight measurements were taken. The meter was oriented upwards at 45° against the horizontal. The diffusor was not utilized; total and spectrally resolved illuminance measurements were taken.

Table 1. Characteristics of Interference Filters Used for the Spectrally Resolved Illuminance Data

wavelength nm	band width nm	transmission %
370	10	43.8
440	8	55.4
530	9	54.4
640	11	61.8

The illuminance is recorded in the photometric unit of Exposure Value (EV). This unit represents the relative luminous flux on a logarithmic scale to base 2, that is, an increment of 1 corresponds to a factor of two in illuminance. The use of the diffusor causes the readings of the exposure meter to be reduced by 2.5 in terms of EV units (logarithm to base 2 of 0.17). Table 2 summarizes the data.

RESULTS AND DISCUSSION

The two different series of readings give very consistent values. This indicates that the geographic and atmospheric differences between the measurements were negligible. The scale of the exposure meter can be read with a visual resolution of approximately $^1/_3$ EV which amounts to 26% of the luminous flux associated with the reading. Thus the data presented here carry an uncertainty of ±13%. It is within these limits that differences between experimental conditions, and differences due to spectral effects, can be detected. While fairly large, this error margin is considered adequate for the present purpose of establishing an estimate of the acceleration that sunlight causes compared to the blue sky.

Table 2. Luminous Flux at the North and the South Side of a Building (given in the logarithmic unit of Exposure Values, EV)

Series	"South" (direct global) total	spectral (filter wavelength in nm)				"North" (diffuse sky) total	spectral (filter wavelength in nm)			
	note*)	370	440	530	640		370	440	530	640
I	16.5	11.3	12.0	-	-	14.5	7.5	7.5	-	-
II	16.7	11.7	12.5	15.5	15.7	14.5	7.7	8.0	9.5	8.7

*) these readings taken with diffusor (2.5 EV)

Taking an average value of 16.6 for the reading in total sunlight, the equivalent reading without the diffusor in front of the meter is calculated as 19.1; the difference to the blue sky reading is 4.6 which gives a ratio of 24. Thus the light from the blue sky has about 4.1% the illuminance of direct global sunlight. (The lower value from series I gives 4.7% for this case as reported in the abstract.)

For the spectrally resolved illuminance data, again taking averages where applicable, the differences between "South" and "North" side, in EV units, and the corresponding spectral illuminance ratios and percentages, are compiled below:

wavelength	[nm]	370	440	530	640
S-N diff.	[EV]	3.9	4.5	6.0	7.0
ratio		15	27	64	128
North/South	[%]	6.7	3.7	1.6	0.78

(The values from series II at 370 nm give a value of 6.3% as quoted in the abstract.) The relatively higher content of the diffuse radiation from the sky in the blue part of the spectrum with respect to direct sunlight is clearly evident. This confirms the spectral dependence of scattered light: The least-squares regression of the -(South-North) data versus the $\log_2$ of wavelength gives a slope of -4.1 (correlation coefficient of 0.991) which is very close to the value expected for Rayleigh scattering, see Figure 2.

While these spectrally resolved data are sufficient to estimate acceleration factors for degradation in direct sunlight compared to diffuse light from the blue sky they are inappropriate to calculate absolute values from the total solar irradiance without spectrally resolving the sunlight incident on the surface of the earth. The total irradiance received from the sun outside of the atmosphere, the "solar constant", is 1.35 kW/m^2 [1]; after passing through the atmosphere the sunlight has left between 900 and 1000 W/m^2 (at Air Mass 1, i.e., at approximately normal incidence). From a spectral distribution of sunlight [2] absolute irradiances at 370 and 440 nm are 0.5 and 1.5 $W/(m^2\ nm)$, respectively.

The practical side of these findings is that the durability of sealants and coatings has to be specified by the manufacturer as to north or south side exposure. One has to ask when samples of a certain product are reported to last 10 or 15 years, does this mean 15 years on the north side of a building or just one year on the south side? Unless a specific product is tested and studied, the exact acceleration factor will not be known. However, since higher temperature (and temperature cycling) at southern exposure tends to promote weathering in general, a factor between 15 and 27 based only on spectral irradiance ratios (near-UV and blue) can be assumed to be a lower limit on durability.

CONCLUSION

From measurements of the spectrally resolved luminous flux of direct global sunlight and of the blue sky radiation, on the south and north side of a building, respectively, the relative light exposure of a structural wall (building) could be determined in four spectral bands. For blue light, i.e., for wavelengths below 440 nm, the north side of a building receives between 3.7 and 6.7 % of direct sunlight radiation. Assuming a linear relationship between light intensity and photo-degradation of materials in this spectral range, an acceleration factor between 27 and 15, or ≈20 as an average, is calculated from this ratio.

REFERENCES

[1] M.P. Thekaekara, R. Kruger, C.H. Duncan, "Solar Irradiance Measurements from a Research Aircraft", Applied Optics 8(8), 1713 (1969)

[2] W.G. Driscoll, W. Vaughan (eds), "Handbook of Optics" (ch.3); McGraw-Hill, New York 1978.

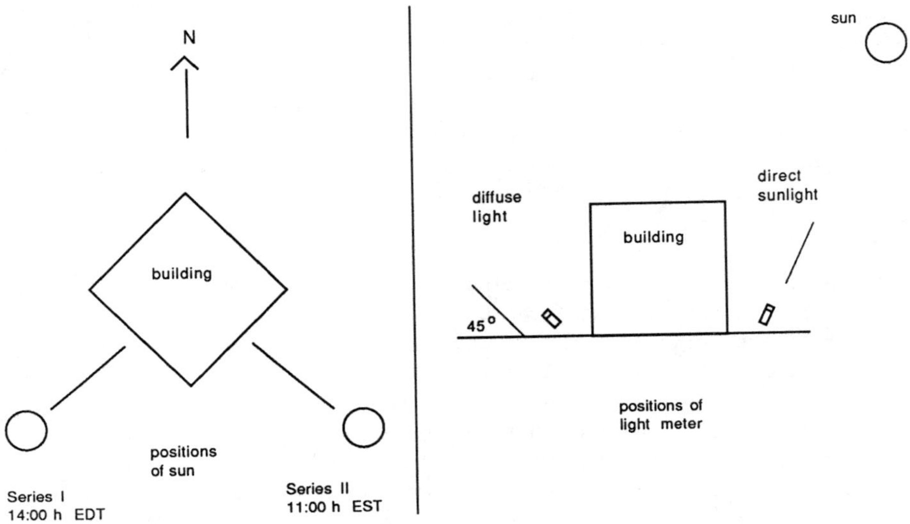

Figure 1

Left panel: Top view for the two series of measurements.

Right panel: Elevation of experimental arrangement; the small rectangles represent the light meter.

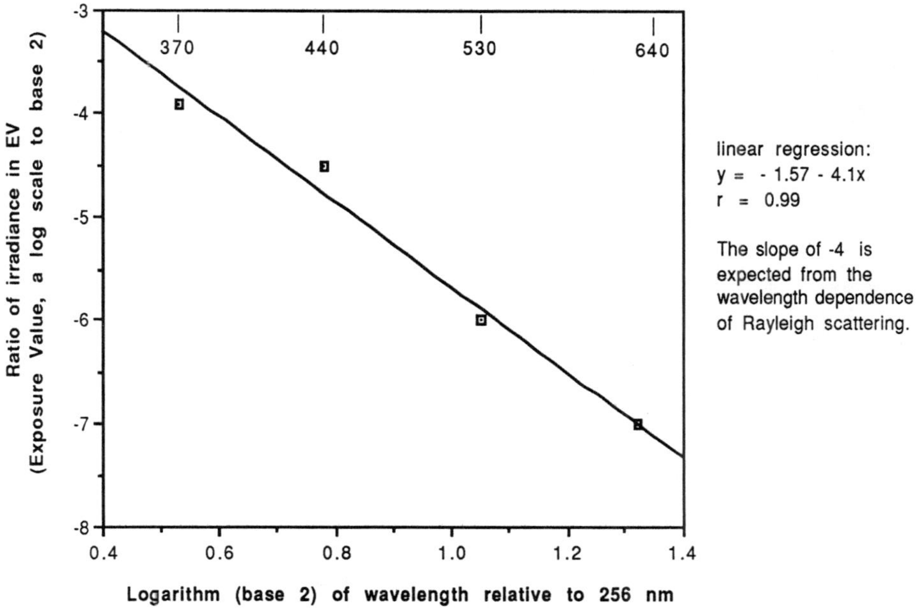

Figure 2

Ratio of light from clear blue sky (Rayleigh scattering) to direct solar irradiance, as a function of wavelength. Data taken with narrow-band interference filters at specified wavelengths (in nm).

Both scales are logarithmic to base 2; wavelengths are relative to (the arbitrary value of) 256 nm. The straight line represents the linear least-squares regression through the data; the slope is -4.1; the correlation factor, 0.991.

Author Index

Subject Index

S

T

U

V

W

X